Lecture Notes in Computer Science   14511

Founding Editors

Gerhard Goos
Juris Hartmanis

The series Lecture Notes in Computer Science (LNCS), including its subseries Lecture Notes in Artificial Intelligence (LNAI) and Lecture Notes in Bioinformatics (LNBI), has established itself as a medium for the publication of new developments in computer science and information technology research, teaching, and education.

LNCS enjoys close cooperation with the computer science R & D community, the series counts many renowned academics among its volume editors and paper authors, and collaborates with prestigious societies. Its mission is to serve this international community by providing an invaluable service, mainly focused on the publication of conference and workshop proceedings and postproceedings. LNCS commenced publication in 1973.

Bin Ma · Jian Li · Qi Li

Editors

Digital Forensics and Watermarking

22nd International Workshop, IWDW 2023
Jinan, China, November 25–26, 2023
Revised Selected Papers

 Springer

Editors
Bin Ma
Qilu University of Technology
Jinan, China

Jian Li
Qilu University of Technology
Jinan, China

Qi Li
Qilu University of Technology
Jinan, China

ISSN 0302-9743   ISSN 1611-3349 (electronic)
Lecture Notes in Computer Science
ISBN 978-981-97-2584-7   ISBN 978-981-97-2585-4 (eBook)
https://doi.org/10.1007/978-981-97-2585-4

Preface

The International Workshop on Digital Forensics and Watermarking (IWDW) is a premier forum for researchers and practitioners working on novel research, development, and application of digital watermarking, data hiding, and forensic techniques for multimedia security. The 22nd International Workshop on Digital Forensics and Watermarking (IWDW 2023) was held in Jinan, China, during November 25–26, 2023 by Qilu University of Technology (Shandong Academy of Sciences) and sponsored by the Institute of Information Engineering, Chinese Academy of Sciences.

IWDW 2023 received 48 submissions. Each of them was assigned to at least two members of the Technical Program Committee for review. The decisions were made on a highly competitive basis. Only 22 submissions were accepted according to the average score ranking. The accepted papers cover important topics in current research on multimedia security, and the presentations were organized into two "Digital Forensics and Security" sessions, and two "Data Hiding" sessions. There were three invited keynotes: "Advancements in Multimedia Intelligent Security Research" by Zhenxing Qian from Fudan University, "Temporal Image Forensics: What do we actually learn in data driven approaches?" by Andreas Uhl from Univ. of Salzburg, Austria, and "Social Media Steganography and Steganalysis from the Perspective of Cognitive Confrontation" by Zhongliang Yang from Beijing University of Posts and Telecommunications.

November 2023

Bin Ma
Xianfeng Zhao
Alessandro Piva
Pedro Comesaña Alfaro

Organization

Honorary Chairs

Yun-Qing Shi   New Jersey Institute of Technology, USA
Hyoung Joong Kim   Korea University, South Korea

Conference Chairs

Meihong Yang   Qilu University of Technology (Shandong Academy of Sciences), China
Xiaoming Wu   Qilu University of Technology (Shandong Academy of Sciences), China
Linna Zhou   Beijing University of Posts and Telecommunications, China

Technical Program Chairs

Bin Ma   Qilu University of Technology (Shandong Academy of Sciences), China
Xianfeng Zhao   Chinese Academy of Sciences, China
Alessandro Piva   University of Florence, Italy
Pedro Comesaña-Alfaro   University of Vigo, Spain

Steering Committee

Alessandro Piva   University of Florence, Italy
Jiwu Huang   Shenzhen University, China
Pedro Comesaña-Alfaro   University of Vigo, Spain
Wojciech Mazurczyk   Warsaw University of Technology, Poland
Xianfeng Zhao   Chinese Academy of Sciences, China

Technical Program Committee

Alessandro Piva   University of Florence, Italy
Xiangui Kang   Sun Yat-sen University, China
Yongfeng Huang   Tsinghua University, China

Isao Echizen	National Institute of Informatics, Japan
Kai Wang	CNRS, Grenoble Alpes University, France
Akira Nishimura	Tokyo University of Information Sciences, Japan
Andreas Westfeld	HTW Dresden, University of Applied Sciences, Germany
Minoru Kuribayashi	Tohoku University, Japan
Andreas Uhl	University of Salzburg, Austria
Dinu Coltuc	Valahia University of Targoviste, Romania
Hongxia Wang	Sichuan University, China
Wojciech Mazurczyk	Warsaw University of Technology, Poland
Koksheik Wong	Monash University, Australia
Xinpeng Zhang	Fudan University, China
Jiwu Huang	Shenzhen University, China
Yongjian Hu	South China University of Technology, China
Qingzhong Liu	Sam Houston State University, USA
Sang-ug Kang	Sangmyung University, South Korea
Christian Kraetzer	Otto-von-Guericke University Magdeburg, Germany
Pascal Schöttle	Management Center Innsbruck, Austria
William Puech	LIRMM - CNRS, France
Anthony T. S. Ho	University of Surrey, England
Tomas Pevny	University of Prague, Czech Republic
Xiaolong Li	Beijing Jiaotong University, China
Chunpeng Wang	Qilu University of Technology, China
Jinwei Wang	Nanjing University of Information Science and Technology, China
Yongjin Xian	Qilu University of Technology, China
Hong Zhang	Institute of Information Engineering, CAS, China
Chuan Qin	University of Shanghai for Science and Technology, China
Peijia Zheng	Sun Yat-sen University, China
Weiqi Luo	Sun Yat-sen University, China

Organization Committee Chairs

Jian Li	Qilu University of Technology (Shandong Academy of Sciences), China
Chunpeng Wang	Qilu University of Technology (Shandong Academy of Sciences), China
Qi Li	Qilu University of Technology (Shandong Academy of Sciences), China

Organization Committee

Xiaoyu Wang   Qilu University of Technology (Shandong
                               Academy of Sciences), China
Yongjin Xian   Qilu University of Technology (Shandong
                               Academy of Sciences), China

Sponsors

Chinese Academy of Sciences

Qilu University of Technology (Shandong Academy of Sciences)

New Jersey Institute of Technology

Springer

Contents

Data Hiding

Digital Forensics and Security

Image Encryption Scheme Based on New 1D Chaotic System and Blockchain

Yongjin Xian[1,2] , Ruihe Ma[3] , Pengyu Liu[1,2] , and Linna Zhou[4(✉)]

[1] Key Laboratory of Computing Power Network and Information Security, Ministry of Education, Shandong Computer Science Center, Qilu University of Technology (Shandong Academy of Sciences), Jinan 250353, China
[2] Shandong Provincial Key Laboratory of Computer Networks, Shandong Fundamental Research Center for Computer Science, Jinan 250353, China
[3] School of Economics, Jilin University, Jilin 130012, China
[4] School of Cyberspace Security, Beijing University of Posts and Telecommunications, Beijing 100876, China
zhoulinna@tsinghua.edu.cn

Abstract. With the increasing need for secure image transmission and storage, researchers have focused on developing advanced encryption algorithms. This article highlights the significance of a newly constructed 1D chaotic system and its application in constructing a secure image encryption algorithm over blockchain. The newly proposed 1D logistic cosine tangent chaotic system consists of the cosine function and tangent function replacing the two variables in the logistic map, respectively, thus improving the dynamics of the new 1D chaotic map. The image encryption scheme constructed based on the newly proposed 1d chaotic system relies on the privilege management function of the blockchain's smart contract to design the key transfer scheme, which simultaneously achieves the security of the cryptographic algorithm and the reliability of the key transfer. The experimental results provide strong evidence that the proposed cryptographic system ensures robust security. This research provides a novel and effective solution for ensuring the confidentiality and integrity of encrypted images, addressing the growing concerns of image security in the digital era.

Keywords: Chaotic System · Image Encryption · Blockchain

1 Introduction

With the improvement of the level of information security requirements, image encryption, as a common method of image information protection, has been extensively studied by scholars in these years [1–5]. The chaos-based image encryption method has become one of the most ordinary design techniques for designing new encryption algorithms [6, 7]. As the image data that needs to be encrypted increases, the security of chaotic image encryption becomes particularly important. How to prevent attackers from obtaining valid information from the ciphertext image or destroying the normal decryption effect has become a question worthy of research [8, 9].

© The Author(s), under exclusive license to Springer Nature Singapore Pte Ltd. 2024
B. Ma et al. (Eds.): IWDW 2023, LNCS 14511, pp. 3–17, 2024.
https://doi.org/10.1007/978-981-97-2585-4_1

Chaotic systems, as the important core of chaotic image encryption algorithms, have had many important research results in recent years. It is mainly reflected in the innovation of one-dimensional chaotic systems [10, 11], two-dimensional chaotic systems [12, 13], high-dimensional chaotic systems [14, 15], and spatiotemporal chaotic systems [16]. One-dimensional chaotic systems may seem to have weak chaotic properties due to the simple algebraic structure, however, they play an important role as the basis for multi-dimensional chaotic research and spatiotemporal chaotic research.

The combination of classical chaotic systems and trigonometric functions innovates the research of one-dimensional chaotic systems. Zhu et al. [17] presented a logistic-tent map by combining a logistic map and a tent map, which combines the good chaotic property of a logistic map and the duality of tent map to make the logistic-tent map better chaotic property. Meng et al. [18] proposed a new one-dimensional cos-cot system with better chaotic properties than the traditional chaotic system, which nests the cosine function and cotangent function to get the new one-dimensional cos-cot system with better chaotic performance than the traditional chaotic system. Hu et al. [19] coupled the sine map and the logistic map and constructed a new one-dimensional chaotic system, in which the sine operation was executed twice to effectively improve the one-dimensional chaotic map. Kafetzis et al. [20] extended the choice of the factorization function for a one-dimensional chaotic map by introducing a hyperbolic tangent function and time lag factor for constructing a one-dimensional chaotic system to improve the nonlinear dynamics performance. Further, Midoun et al. [21] divided two factorization functions of cosine and sine. This new class of chaotic systems was constructed in the form of cosine function over sine function, which makes the designed image encryption algorithm accordingly better security.

Through the above literature, it can be found that the coupling of trigonometric functions as well as hyperbolic trigonometric functions with logistic map and tent map can effectively improve the chaotic properties of one-dimensional chaotic systems, which in turn improves the initial value sensitivity and pseudo-randomness of chaotic systems in their application to image encryption algorithms. The security of the image encryption algorithm designed based on such a chaotic system is trustworthy. Continuing the idea of the above research, this article combines the logistic map, cosine map, and tangent map, to propose a new one-dimensional chaotic system. In this system, the cosine function and tangent function replace the variables of the classical logistic map, respectively, thus enhancing the chaotic dynamics. Further, the designed image encryption algorithm can be verified by enough experiments to have good security.

The remainder of the article is organized as follows. Related methods are given in Sect. 2 to introduce the novel proposed chaotic system logistic cosine tangent map. In Sect. 3, the chaotic encryption method based on a new chaotic system is described and simulated. A series of security analyses are presented in Sect. 4. Section 5 shows the blockchain structure designed for the newly proposed cryptographic algorithm. Finally, this article is concluded in Sect. 6.

2 New Chaotic System

2.1 Logistic Cosine Tangent Map

In this section, we introduce the newly designed chaotic map with the name of logistic cosine tangent map (LCTM).

The logistic map is described as a classical chaotic map as follows

$$y = \mu x(1 - x) \tag{1}$$

Based on the chaotic map, the two terms in Eq. (1) are replaced by the cosine function and tangent function, respectively, to obtain the following equation:

$$y = \mu \cos(x)(1 - \tan(x)) \tag{2}$$

According to the method of constructing a one-dimensional chaotic map in [22], the LCTM can be obtained by applying the optimization Eq. (2) of the remainder operation as follows:

$$y = \mathrm{mod}\,(\mu \cos(x)(1 - \tan(x)), 1) \tag{3}$$

where, mod is the residual function.

The newly proposed chaotic system maintains the properties of the logistic map while the complexity of the map during the iteration process is significantly changed because of the introduction of the cosine function and the tangent function. This change has an enhancing effect on the chaotic nature of the new map. In other words, the newly proposed LCTM can show better chaotic properties.

2.2   The Lyapunov Exponent of LCTM

Lyapunov exponent can be intuitively seen whether a certain system is in chaos or not. The Lyapunov exponent should be greater than 0 for the system when it is in the chaotic phase. The Lyapunov exponent value is calculated as

$$\lambda = \lim_{n\to\infty} \frac{1}{n} \sum_{i=0}^{n-1} \ln|f'(x_i)| \tag{4}$$

then the Lyapunov exponent of the new proposed LCTM is shown in Fig. 1.

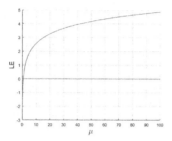

Fig. 1. Lyapunov exponent.

2.3   The Bifurcation Diagrams of LCTM

The bifurcation phenomenon is a qualitative change in the qualitative behavior of a dynamical system as the parameters of the system change. The bifurcation diagram can visualize the effect of the taking of values of the parameters of a mapping on the chaotic behavior of that mapping.

As shown in Fig. 2, the newly proposed chaotic system has a good bifurcation map. It can be found that the parameter μ in the newly proposed LCTM is in a chaotic state. Such a large range of parameter values guarantees the diversity of parameter selection when applying the LCTM. It also shows that the newly designed chaotic system has good chaotic properties. By analyzing Fig. 2(b) and (c), it can be inferred that the LCTM is in a chaotic state when $\mu > 1.53353$.

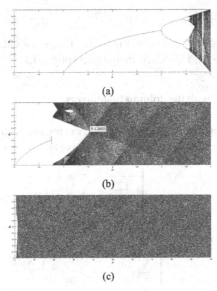

(a)

(b)

(c)

Fig. 2. Bifurcation diagrams: (a) logistic map at μ ∈ (0, 4); (b) LCTM at μ ∈ (0, 4); (c) LCTM at μ ∈ (0, 100).

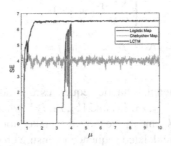

Fig. 3. Shannon entropy.

2.4 The Shannon Entropy of LCTM

As a tool to measure randomness, information entropy can test the randomness of chaotic sequences generated by chaotic systems. The calculation formula of information entropy is shown as Eq. (5) in [23]. As shown in Fig. 3, the Shannon entropy of LCTM is the best as compared to other maps. That is, LCTM shows better randomness. Further, the chaotic property of LCTM is better.

3 Encryption Algorithm

In this section, we introduce the novel chaotic image encryption and decryption algorithms based on the new proposed LCTM.

3.1 Encryption Algorithm

Based on the good chaotic dynamic properties of LCTM, a new chaotic image encryption method is proposed. The LCTM-based image encryption (LCTMIE) is being specified step by step as follows:

Step 1. Input the image P with the column first traversal method, and calculate the width denoted as M and the height denoted as N.

Step 2. Generate a 128-length hexadecimal string Λ using the hash function of SHA-512 from P, and convert Λ into $K = [k_1, k_2, k_3, k_4, k_5, k_6]$ as follows:

$$k_\tau = \begin{cases} \frac{a_\tau}{a_1 + a_2}, & \tau = 1, 2 \\ \frac{a_1 + a_2}{a_\tau}, & \tau = 3, 4 \\ a_1 - a_3, & \tau = 5 \\ a_2 - a_4, & \tau = 6 \end{cases} \tag{5}$$

$$a_i = \begin{cases} \sum_{j=1}^{32 \times i} \Theta_j, & i = 1, 2 \\ \sum_{j=1}^{8} \Theta_{9 \times (j-1)+1}, & i = 3 \\ \sum_{j=1}^{13} \Theta_{5 \times (j-1)+1}, & i = 4 \end{cases} \tag{6}$$

$$\Theta_j = (\Lambda_j \oplus \Lambda_{128-j})_{(10)}, j = 1, 2, \cdots, 64 \tag{7}$$

where, $\oplus$ is the xor symbol.

Step 3. Obtain the $M \times N$ length chaotic sequences C_1 and C_2 by the new proposed LCTM with two key groups $[k_1, k_3, k_5]$ and $[k_2, k_4, k_6]$, where k_1 and k_2 are the initial states of the chaotic system, k_3 and k_4 are the values taken for the μ of the system, while k_5 and k_6 are the length of the deleted sequence to ensure chaotic states of C_1 and C_2.

Step 4. Get the index vector S of the position of the sorted backup element in the original vector from C_1. Rearrange the pixels of the image P by S, and denote the rearranged image as P'.

Step 5. Diffuse P' with second chaotic sequence C_2, and obtain the diffused image P'' as follow:

$$T_i = \mod\left(\lfloor C_{2,i} \times M \times N \rfloor, 256\right) \tag{8}$$

$$P''_i = \begin{cases} P'_i \oplus T_i, & i = 1 \\ P'_i \oplus T_i \oplus P'_{i-1}, & i = 2, 3, \cdots, M \times N \end{cases} \tag{9}$$

where, mod is the remainder function.

Step 6. Output the encrypted image P_E with the column first traversal method of P''.

With the six steps above, the ciphertext image was obtained, while the key sequence consists of $K = [k_1, k_2, k_3, k_4, k_5, k_6]$, which are used to encrypt and decrypt the image.

3.2 Decryption Algorithm

As the proposed encryption algorithm is symmetric and reversible, the decryption process is the reverse of the encryption one. Based on the above proposed LCTM based image encryption algorithm, the LCTM image decryption (LCTMID) can be achieved in the following steps when the key is known.

Step 1. Input the image P_E to be decrypted with the column first traversal method, and calculate the width denoted as M and the height denoted as N.

Step 2. Enter the key group $K = [k_1, k_2, k_3, k_4, k_5, k_6]$.

Step 3. Obtain the $M \times N$ length chaotic sequences C_1 and C_2 by the new proposed LCTM with the key.

Step 4. Inv-diffuse P_E with second chaotic sequence C_2, and obtain the inv-diffused image P'_E as follow:

$$P'_E = \begin{cases} P'_{E,i} \oplus T_i, & i = 1 \\ P_{E,i} \oplus T_i \oplus P'_{E,i-1}, & i = 2, 3, \cdots, M \times N \end{cases} \tag{10}$$

Step 5. Get the index vector S of the position of the sorted backup element in the original vector from C_1. Inv-permute the pixels of the image P'_E by S, and denote the inv-permuted image as P''_E.

Step 6. Output the decrypted image P_D with the column first traversal method of P''_E.

With the six steps above, the decrypted image was obtained with the same key sequence of encryption process which consists of $K = [k_1, k_2, k_3, k_4, k_5, k_6]$.

Comparing the LCTMIE process and LCTMID process it can be found that the proposed cryptosystem designed in this article is symmetric.

3.3 Simulation Results

An example of the proposed algorithm is given below with the data of 4×4 in Fig. 4, hence it would help the reader to go for a simpler understanding of the encryption and decryption process.

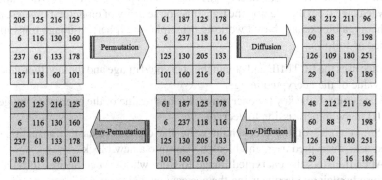

Fig. 4. Example of encryption and decryption process with the data of 4×4.

Fig. 5. Image encryption and decryption results of 3 images: (a) boat.512, (b) Encrypted boat.512, (c) Decrypted boat.512, (d) elaine.512, (e) Encrypted elaine.512, (f) Decrypted elaine.512, (g) gray.512, (h) Encrypted gray.512, (i) Decrypted gray.512.

A qualified encryption algorithm should be able to convert various types and sizes of images into disordered ciphertext images. This section will show the encryption and decryption results of LCTM with different kinds of images. In Fig. 5, the encryption results and decryption results for representative images in the dataset are presented in detail, which demonstrate the effectiveness of the proposed method.

3.4 Blockchain Design for Cryptosystem

The security of smart contracts of blockchain can be used for permission management of image encryption. By storing the encryption key of an image in a smart contract, only users with decryption rights can decrypt the image. This protects image privacy and prevents key leakage from threatening the transmission security of ciphertext images. Based on this theory, a blockchain-based working principle for ciphertext image transmission is designed as follows.

Step 1. Alice uses the LCTMIE and obtain the encrypted image and the key, then calculate the hash value of the encrypted image.
Step 2. Alice uploads the key for encryption, the hash value of the encrypted image, and the user ID the with permission to decrypt.
Step 3. Smart nodes of the blockchain write the ciphertext hash, transaction time, encrypted user, decrypted user, and block header into a new block.
Step 4. Alice transfers the encrypted image to Bob, whom Alice desires to be able to obtain the right plaintext image using the correct key.

Step 5. Bob receives the ciphertext image and then uploads its ID to the smart node for authentication.

Step 6. The smart node of the blockchain sends the key to Bob.

Step 7. Bob decrypts the ciphertext image using LCTMID with the key from the smart node to get the plaintext image.

Remarks: In this key transfer scheme based on the blockchain smart nodes, it needs to be assumed that the ID of each user in the blockchain is unique and cannot be copied.

Based on the above steps, secure blockchain-based ciphertext image transfer can be integrated. The working schematic of the process is shown in Fig. 6.

With such a blockchain workflow design, the leakage of key transmission during ciphertext image transmission is eliminated. The security of the key is guaranteed, and the secure transmission between the sender and receiver of the ciphertext message is maintained. However, the application prospects of the newly proposed LCTMIE and LCTMID are enhanced.

4 Security Analysis

To verify the security and reliability of the algorithm, we have done a lot of experiments on general image sets and representative experimental images. Typical images of the USC-SIPI 'Miscellaneous' dataset were used for experimental testing and analysis as follows.

4.1 Robustness Analysis of Cropping

In the process of information transmission, attackers often use methods such as cropping attacks destroying the integrity of the ciphertext information to prevent the receiver of the ciphertext from successfully decrypting to obtain the correct decryption information. The decryption results of different encrypted images are demonstrated in Fig. 6. So that, the LCTMIE can effectively resist cropping attacks with good robustness.

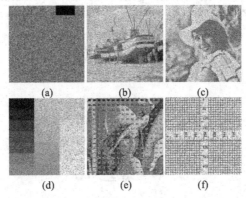

Fig. 6. Decryption results after cropping: (a) 3.125% data cropping, (b) boat.512, (c) elaine.512, (d) gray.512, (e) numbers.512, (f) ruler.512.

4.2 Histogram Analysis

The distributions of pixels in an image can be visually described by histograms, which show the frequency of each grayscale. A qualified encryption algorithm needs to make the histograms of the ciphertext images appear horizontal. The histograms of plaintext images and ciphertext images illustrated in Fig. 7, indicate that the histograms of the encrypted images of LCTMIE are distributed uniformly. Therefore, the proposed algorithm can resist statistical attacks.

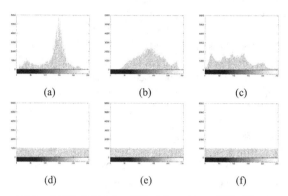

Fig. 7. Histograms: (a) boat.512, (b) elaine.512, (c) numbers.512, (d) encrypted boat.512, (e) encrypted elaine.512, (f) encrypted numbers.512.

4.3 Correlation Analysis

The correlation between adjacent pixels of the ciphertext image limit to 0, makes it very difficult for attackers to obtain any effective information through the association between the pixels of the encrypted image. When the correlation of the ciphertext image is close to 0 to a certain extent, the algorithm has a high enough security to resist statistical analysis attacks. In this article, adjacent pixel correlation coefficients of three adjacent pixel directions (horizontal, vertical, and diagonal) are calculated as follows [6]:

$$r_{xy} = \frac{\mathrm{cov}(x, y)}{\sqrt{D(x)}\sqrt{D(y)}} \tag{11}$$

where,

$$\mathrm{cov}(x, y) = \frac{1}{N} \sum_{i=1}^{N} (x_i - E(x))(y_i - E(y)) \tag{12}$$

$$E(x) = \frac{1}{N} \sum_{i=1}^{N} x_i \tag{13}$$

$$D(x) = \frac{1}{N} \sum_{i+1}^{N} (x_i - E(x))^2 \tag{14}$$

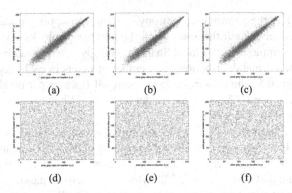

Fig. 8. Adjacent pixel correlation of original and encrypted elaine.512: (a) Horizontal of the original image, (b) Vertical of the original image, (c) Diagonal of the original image, (d) Horizontal of encryption, (e) Vertical of encryption, (f) Diagonal of encryption.

The correlation distributions of elaine.512 before and after encryption are intuitively shown in Fig. 8, which indicates the plaintext images have a strong correlation regardless of the three directions, but the correlations are significantly reduced after encrypted by LCTMIE, and the detailed comparison is shown in Table 1. As the structure and characteristics of the plaintext image are disturbed, anti-statistical attacks cannot achieve the desired result.

Table 1. Correlation coefficients of encrypted Lena

Algorithm	Correlation coefficients		
	Horizontal	Vertical	Diagonal
Mean of LCTMIE	**−0.000688**	**0.000091**	**0.000398**
Ref. [13]	−0.0023	0.0019	−0.0029
Ref. [25]	−0.0008	−0.0013	0.0018
Ref. [26]	−0.0013	−0.0023	0.0025

4.4 Local Shannon Entropy

Local Shannon entropy is a significant and better indicator to reflect the randomness of information. In this section, this indicator is compared with quantitative data to analyze the differences between different algorithms.

Local Shannon entropy can be calculated as [24]:

$$H_{k,T_B}(S, L) = \sum_{i=1}^{k} \frac{H(S_{T_B}, L)}{k} \tag{15}$$

where, $H(S_{T_B}, L)$ is the information entropy of non-overlapping image blocks S_{T_B}. According to the central limit theorem, when the number of blocks k is chosen as 30, which makes the sample mean of local Shannon entropy approximately the minimum requirement of the normal distribution; the number of pixels in each block T_B is set as 1936 so that the local Shannon entropy can capture all L scales per pixel. And the significance α is selected as 0.05. The critical interval of local Shannon entropy conformed to standard is [7.9019014, 7.9030373].

Based on these settings, we can get the quantitative results shown in Table 2. Compared with other algorithms, the Local Shannon entropy of LCTMIE has an average value of 7.902561 which meets the critical interval, a standard deviation of 0.0004 which is the minimum of all, and a pass rate of 27/28 which is close to all passing. In general, the LCTMIE can almost pass Local Shannon entropy tests for all images in the USC-SIPI 'Miscellaneous' dataset, which signifies that LCTMIE has very high security.

Table 2. Comparison of local Shannon entropy

Item	Ref. [6]	Ref. [13]	Ref. [26]	LCTMIE
Mean	7.903029	7.902462	7.902611	7.902561
Std	0.000416	0.000921	0.001169	**0.000400**
Pass Rate	16/28	17/28	20/28	27/28

Table 3. NPCR and UACI values of ciphertext images

Item	NPCR (%)			UACI (%)		
	Ref. [13]	Ref. [28]	LCTMIE	Ref. [13]	Ref. [28]	LCTMIE
Mean	99.6094	99.5966	**99.6091**	33.4476	33.4322	**33.4644**
Std	0.0131	0.0190	**0.0014**	0.0331	0.0312	**0.0210**
Pass Rate	26/28	23/28	**28/28**	28/28	27/28	**28/28**

4.5 Resistance Differential Attack Analysis

The resistance of encryption algorithms against differential attacks can be tested by the number of pixels changing rate (NPCR) and unified averaged changed intensity (UACI) which are acquired as follows [27, 28]:

$$\eta(i, j) = \begin{cases} 0, & c_1(i,j) = c_2(i,j) \\ 1, & c_1(i,j) \neq c_2(i,j) \end{cases} \tag{16}$$

$$NPCR = \frac{\sum_{i,j} \eta(i,j)}{W \times H} \times 100 \tag{17}$$

$$UACI = \frac{1}{M \times N} \left[\sum_{i,j} \frac{|c_1(i,j) - c_2(i,j)|}{255} \right] \times 100 \qquad (18)$$

where, $c_1(i,j)$ and $c_2(i,j)$ are the pixel values of the same position (i,j) in the two different encrypted images, respectively. M and N represent the sizes of the image.

Table 3 compares NPCR and UACI of encrypted images in the dataset. Compared with other algorithms, NPCR of LCTMIE achieves an average value of 99.6091% which is closest to the theoretical value, a standard deviation of 0.0014 which is the minimum of all, and a pass rate of 28/28 which means all values are bigger than the critical value. And UACI of LCTMIE has an average value of 33.4644% which is closest to the theoretical value, a standard deviation of 0.0210 which is the minimum of all, and a pass rate of 28/28 which means all values meet the critical interval. In general, the proposed method can pass the NPCR and UACI tests for all images, implying that LCTMIE has better resistance to differential attacks.

5 Conclusion

This article proposed a new one-dimensional chaotic system named LCTM, which replaces the variable terms in the logistic map of attractions by cosine function and tangent function separately. The new proposed LCTM had shown good Lyapunov exponent performance and bifurcation phenomenon effect as well as pseudo-randomness. The image cryptosystem designed by applying the good chaotic property of LCTM and the key sending scheme of blockchain smart contract effectively improves the security of the cryptographic algorithm and the reliability of key transmission, additionally emphasizing the importance of utilizing blockchain-based key management methods to enhance the security of chaotic image encryption. The security experiments visually analyze the good performance of the cryptosystem in resisting various attacks. Overall, the image encryption and decryption algorithms based on LCTM are secure and reliable, and the blockchain technology can effectively guarantee the security of the cryptosystem and key transmission.

Acknowledgement. This work was supported in part by the National Natural Science Foundation of China (62302248, 62302249, 62272255); National key research and development program of China (2021YFC3340600, 2021YFC3340602); Taishan Scholar Program (tsqn202306251); Shandong Provincial Natural Science Foundation (ZR2020MF054, ZR2023QF018, ZR2023QF032, ZR2022LZH011), Ability Improvement Project of Science and Technology SMES (2022TSGC2485, 2023TSGC0217); "20 Universities"-Project of Jinan Research Leader Studio (2020GXRC056); "New 20 Universities"-Project of Jinan Introducing Innovation Team (202228016); Youth Innovation Team of Colleges and Universities in Shandong Province (2022KJ124); The "Chunhui Plan" Cooperative Scientific Research Project of Ministry of Education (HZKY20220482); Achievement transformation of science, education and production integration pilot project (2023CGZH-05), QLUT First Talent Research Project (2023RCKY131, 2023RCKY143), QLUT Integration Pilot Project of Science Education Industry (2023PX006, 2023PY060, 2023PX071).

References

1. Zhang, Y., Xiao, X., Yang, L.-X., Xiang, Y., Zhong, S.: Secure and efficient outsourcing of PCA-based face recognition. IEEE Trans. Inform. Forensic Secur. **15**, 1683–1695 (2020)
2. Yu, K., et al.: A blockchain-based shamir's threshold cryptography scheme for data protection in industrial Internet of Things settings. IEEE Internet Things J. **9**, 8154–8167 (2022)
3. He, Y., Zhang, C., Wu, B., Yang, Y., Xiao, K., Li, H.: A Cross-chain trusted reputation scheme for a shared charging platform based on blockchain. IEEE Internet Things J. **9**, 7989–8000 (2022)
4. Chuman, T., Sirichotedumrong, W., Kiya, H.: Encryption-then-compression systems using grayscale-based image encryption for JPEG images. IEEE Trans. Inform. Forensic Secur. **14**, 1515–1525 (2019)
5. Feixiang, Z., Mingzhe, L., Kun, W., Hong, Z.: Color image encryption via Hénon-zigzag map and chaotic restricted Boltzmann machine over blockchain. Opt. Laser Technol. **135**, 106610 (2021)
6. Chan, J.C.L., Lee, T.H., Tan, C.P.: Secure communication through a chaotic system and a sliding-mode observer. IEEE Trans. Syst. Man Cybern. Syst. **52**, 1869–1881 (2022)
7. Gao, X., Mou, J., Xiong, L., Sha, Y., Yan, H., Cao, Y.: A fast and efficient multiple images encryption based on single-channel encryption and chaotic system. Nonlinear Dyn. **108**, 613–636 (2022)
8. Xian, Y., Wang, X., Wang, X., Li, Q., Yan, X.: Spiral-transform-based fractal sorting matrix for chaotic image encryption. IEEE Trans. Circ. Syst. I(69), 3320–3327 (2022)
9. Xian, Y., Wang, X., Teng, L.: Double parameters fractal sorting matrix and its application in image encryption. IEEE Trans. Circ. Syst. Video Technol. **32**, 4028–4037 (2022)
10. Panwar, K., Purwar, R.K., Srivastava, G.: A fast encryption scheme suitable for video surveillance applications using SHA-256 hash function and 1D sine-sine chaotic map. Int. J. Image Graph. **21**, 2150022 (2021)
11. Daoui, A., Yamni, M., Chelloug, S.A., Wani, M.A., El-Latif, A.A.A.: Efficient image encryption scheme using novel 1D multiparametric dynamical tent map and parallel computing. Mathematics **11**, 1589 (2023)
12. Hua, Z., Chen, Y., Bao, H., Zhou, Y.: Two-dimensional parametric polynomial chaotic system. IEEE Trans. Syst. Man Cybern. Syst. **52**, 4402–4414 (2022)
13. Hua, Z., Zhou, Y.: Image encryption using 2D Logistic-adjusted-Sine map. Inf. Sci. **339**, 237–253 (2016)
14. Afendee Mohamed, M., et al.: A speech cryptosystem using the new chaotic system with a capsule-shaped equilibrium curve. Comput. Mater. Continua **75**, 5987–6006 (2023)
15. Jing-yu, S., Hong, C., Gang, W., Zi-bo, G., Zhang, H.: FPGA image encryption-steganography using a novel chaotic system with line equilibria. Digit. Sig. Process. **134**, 103889 (2023)
16. Xian, Y., Wang, X., Teng, L., Yan, X., Li, Q., Wang, X.: Cryptographic system based on double parameters fractal sorting vector and new spatiotemporal chaotic system. Inf. Sci. **596**, 304–320 (2022)
17. Zhu, S., Deng, X., Zhang, W., Zhu, C.: A new one-dimensional compound chaotic system and its application in high-speed image encryption. Appl. Sci. **11**, 11206 (2021)
18. Meng, X., Li, J., Di, X., Sheng, Y., Jiang, D.: An encryption algorithm for region of interest in medical DICOM based on one-dimensional e^{λ}-cos-cot map. Entropy **24**, 901 (2022)
19. Hu, Y., Wang, X., Zhang, L.: 1D Sine-Map-Coupling-Logistic-Map for 3D model encryption. Front. Phys. **10**, 1006324 (2022)
20. Kafetzis, I., Moysis, L., Tutueva, A., Butusov, D., Nistazakis, H., Volos, C.: A 1D coupled hyperbolic tangent chaotic map with delay and its application to password generation. Multimedia Tools Appl. **82**, 9303–9322 (2023)

21. Midoun, M.A., Wang, X., Talhaoui, M.Z.: A sensitive dynamic mutual encryption system based on a new 1D chaotic map. Opt. Lasers Eng. **139**, 106485 (2021)
22. Wang, X., Du, X.: Pixel-level and bit-level image encryption method based on Logistic-Chebyshev dynamic coupled map lattices. Chaos Solitons Fractals **155**, 111629 (2022)
23. Wang, X., Guan, N., Liu, P.: A selective image encryption algorithm based on a chaotic model using modular sine arithmetic. Optik **258**, 168955 (2022)
24. Wu, Y., Zhou, Y., Saveriades, G., Agaian, S., Noonan, J.P., Natarajan, P.: Local Shannon entropy measure with statistical tests for image randomness. Inf. Sci. **222**, 323–342 (2013)
25. Cao, W., Mao, Y., Zhou, Y.: Designing a 2D infinite collapse map for image encryption. Sig. Process. **171**, 107457 (2020)
26. Hua, Z., Jin, F., Xu, B., Huang, H.: 2D Logistic-Sine-coupling map for image encryption. Sig. Process. **149**, 148–161 (2018)
27. Zhang, Y.: The unified image encryption algorithm based on chaos and cubic S-Box. Inf. Sci. **450**, 361–377 (2018)
28. Himeur, Y., Boukabou, A.: A robust and secure key-frames based video watermarking system using chaotic encryption. Multimedia Tools Appl. **77**, 8603–8627 (2018)

High-Quality PRNU Anonymous Algorithm for JPEG Images

Jian Li[1,2] (ID), Huanhuan Zhao[1,2] (ID), Bin Ma[1,2(✉)] (ID), Chunpeng Wang[1,2] (ID),
Xiaoming Wu[1,2] (ID), Tao Zuo[1,2] (ID), and Zhengzhong Zhao[1,2] (ID)

[1] Key Laboratory of Computing Power Network and Information Security, Ministry of
Education, Shandong Computer Science Center (National Supercomputer Center in Jinan), Qilu
University of Technology (Shandong Academy of Sciences), Jinan, China
ljian_20@163.com, sddxmb@126.com
[2] Shandong Provincial Key Laboratory of Computer Networks, Shandong Fundamental
Research Center for Computer Science, Jinan, China

Abstract. The utilization of Photo Response Non-Uniformity (PRNU) technology has found extensive application in the field of multimedia forensics, particularly in the authentication of the original camera source of an image. However, this technique has also given rise to significant concerns regarding privacy breaches. For instance, adversaries can exploit publicly available images to generate PRNU and subsequently impersonate the owners of the images. In response to these challenges, we propose an algorithm for achieving source device anonymity in widely used JPEG images. The method combines the discrete cosine transform (DCT) with JPEG compression to process the DCT coefficients of an image after inverse quantization. By ensuring the high quality of the processed image, this approach effectively breaks the link between an image and its source camera. Additionally, a reversible data hiding method is employed, enabling the recovery of traceability if necessary. Our algorithm offers several advantages over existing schemes. It operates within the domain of JPEG image compression, maintaining a low time complexity. Additionally, it effectively preserves the visual quality of images and eliminates the typical traceability effects associated with images.

Keywords: JPEG · Image · PRNU · Discrete Cosine Transform · Anonymous

1 Introduction

Source camera identification technology based on PRNU is one vital focus of the research on multimedia forensics [1–7]. However, its availability has raised concerns about personal privacy security [8]. Attackers are able to perform illegal identity tracking of shared images across different social networks by analyzing and combining information obtained from one or more images. To protect photographers' privacy rights and ensure anonymity, it is essential to enable secure and anonymous image communication [9, 10]. Lukas et al. [1] investigated the robustness of PRNU fingerprints to gamma correction and JPEG compression (with various quality factors) and showed that PRNU fingerprints can still be identified from compressed images with a JPEG quality factor of 50.

B. Ma et al. (Eds.): IWDW 2023, LNCS 14511, pp. 18–32, 2024.
https://doi.org/10.1007/978-981-97-2585-4_2

Subsequently, Rosenfeld et al. [11] studied the use of multiple denoising steps at the expense of image quality, but it was still insufficient to cut off the connection between the image and the source device [12, 13].

To solve the above problems, an anonymization method that can remove or weaken PRNU from images has been proposed in the literature. Lameri et al. [14] proposed a patching-based method that successfully achieved anonymization of the source device in the final image by deleting and reconstructing image pixels. Dirik et al. [15] proposed an adaptive denoising method to remove PRNU fingerprints. Bonettini et al. [16] utilized Convolutional Neural Networks (CNN) to impede the traceability of the image under examination through the source device PRNU. Memon et al. [17] investigated the performance bounds of source camera anonymity in seam engraving and proposed a content-aware sizing method. Zeng et al. [18] introduced a method for fingerprint attack removal and division, which estimates the optimal subtraction weights directly in the null field. Kirchner et al. [19] utilized image self-similarity and a block-matching technique to disturb the pixel positions of the image. Villalba et al. [20] proposed two anti-forensics algorithms based on sensor and wavelet transform to undermine the accuracy of image source camera recognition technology. Sudipta et al. [21] conducted the Discrete Cosine Transform (DCT) on a given image by zeroing its high-frequency coefficients and preserving the low frequencies, but it resulted in reduced image quality. Francesco et al. [22] integrated the source device anonymization algorithm into a Deep Image Prior (DIP) framework. However, this method is characterized by high computational time, cost, and hardware requirements.

We present a novel approach to address prior shortcomings in PRNU anonymization. The method involves using JPEG compression and inverse quantization in DCT coefficients to attenuate PRNU strength for enhanced privacy protection. Additionally, reversible data hiding technology is employed to preserve image traceability, making it suitable for legal purposes. Importantly, unlike conventional deep learning techniques, the proposed algorithm streamlines the process by solely relying on the PRNU data of the test image and knowledge of the source device, eliminating the need for extensive neural network training. This algorithm validates the results in the classical natural image dataset Dresden [23] using 2731 color images from 10 devices and 2559 images from the MICHE-I dataset [24]. The results show that this algorithm can successfully hide the test image and maintain the high quality of the image.

This paper mainly makes the following two contributions:

1. We propose a novel DCT-based PRNU anonymity technique specifically designed for JPEG compression, which effectively anonymizes natural images. Our experimental results using the Dresden dataset demonstrate that the proposed technique successfully achieves anonymity.
2. We introduce a technology that combines image anonymization with reversible data hiding, enabling the embedding of traceability information. Our technique allows for the recovery of most PRNU information while preserving image biometrics. To validate our approach, we conducted experiments on the MICHE-I dataset.

The rest of this paper is organized as follows: Sect. 2 introduces the knowledge related to the reference PRNU, JPEG compression, and the reversible data hiding technology used in the extraction source camera. Section 3 describes the new method proposed in this paper. Section 4 describes the dataset used in the algorithm, the algorithm process, and the final algorithm results. We conclude the whole paper in Sect. 5.

2   Related Works

To make our paper more complete, this section introduces the source device identification method, the JPEG compression process involved in the proposed scheme, and the reversible data hiding technology based on code division multiplexing, to help readers understand the other parts of this paper.

2.1   PRNU-Based Source Camera Identification

Image capturing devices use a light-sensitive imaging chip (made of CCD or CMOS), which consists of several light-sensitive crystals, each of which corresponds to a pixel in the image. However, due to the limitations of the manufacturing process, the light-sensing ability of the crystals can vary slightly, thus leaving a pattern of noise in the image, known as Photo Response Non-Uniformity (PRNU) [25]. The classical PRNU model is obtained by analyzing the composition of the media signal:

$$I = I_0 + I_0 K + \theta \tag{1}$$

I_0 represents an ideal sensor output image without any sensor noise or optical distortion, K is the multiplicative PRNU factor (camera fingerprint), and θ is a combination of dark current, shot noise, readout noise, quantization noise, etc.

Given the noise residuals of the image to be measured and an estimated camera fingerprint, the set Peak to Correlation Energy (PCE) is used to determine whether the image is attributed to the camera. The formula is as follows:

$$\text{PCE}\left(Y, \widehat{F}\right) = \frac{\max(\rho)^2}{\frac{1}{L-\varepsilon} \sum_{l=1,\ldots,L \notin \varepsilon} \rho[l]^2} > \gamma \tag{2}$$

where γ represents the threshold, greater than the threshold value indicates that it can be successfully matched with the camera fingerprint $\widehat{F}$. ε is represents the peak neighborhood. L represents the pixel and ρ represents the cross-correlation value [26–28].

2.2   JPEG Compression

To reduce the burden of image storage and transmission, JPEG compression is extensively utilized in digital cameras, image editing software, and social networking platforms. The primary steps involved in JPEG compression include DCT transformation, quantization, and coding, as presented below:

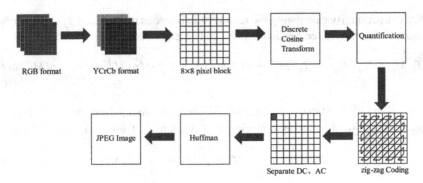

Fig. 1. JPEG compression flow chart

We primary focus revolves around the meticulous design of the discrete cosine transform (DCT) and the quantization process. DCT, a commonly employed image processing technique, transforms a discrete signal into a set of discrete cosine coefficients, facilitating representation in the frequency domain. In the context of JPEG compression, quantization, an irreversible lossy procedure, discards non-essential information for visual perception. To initiate quantization, the creation of a quantization table is imperative, followed by division of the DCT coefficients by this table, resulting in the derivation of "quantized DCT coefficients".

2.3 Reversible Data Hiding

We need to embed a large amount of information, and ordinary embedding algorithms may not be able to fulfill the requirements. Reversible Data Hiding based on Code Division Multiplexing (RDH-CDM) provides a viable solution to this problem [29]. The method can be divided into four steps as follows:

1. Data embedding. Assuming that $\Gamma_0 = [\gamma_1, \gamma_2, \cdots, \gamma_n](\gamma_z, \in \{1, 0\}, z \in \{1, 2, \cdots, n\})$ is the data to be embedded, it is modified to the corresponding bit by the following formula:

$$\beta_e = \begin{cases} 1 & if \ \gamma_z = 1 \\ -1 & if \ \gamma_z = 0 \end{cases} \tag{3}$$

The modified data is $\Gamma_c = [\beta_1, \beta_2, \cdots, \beta_n]$, where $\beta_e \in \{-1, 1\}, e \in \{1, 2, \cdots, n\}$, Select k orthogonal sequences $M_i = \{q_1, q_2 \cdots, q_i\}(i \in \{1, 2, \cdots, k\})$ from the walsh hadamard matrix, The length of each sequence l is even, and the number of "1" and "−1" is the same, which guarantees the zero mean and orthogonality of the spread spectrum sequence. Select adjacent pixels $T_j = [p_1, p_2, \cdots, p_l]$ from the original image I to form a vector, which has the same length as M_i. The bit embedding formula is:

$$\widehat{T}_j = T_j + \alpha[\beta_1 M_1 + \beta_2 + \cdots + \beta_k M_k] \tag{4}$$

where, k represents the number of bits to be embedded, and a represents the embedding strength.

2. Data extraction. By calculating the relationship between mark vector T_j and extended sequence M_i, the j-th embedded data bit can be extracted:

$$\langle \widehat{T_j}, M_i \rangle = \widehat{T_j} \cdot M_i^T = T_j \cdot M_i^T + \alpha \left[\beta_1 M_1 \cdot M_i^T + \beta_2 M_2 \cdot M_i^T + \cdots + \beta_k M_k \cdot M_i^T \right] \tag{5}$$

3. Image restoration. After extracting the embedded information from the marked image, restore the original image by replacing all $\widehat{T_j}$ back to T_j, the formula is as follows:

$$T_j = \widehat{T_j} - \alpha [\beta_1 M_1 + \beta_2 M_2 + \cdots + \beta_k M_k] \tag{6}$$

3   The Proposed Method

The object processed by this algorithm is the JPEG-format image. To illustrate the use of this work, this section introduces the detailed steps of the algorithm and the feasibility analysis of the technology involved.

3.1   Preliminary

JPEG compression's quantization step sets some high-frequency data to 0. This results in the discarding of a portion of high-frequency information. However, the fingerprint information of the image remains widely present. Therefore, to weaken the image traceability, this algorithm starts with the DCT coefficient after inverse quantization and further attenuates the noise residual information.

The AC component carries high-frequency information, including image texture and detail. Processing AC components enable denoising, compression, and other signal operations. Therefore, this article uses zigzag to deal only with the AC part.

Whether the image is connected to the source camera depends on the correlation between the image noise residuals and the reference device PRNU. To achieve this, we extract noise residuals from the image and applies a discrete cosine transform. We refer to the parameters in [22] to control the number range of AC components. In order to continuously weaken the traceability of the image, the DCT coefficient φ of the inverse-quantized JPEG image is subtracted from the transformed noise residuals ξ: $\varphi' = \varphi - \xi$. Where, $\varphi = DCT(f)$, $\xi = DCT(\kappa)$, κ indicates the image noise residual AC part. The difference results are quantized according to the quantization matrix, subsequent JPEG compression operations are performed, and finally, the initial anonymous image is saved.

Scheme involves selecting two sets of images from each camera. Each set must have more than 50 images. Reference PRNU is calculated for both sets of images. One reference PRNU is used to control the anonymization process end time (*ref*). The other reference PRNU is used to test anonymization success (*ref* 1). When initially calculating the PCE value for the image and the reference PRNU, we discovered that doubling the residual weakened the traceability, but it wasn't completely eliminated. To address this, the article introduced a parameter with twice the strength, resulting in a decrease in PCE value, albeit not by much. However, when introducing ten times the intensity, it significantly reduced the image quality, as follows:

Table 1. Comparison of results of strength coefficient

Strength Coefficient	1	2	10
PSNR	54.3456	46.1560	32.1774
PCE	442.45	189.78	−106.73

In order to solve the strength coefficient problem, reference PRNU was used to control the experimental progress. The optimal intensity coefficient (τ) is searched automatically to achieve the expected results of the experiment. In this case, the two parts are subtracted: $\varphi' = \varphi - \tau\xi$. Through this operation, the algorithm can improve the efficiency and accuracy of the algorithm, while protecting the privacy of the photographer. The anonymous flow chart is as follows:

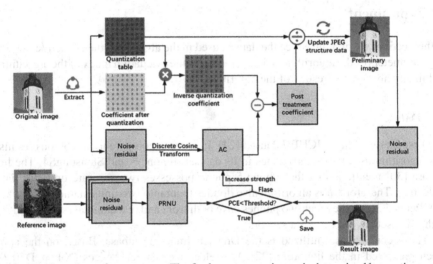

Fig. 2. The flow chart of anonymous. The final anonymous image is determined by continuously decaying the PRNU intensity.

3.2 Processing

We round the noise residual information with the addition of the intensity factor to satisfy the reversibility of the method. The method incorporates noise residual information and intensity coefficients, rounds them, and utilizes code division multiplexing for reversible data hiding, allowing for large capacity storage and multiple embeddings.

To make data suitable for embedding, we employ Huffman encoding, which reduces data storage needs and enhances data transfer efficiency, and modifies specific bits of the encoded information. It then combines them with a chosen orthogonal sequence from the walsh-hadamard matrix and embeds them into the anonymous image. This concealed information relates to traceability but cannot be extracted from the image's noise, ensuring the privacy of the photographer's source device.

All the embedded bit information can be extracted from the image using the marker vector T_j. Next, the original image before the embedded information is recovered by replacing the marker vector T_j using Eq. (6). To enhance the recovery of image traceability, the scheme maintains consistency with the anonymous process when operating on the DCT coefficients. The difference is that this scheme uses the opposite operation, namely summing instead of differentiating: $\varphi' = \varphi + \tau\xi$. After quantization, the final image is saved.

It should be noted that due to the loss of Information in JPEG compression itself, the recovered image cannot guarantee an exact match with the original image. However, the results demonstrate that the algorithm can successfully recover the majority of the PRNU information.

4 Experiment

In this section, we first introduce the dataset used in the algorithm, then describe the setting parameters of the algorithm in detail, then give the evaluation index of the algorithm, and finally analyze the results of the algorithm.

4.1 Dataset

The first dataset is the MICHE-I dataset [24]. In order to compare more obvious results, this algorithm uses two smartphones in its dataset: iPhone 5 and SamsungS4. The literature [30] mentions that the iPhone 5 phone possesses two separate units for data collection. Therefore, this algorithm also divides them into two units, named iPhone5 1 and iPhone5 2, and the other two phones have both front and rear cameras, so the devices in this dataset have six camera sensors.

The second dataset utilized is the Dresden Image Database. Based on the approach presented in the literature [22], We select six distinct devices (Nikon_D70_0, Nikon_D70_1, Nikon_D70s_0, Nikon_D70s_1, Nikon_D200_0, Nikon_D200_1). To verify the feasibility of the experiment, four more cameras are randomly selected from the Dresden dataset for testing (Agfa_DC-733s_0, Kodak_M1063_0, Samsung_L74wide_1, Sony_DSC-H50_0).

4.2 Setup

The PRNU noise extraction algorithm employed in this paper is the MLE PRNU algorithm. Additionally, the Wiener filter and zero-mean algorithm are utilized to remove artifacts in images [31]. During the anonymous stage, we observed that the PCE value of the image and the non-source camera were both close to 0. Consequently, we set the PCE value to 0.2 as the transformation cutoff threshold. In addition, because of the past techniques in the field of natural image anonymity [22], this solution crops the center of a full-size JPEG format image to 512×512.

This algorithm is implemented and tested on MATLAB 2021b without the need for extensive equipment. We compare the proposed algorithm with two existing PRNU anonymization techniques: (1) For biometric images, the approach described in reference [21] is directly applied to JPEG images, setting the high-frequency coefficient to 0 after DCT. (2) For natural images, the depth-first framework presented in the literature [22] is utilized. In this study, the termination condition for image generation was set to 2500 iterations. The PRNU threshold was set to 38 dB, and the number of blocks used was 64.

4.3 Evaluation Criteria

PSNR is often used to represent the change of an anonymized image relative to the original image, with higher values indicating less distortion. The calculation formula is as follows:

$$PSNR = 10 \times \log_{10}\left(\frac{(2^n-1)^2}{MSE}\right) \tag{7}$$

where, n represents the number of bits of each sample value, and MSE represents the mean square error between the original image and the processed image.

The proposed algorithm incorporates the SSIM metric to assess its performance. SSIM provides a perceptual model that comprehensively evaluates the brightness, contrast, and structure of the image:

$$SSIM(i,j) = \frac{(2\mu_i\mu_j+c_1)+(2\sigma_{ij}+c_2)}{\left(\mu_i^2+\mu_j^2+c_1\right)\left(\sigma_i^2+\sigma_j^2+c_1\right)} \tag{8}$$

where μ_i is the mean of i, μ_j is the mean of j, σ_i is the variance of i, σ_j is the variance of j, σ_{ij} is the covariance of ij, $c_1 = (k_1L)^2$, $c_2 = (k_2L)^2$ is the constant used to maintain stability, $k_1 = 0.01$, $k_2 = 0.03$, L is the dynamic range of the pixel.

We use cosine similarity to measure the utility of anonymized biometric images [32–34]:

$$y = \cos(\theta) = \frac{A \cdot B}{|A||B|} = \frac{\sum_{i=1}^{n} A_i \times B_i}{\sqrt{\sum_{i=1}^{n}(A_i)^2} \times \sqrt{\sum_{i=1}^{n}(B_i)^2}} \tag{9}$$

where A_i, B_i represents the vectors of A and B, respectively.

ROC curve is commonly used in source camera recognition. Scheme uses PCE value from source device image and reference PRNU as positive samples. Images from other cameras and matched values used as negative samples. The purpose of this paper is to reduce the AUC under the ROC curve to prevent the source camera from being identified.

4.4 Results

This section shows the results of each evaluation metric for anonymous images. For the convenience of graphing experimental results presentation, we denote the six devices in the MICHE-I dataset as ID1, IU1, ID2, IU2, SD1, SU1. The 10 devices in the Dresden dataset are noted as N_D70_0, N_D70_1, N_D70s_0, N_D70s_1, N_D200_0,

N_D200_1, Agfa, Kodak, Samsung, Sony. In the legend shown in the following content, Origin denotes the original and untouched image set, DCT denotes the image set processed by the algorithm of literature [21], DIP denotes the image set processed by the algorithm of literature [22], J_ANO denotes the image set after anonymization by the algorithm, and J_EM denotes the image set after embedding information by the algorithm.

Analysis of Biometric Image Results
Considering that one of the criteria for the success of anonymization is the visual quality of the image, we calculated the average PSNR of the three algorithms on the MICHE-I dataset. Figure 3 shows that the algorithms proposed are significantly better than DCT, in which the IU2 device improves by 4.31 dB. Compared with DIP, only the SU1 device has a slightly lower average, and it is found by observing the dataset that the resolution of the images in SU1 is higher, so it may contain some artifacts inside that interfere with the PRNU estimation. In embedding processing, the average PSNR value of our method, on SD1 and SU1 devices is slightly lower, the reason is that the embedding information in the image is to bring a small damage to the image quality, but the final experimental results of the PSNR value are all above 37 dB. Through observation, it is found that the human eye cannot perceive a significant difference in the images at this level. The results are shown in Fig. 4. Therefore, the results verify that the algorithm can effectively retain the visual quality of the images.

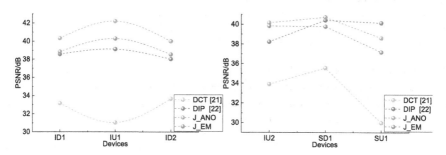

Fig. 3. The plot of PSNR results for the MICHE-I dataset

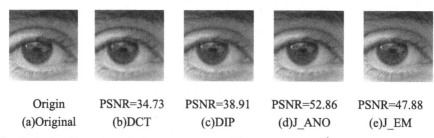

| Origin | PSNR=34.73 | PSNR=38.91 | PSNR=52.86 | PSNR=47.88 |
| (a)Original | (b)DCT | (c)DIP | (d)J_ANO | (e)J_EM |

Fig. 4. Comparison of processing results of the MICHE-I dataset. The first one is the original image and the remaining four are the images after processing by different methods.

In the subsequent phase, we performed biometric retention tests on face images in the MICHE-I dataset. The cosine similarity was tested across six devices, and the DCT and DIP methods, which were used in the literature, were compared. As depicted in Fig. 5, the results of this algorithm outperform the other two comparison experiments significantly, with some images reaching a cosine similarity of 0.999. This indicates that the proposed algorithm can achieve superior image anonymity on biometric datasets while preserving biometric utility.

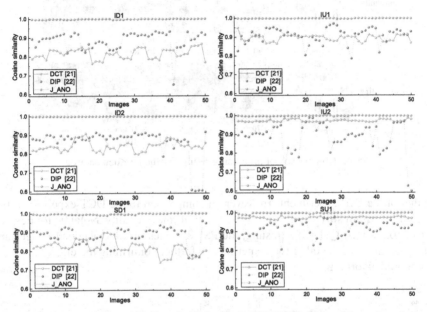

Fig. 5. Comparison results of Cosine similarity in the MICHE-I dataset. The value close to 1 indicates better performance.

Natural Image Result Analysis

To demonstrate the algorithm's extensibility, we further tested it on 10 devices from the Dresden dataset. Figure 6 presents the results of this study by randomly selecting a picture from 5 devices. It is evident from the figure that the visual differences in the content of the images, whether anonymous or embedded, are imperceptible to the human eye.

Figure 7 illustrates that the proposed algorithm outperforms the other two algorithms in terms of anonymized results, with an average PSNR value exceeding 38.8 dB. Among them, the N_D70_0 device with the lowest PSNR value is 11.69 dB higher than the DCT experiment, and the Kodak device with the highest PSNR value is 5.13 dB higher than the DIP experiment. When comparing the image quality after anonymity with the image quality after embedding, a slight decrease in quality can be observed in the embedded images due to the injection of additional information. Specifically, the PSNR value of the Sony device is 1.17 dB lower than its value before embedding and 0.89 dB lower

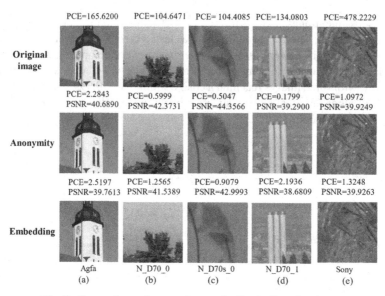

Fig. 6. Comparison of processing results for the Dresden dataset

than that of the DIP experiment. However, in comparison to the DCT experiment, the proposed algorithm still maintains a higher level of image quality. Notably, for devices N_D200_0, N_D200_1, and Samsung, the PSNR values of the embedded images are higher than those of the other two experiments, further confirming the effectiveness of the proposed algorithm.

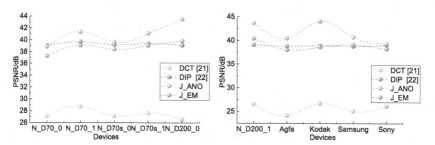

Fig. 7. PSNR result graph for the Dresden dataset

Analysis of Restored Image Results

We continue to process the embedded image to meet the requirements of certification bodies for higher-quality images with authentication information, building upon the successful achievement of the expected anonymous effect. By utilizing the reversible data hiding method based on code division multiplexing, the information is extracted from the image after embedding, and the same processing steps used in the anonymous stage are applied to inject the information into the coefficients after inverse quantization in the JPEG compression domain, enabling incomplete recovery of the image. Figure 8. Demonstrates that the processed images exhibit a PSNR of approximately 49 and an SSIM above 0.99, suggesting that the current algorithm effectively recovers the majority of anonymized information in the images.

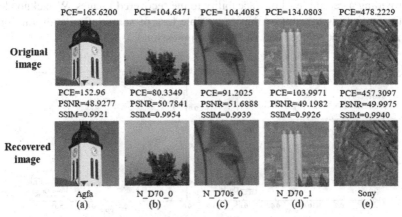

Fig. 8. Comparison of processing results on the Dresden dataset. The PCE value in the first row indicates the original image information and the other values indicate the recovered image information. As the SSIM value approaches 1, it indicates a higher level of structural similarity between the original image and the processed image.

Analysis of AUC and ROC Curves

AUC is the evaluation metric used to evaluate the binary classification model. Our experiment is to get the AUC value close to 0.5 to satisfy our goal of anonymization. From Table 2, it can be seen that the DCT experiment performs well in terms of anonymization ability. Nevertheless, when considering image quality (PSNR), the DCT approach exhibits subpar performance. Conversely, comparing the DIP experiments, this algorithm demonstrates exceptional results in both datasets for both anonymization and embedding. In conclusion, the present algorithm successfully achieves effective anonymization.

Table 2. AUC comparison results

AUC	MICHE-I dataset	Dresden dataset
Origin	0.9273	0.9677
DCT [21]	0.4958	0.5408
DIP [22]	0.7408	0.8201
J_ANO	0.5155	0.5976
J_EM	0.5094	0.6010

Figure 9 plots the ROC curves of the algorithm on the MICHE-I dataset (a) and the Dresden dataset (b), the curves compare the original and restored images. The results show that while this algorithm is unable to fully restore the original traceability effect of JPEG images, it has achieved a high level of success. This indicates that the extracted noise residuals from the processed images can match the camera of the original source, thus providing a certain level of traceability to the recovered images. We acknowledge that this outcome is a result of the irreversible loss of some information during the quantization and encoding stages of JPEG compression.

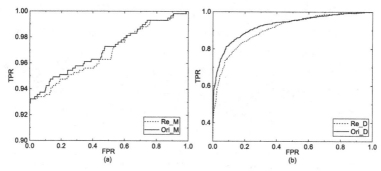

Fig. 9. Comparison of ROC results. (a) Ori_M and Re_M denote the ROC curves of the original and restored images of the MICHE-I dataset, respectively. (b) Ori_D and Re_D represent ROC curves of the original and recovered images of the Dresden dataset, respectively.

5   Conclusion

We present a novel anonymity algorithm for JPEG image compression, achieving the following: (a) Dynamic removal of noise residual traces in the JPEG compression domain using discrete cosine transform, effectively severing the connection with the source device. (b) Enabling traceability-free images to embed lost information and regain strong traceability upon extraction. Experimental results demonstrate that the algorithm successfully severs the link between images and their source devices while maintaining high image quality. The proposed algorithm holds significant research and practical value by providing an efficient and reliable method for source camera anonymization and privacy protection in digital image domains. Future work will involve evaluating the algorithm's feasibility across diverse datasets.

Acknowledgements. This work was supported by National Natural Science Foundation of China (62272255, 62302248, 62302249); National key research and development program of China (2021YFC3340600, 2021YFC3340602); Taishan Scholar Program of Shandong (tsqn202306251); Jinan "New 20 Universities"-Project of Introducing Innovation Team (202228016); The "Chunhui Plan" Cooperative Scientific Research Project of Ministry of Education (HZKY20220482); First Talent Research Project under Grant (2023RCKY131, 2023RCKY143), Integration Pilot Project of Science Education Industry under Grant (2023PX006, 2023PY060, 2023PX071).

References

1. Lukas, J., Fridrich, J., Goljan, M.: Digital camera identification from sensor pattern noise. IEEE Trans. Inf. Forensics Secur. **1**, 205–214 (2006)
2. Chen, S.-H., Hsu, C.-T.: Source camera identification based on camera gain histogram. In: 2007 IEEE International Conference on Image Processing, pp. IV-429–IV-432 (2007)
3. Bayram, S., et al.: Source camera identification based on CFA interpolation. In: IEEE International Conference on Image Processing 2005, vol. 3. IEEE (2005)
4. Zhao, X., Stamm, M.C.: Computationally efficient demosaicing filter estimation for forensic camera model identification. In: 2016 IEEE International Conference on Image Processing (ICIP), pp. 151–155 (2016)
5. Kirchner, M., Gloe, T.: Forensic camera model identification. In: Handbook of Digital Forensics of Multimedia Data and Devices, pp. 329–374. Wiley (2015)
6. Goljan, M., Fridrich, J., Mo, C.: Defending against fingerprint-copy attack in sensor-based camera identification. IEEE Trans. Inf. Forensics Secur. **6**, 227–236 (2011)
7. Goljan, M., Fridrich, J.J., Mo, C.: Sensor Noise Camera Identification: Countering Counter-Forensics. International Society for Optics and Photonics (2010)
8. Nagaraja, S., Schaffer, P., Aouada, D.: Who clicks there!: anonymising the photographer in a camera saturated society. In: Proceedings of the 10th Annual ACM Workshop on Privacy in the Electronic Society - WPES 2011, Chicago, Illinois, USA, p. 13. ACM Press (2011)
9. Böhme, R., Kirchner, M.: Counter-forensics: attacking image forensics. In: Sencar, H., Memon, N. (eds.) Digital Image Forensics, pp. 327–366. Springer, New York (2013). https://doi.org/10.1007/978-1-4614-0757-7_12
10. Pfitzmann, A., Hansen, M.: A terminology for talking about privacy by data minimization: anonymity, unlinkability, undetectability, unobservability, pseudonymity, and identity management (2010)
11. Rosenfeld, K., Sencar, H.T.: A study of the robustness of PRNU-based camera identification. In: Media Forensics and Security, pp. 213–219. SPIE (2009)
12. Bernacki, J.: On robustness of camera identification algorithms. Multimedia Tools Appl. **80**, 921–942 (2021)
13. Chen, M., Fridrich, J., Goljan, M.: Digital imaging sensor identification (further study). In: Security, Steganography, and Watermarking of Multimedia Contents IX, pp. 258–270. SPIE (2007)
14. Mandelli, S., Bondi, L., Lameri, S., Lipari, V., Bestagini, P., Tubaro, S.: Inpainting-based camera anonymization. In: 2017 IEEE International Conference on Image Processing (ICIP), pp. 1522–1526 (2017)
15. Dirik, A.E., Karaküçük, A.: Forensic use of photo response non-uniformity of imaging sensors and a counter method. Opt. Express **22**, 470–482 (2014)
16. Bonettini, N., et al.: Fooling PRNU-based detectors through convolutional neural networks. In: 2018 26th European Signal Processing Conference (EUSIPCO), pp. 957–961 (2018)
17. Dirik, A.E., Sencar, H.T., Memon, N.: Analysis of seam-carving-based anonymization of images against PRNU noise pattern-based source attribution. IEEE Trans. Inf. Forensics Secur. **9**, 2277–2290 (2014)
18. Zeng, H., Chen, J., Kang, X., Zeng, W.: Removing camera fingerprint to disguise photograph source. In: 2015 IEEE International Conference on Image Processing (ICIP), pp. 1687–1691 (2015)
19. Entrieri, J., Kirchner, M.: Patch-based desynchronization of digital camera sensor fingerprints. Electron. Imaging **28**, 1–9 (2016)
20. García Villalba, L.J., Sandoval Orozco, A.L., Rosales Corripio, J., Hernandez-Castro, J.: A PRNU-based counter-forensic method to manipulate smartphone image source identification techniques. Future Gener. Comput. Syst. **76**, 418–427 (2017)

21. Banerjee, S., Ross, A.: Smartphone camera de-identification while preserving biometric utility. In: 2019 IEEE 10th International Conference on Biometrics Theory, Applications and Systems (BTAS), pp. 1–10 (2019)
22. Picetti, F., Mandelli, S., Bestagini, P., Lipari, V., Tubaro, S.: DIPPAS: a deep image prior PRNU anonymization scheme. EURASIP J. Inf. Secur. **2022**, 2 (2022)
23. Gloe, T., Böhme, R.: The "Dresden Image Database" for benchmarking digital image forensics. In: Proceedings of the 2010 ACM Symposium on Applied Computing, pp. 1584–1590. Association for Computing Machinery, New York (2010)
24. De Marsico, M., Nappi, M., Narducci, F., Proença, H.: Insights into the results of MICHE I - mobile iris challenge evaluation. Pattern Recogn. **74**, 286–304 (2017)
25. Chen, M., Fridrich, J., Goljan, M., Lukas, J.: Determining image origin and integrity using sensor noise. IEEE Trans. Inf. Forensics Secur. **3**, 74–90 (2008)
26. Goljan, M., Fridrich, J., Filler, T.: Large scale test of sensor fingerprint camera identification. In: Media Forensics and Security, pp. 170–181. SPIE (2009)
27. Debiasi, L., Uhl, A.: Techniques for a forensic analysis of the CASIA-IRIS V4 database. In: 3rd International Workshop on Biometrics and Forensics (IWBF 2015), pp. 1–6 (2015)
28. Debiasi, L., Uhl, A., Sun, Z.: Generation of iris sensor PRNU fingerprints from uncorrelated data. In: 2nd International Workshop on Biometrics and Forensics, pp. 1–6 (2014)
29. Ma, B., Shi, Y.Q.: A reversible data hiding scheme based on code division multiplexing. IEEE Trans. Inf. Forensics Secur. **11**, 1914–1927 (2016)
30. Galdi, C., Nappi, M., Dugelay, J.-L.: Multimodal authentication on smartphones: combining iris and sensor recognition for a double check of user identity. Pattern Recogn. Lett. **82**, 144–153 (2016)
31. Kang, X., Li, Y., Qu, Z., Huang, J.: Enhancing source camera identification performance with a camera reference phase sensor pattern noise. IEEE Trans. Inf. Forensics Secur. **7**, 393–402 (2012)
32. ImageNet: A large-scale hierarchical image database. IEEE Conference Publication. IEEE Xplore. https://ieeexplore.ieee.org/document/5206848
33. He, K., Zhang, X., Ren, S., Sun, J.: Deep residual learning for image recognition. Presented at the Proceedings of the IEEE Conference on Computer Vision and Pattern Recognition (2016)
34. Hernandez-Diaz, K., Alonso-Fernandez, F., Bigun, J.: Periocular recognition using CNN features off-the-shelf. In: 2018 International Conference of the Biometrics Special Interest Group (BIOSIG), pp. 1–5 (2018)

Limiting Factors in Smartphone-Based Cross-Sensor Microstructure Material Classification

Johannes Schuiki(✉)[ID], Christof Kauba[ID], Heinz Hofbauer[ID],
and Andreas Uhl[ID]

Department of Artificial Intelligence and Human Interfaces, University of Salzburg,
Salzburg, Austria
{jschuiki,ckauba,hofbauer,uhl}@cs.sbg.ac.at

Abstract. Intrinsic, non-invasive product authentication is the preferred way of detecting counterfeit products as it does not generate
additional costs during the production process. Previous works achieved
promising results for smartphone-based product authentication. However, while promising, the methods fail when enrollment and authentication are performed on different devices (cross-device). This work investigates the underlying reasons for the limitations in the practical application of cross-device intrinsic surface structure-based product authentication. In particular by utilising micro-texture classification approaches
applied on images of zircon oxide blocks (dental implants) captured using
a commodity smartphone device. The main result is that the device-
specific artefacts (image sensor as well as image processing-specific ones)
are so strong that they obfuscate the material microstructure. To be more
precise, the device's intrinsic signal makes device identification easier to
perform than the material authentication.

Keywords: intrinsic product authentication · material classification ·
microstructure texture features · dental ceramic blocks · camera source
classification

1 Introduction

The wide-spread use and availability of mobile smartphone devices with built-in
high quality cameras opened new possibilities for mobile applications such as
classification of paving materials in urban environments [10], wood type identification [25], or personal authentication using biometrics [7]. Another recent
application employing non-modified commodity smartphone devices is to verify
a product's origin or to assure that a product stems from a certain manufacturer

This research was partially funded by the Salzburg State Government within the Science and Innovation Strategy Salzburg 2025 (WISS 2025) under the project AIIV-
Salzburg (Artificial Intelligence in Industrial Vision), project no 20102-F2100737-FPR.

[3, 8, 24, 26], denoted as product authentication. Counterfeit products do not only cause economic damage to the original manufacturers especially counterfeit medical and health related products can directly harm the patients' health. Hence, manufacturers strive to implement ways to reliably detect non-genuine products, with several available commercial solutions for extrinsic, e.g. AuthenticVision, QLIKTag using NFC Tags as well as intrinsic mobile product authentication e.g. AlpVision, Bosch and Sepio's Logitrak.

Our previous work [22] addressed the practical applicability of using commodity off-the-shelf smartphone devices in combination with a clip-on macro lens to establish the authenticity of zircon oxide blocks, commonly used for dental ceramics in an intrinsic, non-intrusive way based on their surface's micro-texture. Material classification worked well in an intra-sensor ("one device") scenario, but the tested classifiers faced problems and limitations in the inter-sensor (cross-device) one. While a cross-device application is not an issue in many application settings (e.g. in biometrics) it is well-known to cause problems in other settings, e.g. in sensor forensics.

The main aim of this study is to investigate and understand the limiting factors for the cross-device intrinsic product authentication performance, their underlying reasons and in particular what happens during the image acquisition (i.e. the influence of the image signal processing tool-chain). This is an important step prior to developing appropriate countermeasures (i.e. training classifiers that work for smartphone devices unavailable during training) and can only be done in an explorative manner. Therefore, a second version of the zircon oxide block data set is acquired with more smartphone devices (7 in total), additionally capturing raw images as well. Furthermore, sensor identification experiments are performed as it turned out that sensor identification works better than the cross-device material classification. Sensor identification deals with establishing the origin of a digital image, i.e. linking an image to an image sensor/camera and can be done on two levels: manufacturer/type level or device/single instance level with the pixel response non-uniformity (PRNU) [14] being the most common approach for single instance sensor identification. There are other artefacts stemming from the image sensor, e.g. optical lens distortions, chromatic aberration, the layout of the physical pixel pattern on the sensor surface (Bayer pattern) as well as device-inherent processing steps from the image signal processor (ISP) which were successfully utilised as well [1, 4]. The signal of these imaging artefacts might overlay the micro-structure texture features, impacting the cross-sensor material classification accuracy. This influence is evaluated utilising raw images, several variants of raw to RGB conversion and by applying an image denoising filter.

The rest of this work is organised as follows: Sect. 2 gives an overview on previous works about product authentication with a focus on smartphone-based mobile solutions. Section 3 introduces the dental ceramic data set, outlines the micro-texture classification approaches and describes the experimental set-up. Section 4 lists and discusses the experimental results. Finally, Sect. 5 concludes this work and gives an outlook on future work.

2 Related Work

Product authentication can be categorised into intrinsic and extrinsic methods, where the latter ones rely on external properties added to the product. This can be as simple as a signed physical document, which can be easily forged, up to individual, hard-to-forge features that are embedded in each single product. These so called copy detection patterns (CDP) are usually a high resolution QR or 2D code printed with industrial printers on the product or its packaging. With the advancements in digital scanning technology and home printer equipment, CDPs are more often subject to successful illegal reproduction. Taran et al. [26,27] proposed a machine-learning framework (based on a one-class support vector machine) to explore the resistance of CDPs to illegal reproduction. They established a publicly available CDP dataset, tailored to real life conditions, named the "Indigo mobile dataset", captured using commodity smartphones (iPhone XS) under regular light conditions. Their reported results suggest that modern mobile phones allow to reliably authenticate CDPs under the considered classes of fakes. Cai et al. [3] proposed a deep learning product authentication approach for leather products based on texture features extracted from the material which are encoded into a non-detachable 2D barcode, which can be easily scanned and verified by consumers. Yan et al. [30] presented a framework that combines micro-texture features in combination with a QR code for anti-counterfeiting. Visual features and the QR code are registered on the assembly line and stored in a cloud and are then compared to images captured using a mobile phone for verification. The EU implemented the Falsified Medicines Directive (FMD) 2011/62/EU, based on product serialisation, i.e. a unique identifier in the form of a 2D barcode is added to each product, enabling to track and identify each medical package along the supply chain. It requires a central database and the packages to be equipped with safety features in order to avoid tampering.

Any form of external embedded features can be cumbersome to implement into a running production chain, require complex processing techniques, might add the demand for a back-end database infrastructure that needs to be administrated and increase the production costs. Thus, manufacturers prefer intrinsic methods, which are based solely on constant but discriminative intrinsic features of the product or its packaging material for product authentication. In practice, a further requirement is that the authentication can be done in a non-invasive way, i.e. without altering the product. The concept of surface micro-structures or so called physical non-cloneable functions (PUFs) is widely used in intrinsic product/material authentication, especially for paper based materials [11]. Paper PUFs use the fiber structure of paper as physical/intrinsic characteristic. In 2012, Voloshynovskiy et al. established the publicly available FAMOS data set [28], captured under different illumination conditions and with two different cameras and derived statistical authentication frameworks which achieved promising results. To overcome the main drawback of using micro-structures for product authentication - the large storage space required for the extracted feature information - the authors [2] suggested to use digital content fingerprints as a short and robust representation. In [9] the authors extended their previous work and showed that it is feasible to uniquely identify packages based on micro-structure images acquired using an unmodified consumer smartphone without any special lighting or adaptation. They pointed out that the smartphone captured images suffer from non-linear distortions (lens distortions), geometrical distortions introduced by the user holding the phone in different manners and non-even, varying lighting conditions, all affecting the classification accuracy. However, no cross-device experiments between the smartphone and the handheld cameras have been performed. In a follow-up work [8] they utilised "SketchPrint", an approach previously introduced by the authors [29], which should provide reasonable invariance to varying lightning conditions and geometric distortions. They acquired a data set of 50 paper sheets with a commodity non-modified smartphone and achieved a reasonable performance. Schraml et al. [20,21] used microtexture images of drug package material for product authentication in an open set scenario, which worked well in the intra-sensor scenario. Their later results in [21] showed that the classification capability is greatly reduced in a cross-sensor/device scenario with scaling issues (different dpi resolutions) as well as artefacts introduced during the image processing pipeline mentioned as reasons for the non-satisfactory performance. Sun et al. [24] utilised an efficient micro-structure orientation estimation technique, which models the entire propagation path of the light, to establish the authenticity of paper sheets using images captured with a mobile camera and its built-in flash. Sun et al. proposed the so called "LiquidHash" approach [23] for the detection of counterfeit liquid food products, based on the characteristics of air bubbles formed if the bottle is flipped. They utilised commodity off-the-shelf smartphone devices to capture a video stream of the bubble movement in combination with computer vision techniques to detect adulterated liquid products on a small data set (3 authentic and 8 counterfeit products).

While a sensor-aware classifier, specifically trained for each of the employed sensors could definitely improve the inter-sensor performance, the actual aim is to train a classifier that also works on smartphone devices not available during training. In practical applications the classification model is trained on a few particular devices, while the authentication should not be restricted to those devices but be possible on any modern off-the-shelf smartphone device so that even an end user can easily check if a product is genuine or counterfeit. The above works show that cross-device extrinsic product authentication as well as intrinsic micro-structure based material classification is feasible in practical applications using commodity smartphone devices. However, cross-device intrinsic micro-structure based product authentication has hardly been investigated. There is at least some evidence that the cross-device authentication performance is drastically reduced [21].

3   Experimental Setup

In the following, the database used within this study is explained in detail. Afterwards the employed texture classification tool-chain for the experiments Sect. 4 is described.

3.1   Database

Fig. 1. Unprocessed Zircon Oxide Ceramic Blocks.

The micro-texture database comprises of images acquired from the top side of zircon oxide blocks produced by three different manufacturers: 10 blocks produced by Ivoclar Vivadent, 6 by Dentsply and 16 by 3M. The different number of zircon oxide blocks is due to their limited availability. See Fig. 1 for examples of such blocks. The acquisition setup is depicted in Fig. 2. Four images, one on every top-side corner, were captured from every block using a macro lens *Agritix WIDK-24X01 Xylorix Wood Identification Tool* clipped onto the smartphone's camera with seven different smartphones (40, 24 and 64 samples per smartphone for *Ivoclar Vivadent, Dentsply Sirona* and *3M*, respectively). The distance between the block surface and the macro lens surface was 8 mm (This distance was best for successful auto-focusing). The macro lens has a circular illumination ring which was used on the brightest of the three settings during the acquisition to suppress the influence of varying lighting conditions and enhance the visibility of the surface micro-structure.

Checkerboard pattern images revealed that the fields of view between the various devices are similar but the resolutions are only coarsely matching. Therefore, a resolution normalization is applied by down-scaling the images to a fixed resolution of 2074×2765 using bilinear interpolation. Sensor resolutions and the scaling factors for each smartphone are listed in Table 1. Scaling the images with different factors naturally introduces resolution specific artefacts. Hence, one experiment in Sect. 4 deals with images without prior scale adjustments. The next step is to extract multiple patches by at first rotating the images to landscape orientation. Afterwards, the images are converted to gray-scale and nine patches of size 512×512 are cropped from the center as shown in Fig. 3,

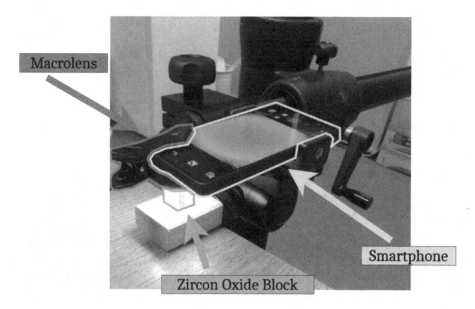

Fig. 2. Acquisition setup for the zircon oxide blocks, smartphone with clip-on macro lens.

resulting in 576, 216 and 360 patches for *Ivoclar Vivadent*, *Dentsply Sirona* and *3M*, respectively. This patching and down-scaling strategy avoids distortion (at the image boundaries) and black area (macro lens not perfectly aligned with the smartphone's lens) artefacts introduced by the macro lens. No further pre-processing (e.g. contrast enhancement) is employed to best preserve the micro-structures of the ceramic material and artefacts due to the sensor and ISP pipeline.

Table 1. Smartphones and their imaging sensor resolution.

Smartphone	Image Resolution	Scaling Factor
Google Pixel 4a (GP)	3024 × 4032	0.686
Huawei P20 Lite (H20)	3456 × 4608	0.600
Huawei P30 Pro (H30)	2736 × 3648	0.758
iPhone 11 (i11)	3024 × 4032	0.686
iPhone 13 Pro (i13)	3024 × 4032	0.686
Samsung Galaxy A52 (SG)	3468 × 4624	0.598
Xiaomi Mi A3 (XM)	3000 × 4000	0.691

Results from previous experiments [22] suggested that the smartphone inherent ISP greatly influences the cross-sensor performance. Hence, all images were captured in raw in addition to the "normal" (ISP) image. Usually, the standard smartphone camera application does not allow to capture raw images. Hence, suitable camera applications were employed: OpenCamera for devices running Android and Halide Mark II for the iPhone devices. Two applications are used for conversion from the raw data, i.e. the Bayer pattern image, to an RGB image: Darktable (DT) and dcraw. Dcraw offers the possibility to set additional parameters such as -a (DCA) to average the whole image for white balance or -d to omit demosaicing, denoted as CFA (for color filter array) in the later experiments.

In order to remove PRNU and other sensor artefacts, images additionally underwent denoising by application of the BM3D [6] denoising filter. Results in Sect. 4 generated using denoised images are denoted with DN.

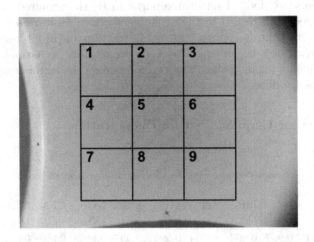

Fig. 3. Patch annotations in a zircon oxide block image.

3.2   Texture Classification Tool-Chain

Six distinct feature extraction techniques are employed in this study: Dense SIFT [13] (SIFT descriptors applied on a fixed-spaced grid), Dense Micro-block Difference (DMD) [15], Local Binary Pattern Histograms (LBP) [17], Local Phase Quantization (LPQ) [18] and Weber Pattern (WP) [16]. In addition to the aforementioned standard LBP variant, a second LBP feature descriptor is employed. LBP can be made rotational invariant by using a number of rotational shifts of the pattern equal to the number of points and taking the minimum. This method will be denoted as *ror* (n.b. this is not uniform LBP). Varying the radius of the LBPs can make it use differently sized structures and this will be denoted as Rx where x is the radius in pixels. The extracted feature vectors (except the ones from the ror LBP) are subsequently encoded using improved Fisher vector encoding [19] in a similar way as originally proposed in [5]. They undergo a soft-quantization using a Gaussian mixture model and dimensionality reduction using principle component analysis. A support vector machine with linear kernel is utilized to classify the encoded features. For more details the interested reader is referred to our previous work [12].

The accuracy, defined as the number of correct classifications divided by the number of total classifications, is used as an evaluation metric to quantify the classification performance. The reported value in the experiments in Sect. 4 is the average accuracy, i.e. the arithmetic mean over all accuracies for a particular experiment.

4   Experimental Results

The experiments are divided into three parts: in (i) the acquired ceramic samples are classified with regard to their manufacturer in an intra-sensor scenario to verify that they contain enough texture information, in (ii) inter-sensor experiments are carried out, which constitutes a more realistic scenario, while in (iii) further experiments (camera classification) concerning the factors responsible for the bad results are done.

4.1   Intra-sensor Ceramic Texture Classification

Table 2. Average ceramic classification accuracy in intra-sensor setup.

	ISP	ISP DN	DT	DT DN	DCA	DCA DN	CFA	CFA DN
SIFT	0.997	0.993	0.994	0.995	0.997	0.997	0.996	0.997
DMD	0.988	0.986	0.987	0.982	0.990	0.991	0.981	0.981
LBP	0.720	0.791	0.656	0.795	0.587	0.809	0.399	0.398
WP	0.760	0.770	0.752	0.787	0.760	0.804	0.670	0.675
LPQ	0.519	0.767	0.428	0.819	0.378	0.757	0.422	0.646

A random 50/50 training/evaluation data split, training using images from the same camera only, was performed. The average accuracy over all cameras for each imaging modality and feature is reported in Table 2. Obviously there is a signal strong enough for classification in all imaging modalities, e.g. SIFT reaches 99% classification accuracy for each modality.

Raw with denoising (DT DN, DCA DN, CFA DN) performs better than regular (ISP) images which include artefacts from the image processing pipeline. As the samples' patches in Fig. 4 show, the surface micro-structure texture components are hardly discernible. Thus, the imaging pipeline's artefacts clearly impact the ceramic sample image patches, lowering the classification accuracy and making raw the preferred choice.

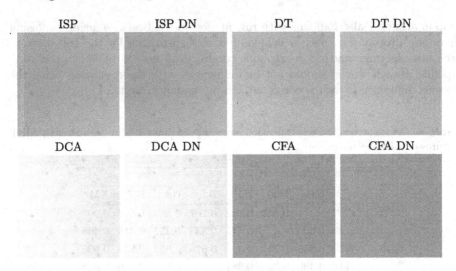

ISP   ISP DN   DT   DT DN

DCA   DCA DN   CFA   CFA DN

Fig. 4. Samples of ceramic images per imaging modality.

4.2 Practical Consideration: Inter-camera Classification

Table 3. Average inter-sensor ceramic classification accuracy.

	ISP	ISP DN	DT	DT DN	DCA	DCA DN	CFA	CFA DN
SIFT	0.892	0.899	0.936	0.945	0.899	0.936	0.701	0.762
DMD	0.774	0.825	0.735	0.826	0.720	0.853	0.483	0.492
LBP	0.389	0.619	0.390	0.720	0.376	0.610	0.332	0.373
WP	0.484	0.676	0.595	0.698	0.593	0.638	0.362	0.366
LPQ	0.380	0.438	0.338	0.456	0.348	0.361	0.296	0.315
LBP R1	0.587	0.758	0.644	0.910	0.643	0.790	0.505	0.510
LBP R1 ror	0.567	0.769	0.966	0.911	0.672	0.776	0.500	0.517
LBP R4 ror	0.847	0.870	0.942	0.942	0.925	0.915	0.602	0.606

The results confirm that ceramic material classification is possible in the intra-device scenario while in practice it needs to be possible across different devices as well. To evaluate this cross-device scenario, a leave-one out training regime is used, i.e. train on 6 out of 7 ceramics and test on the 7th one, performed 7 times (for each ceramic). The (averaged) end results are given in Table 3.

The drop in classification accuracy compared to the intra-sensor scenario is quite substantial, especially for ISP. In most cases denoising improves the classification results, with a best accuracy of 94%, but less than 70% in many other cases, which is not satisfactory. An interesting observation is that the drop in accuracy also happens with raw modes, in the best case using denoising there is a drop of 5%. Such a behaviour is not surprising for the ISP case, as each manufacturer and/or smartphone combination uses a different processing pipeline. Raw however, should not be influenced, with the resolution being the obvious difference (which is corrected, c.f. discussion in Sect. 3.1).

Table 4. Average ceramic classification accuracy in inter-sensor setup using SIFT features and DCA images.

		Test						
		GP	H20	H30	i11	i13	SG	XM
Train	GP	–	0.837	0.852	0.707	0.832	0.607	0.997
	H20	0.605	–	0.515	0.333	0.352	0.455	0.465
	H30	0.846	0.716	–	0.676	0.854	0.751	0.906
	i11	0.480	0.663	0.381	–	0.864	0.380	0.613
	i13	0.813	0.579	0.362	0.819	–	0.344	0.552
	SG	0.831	0.668	0.603	0.675	0.963	–	0.943
	XM	0.982	0.804	0.451	0.787	0.903	0.358	–

To validate if this behaviour is manufacturer or camera specific, an experiment where training is done on images from one camera and evaluated on images from another one, for all ISP/RAW/DN types and features, is conducted. The general trend is the same for all those combinations, so only SIFT on DCA is shown in Table 4 for reasons of brevity. Unfortunately our database is rather limited in terms of multiple smartphone models from the same manufacturer. However, for the Huawei smartphone cameras it is clearly a camera specific issue: if trained on Huawei H30, the H20 has the second worst performance of all cameras (highlighted in red colour in Table 4). On the other hand, the iPhones (i11 vs. i13 - highlighted in green in Table 4) work well in cross-training, still being inferior to many of the other tested combinations. Hence, in general it cannot be considered as a manufacturer specific effect but more a camera specific one. Also note that this is not reciprocal, i.e. if training on A performs well on B, training on B is not guaranteed to work well on A, e.g. for SG and XM (highlighted in orange in Table 4) - training on SG results in an accuracy of 94%, but training on XM only achieves 35% accuracy on SG.

4.3 Camera Classification

The signal of the camera/ISP is strong enough to obfuscate the micro-structure texture information used for material classification to a (sometimes strong) degree, raising the question if this can be used for camera classification. To test this, a leave-one material out cross-validation for camera classification is performed. The results are given in Table 5. Camera classification works well with

Table 5. Average sensor identification accuracy.

	ISP	ISP DN	DT	DT DN	DCA	DCA DN	CFA	CFA DN
SIFT	0.966	0.870	0.880	0.725	0.894	0.770	0.977	0.995
DMD	0.934	0.923	0.879	0.750	0.863	0.716	0.990	0.996
LBP	0.910	0.559	0.823	0.490	0.842	0.580	0.983	0.969
WP	0.796	0.497	0.393	0.375	0.443	0.480	0.962	0.947
LPQ	0.655	0.373	0.694	0.335	0.529	0.390	0.898	0.914
LBP R1	0.935	0.879	0.859	0.606	0.862	0.712	0.999	0.997
LBP R1 ror	0.981	0.880	0.527	0.563	0.883	0.690	0.999	0.996
LBP R4 ror	0.865	0.805	0.692	0.549	0.703	0.615	0.990	0.986

regular images processed by the ISP (98% accuracy) while raw ones perform worse in most cases. However, if the Bayer pattern is kept visible, the classification rate even reaches 99% accuracy. So apparently there are two signals able to identify the camera: the Bayer pattern and the image processing pipeline. What is more, the color filter arrays between models from the same manufacturer are different enough to differentiate between the devices.

Table 6. Classification accuracy results CFA unscaled.

	Texture Classification		Sensor Identification	
	CFA	CFA DN	CFA	CFA DN
SIFT	0.453	0.736	1.000	1.000
DMD	0.333	0.333	1.000	1.000
LBP	0.344	0.277	0.973	0.981
WP	0.539	0.755	0.666	0.399
LPQ	0.325	0.445	0.842	0.616

To rule out the influence of scaling (as a result of unifying the resolution), the camera identification experiment was carried out a second time using CFA samples without prior scaling. Figure 5 shows examples of unscaled CFA patches. The results are given in Table 6. While the texture classification performance is drastically reduced, sensor classification accuracy even reaches 100%, hereby allowing to deduce that scaling artefacts are not the reason for such a high accuracy.

4.4   Discussion of Results

For material classification the camera intrinsic signal has a detrimental effect on the classification as it obfuscates the material intrinsic one to a significant

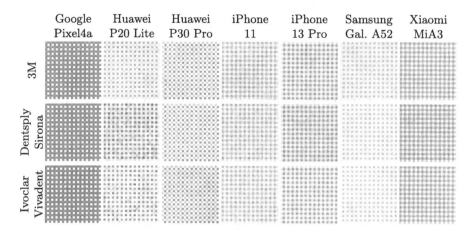

Fig. 5. Ceramic patches of size 26 × 26 converted with "dcraw -d". No down-scaling or denoising involved.

extent. This can partly be counteracted by capturing the images in raw format and denoising the images, as the camera specific signal is noise-like while the material specific signal is more coarse. This leads to a promising accuracy of 94% even in the cross-device scenario. However, it excludes a large number of devices, especially older ones not able to capture in raw mode. Hence, if they should be included, only the ISP images can be used, leading to a drop in accuracy to 87% (with denoising).

For device classification it became apparent that there are in reality two, probably not entirely independent, signals which can be used for camera identification. One is based on the sensor's intrinsics and image processing pipeline and the other one based on the Bayer-pattern and also the scaling done for resolution compensation. By using the second one (Bayer pattern), an incredibly high accuracy (99%) per patch can be achieved, but in practice it is still of little use as typical camera identification tasks are performed on regular images processed by the ISP rather than on raw ones, leaving us with the first one. Based on the first one (signal processing pipeline) the achievable accuracy is also high (98% per patch).

5   Conclusion

This work investigated the underlying reasons for the limitations in using unmodified off-the-shelf smartphone devices for intrinsic product authentication for ceramic materials based on their surface micro-structure. The experiments confirmed that there is an intrinsic signal from the material texture as well as one stemming from the camera itself which can be used to identify either one. The camera intrinsic signal, however, obfuscates the material intrinsic one to a significant extent, lowering the material classification accuracy, which can partly be counteracted by capturing the images in raw mode. In reality there are two, probably not entirely independent, signals stemming from the imaging device which can be used for camera identification, one based on the sensor's intrinsics and image processing pipeline and the other one based on the Bayer-pattern. Even using the latter one an accuracy of 99% can be achieved.

Future work will include deep learning methods for classification and feature extraction. It is worth further investigating what constitutes the signal/texture/feature used for device identification. This is related to the topic of "cover source mismatch" in steganalysis. Hence, ideas from steganalysis to mitigate the cover source mismatch will be investigated too. With more precise information about the signal it can be removed to strengthen the material classification performance. Furthermore, we will acquire additional samples with different entities of the same smartphone model in order to identify the contribution of the ISP and the sensor specific artefacts.

References

1. Bayram, S., Sencar, H., Memon, N., Avcibas, I.: Source camera identification based on CFA interpolation. In: IEEE International Conference on Image Processing 2005, vol. 3, pp. III–69 (2005). https://doi.org/10.1109/ICIP.2005.1530330
2. Beekhof, F.P., Voloshynovskiy, S., Diephuis, M., Farhadzadeh, F.: Physical object authentication with correlated camera noise. In: Saake, G., Henrich, A., Lehner, W., Neumann, T., Köppen, V. (eds.) Datenbanksysteme für Business, Technologie und Web (BTW) 2013 - Workshopband, pp. 65–74. Gesellschaft für Informatik e.V, Bonn (2013)
3. Cai, S., Zhao, L., Chen, C.: Open-set product authentication based on deep texture verification. In: Peng, Y., Hu, S.M., Gabbouj, M., Zhou, K., Elad, M., Xu, K. (eds.) Image and Graphics, vol. 12888, pp. 114–125. Springer, Cham (2021). https://doi.org/10.1007/978-3-030-87355-4_10
4. Choi, K.S., Lam, E.Y., Wong, K.K.Y.: Automatic source camera identification using the intrinsic lens radial distortion. Opt. Express 14(24), 11551–11565 (2006). https://doi.org/10.1364/OE.14.011551
5. Cimpoi, M., Maji, S., Kokkinos, I., Mohamed, S., Vedaldi, A.: Describing textures in the wild. In: 2014 IEEE Conference on Computer Vision and Pattern Recognition, pp. 3606–3613 (2014). https://doi.org/10.1109/CVPR.2014.461
6. Dabov, K., Foi, A., Katkovnik, V., Egiazarian, K.: Image denoising by sparse 3-d transform-domain collaborative filtering. IEEE Trans. Image Process. 16(8), 2080–2095 (2007). https://doi.org/10.1109/TIP.2007.901238
7. Das, A., Galdi, C., Han, H., Ramachandra, R., Dugelay, J.L., Dantcheva, A.: Recent advances in biometric technology for mobile devices. In: 2018 IEEE 9th International Conference on Biometrics Theory, Applications and Systems (BTAS), pp. 1–11 (2018). https://doi.org/10.1109/BTAS.2018.8698587
8. Diephuis, M., Voloshynovskiy, S., Holotyak, T.: Sketchprint: physical object microstructure identification using mobile phones. In: 2015 23rd European Signal Processing Conference (EUSIPCO), pp. 834–838 (2015).https://doi.org/10.1109/EUSIPCO.2015.7362500
9. Diephuis, M., Voloshynovskiy, S., Holotyak, T., Stendardo, N., Keel, B.: A framework for fast and secure packaging identification on mobile phones. In: Alattar,

A.M., Memon, N.D., Heitzenrater, C.D. (eds.) Media Watermarking, Security, and Forensics 2014, vol. 9028, p. 90280T. International Society for Optics and Photonics, SPIE (2014). https://doi.org/10.1117/12.2039638

10. Jain, S., Gruteser, M.: Recognizing textures with mobile cameras for pedestrian safety applications. IEEE Trans. Mob. Comput. **18**(8), 1911–1923 (2019). https://doi.org/10.1109/TMC.2018.2868659

11. JD, R.B., et al.: Forgery:'fingerprinting' documents and packaging. Nat. Brief Commun. **436**, 475 (2005)

12. Kauba, C., Debiasi, L., Schraml, R., Uhl, A.: Towards drug counterfeit detection using package paperboard classification. In: Chen, E., Gong, Y., Tie, Y. (eds.) Advances in Multimedia Information Processing – Proceedings of the 17th Pacific-Rim Conference on Multimedia (PCM 2016), LNCS, Xi'an, China, vol. 9917, pp. 136–146. Springer, Heidelberg (2016). https://doi.org/10.1007/978-3-319-48896-7_14

13. Lowe, D.G.: Distinctive image features from scale-invariant keypoints. Int. J. Comput. Vision **60**(2), 91–110 (2004). https://doi.org/10.1023/B:VISI.0000029664.99615.94

14. Lukas, J., Fridrich, J.J., Goljan, M.: Digital camera identification from sensor pattern noise. IEEE Trans. Inf. Forensics Secur. **1**(2), 205–214 (2006)

15. Mehta, R., Egiazarian, K.: Texture classification using dense micro-block difference (DMD). In: Cremers, D., Reid, I., Saito, H., Yang, MH. (eds.) ACCV 2014, LNCS, vol. 9004, pp. 643–658. Springer, Heidelberg (2015). https://doi.org/10.1007/978-3-319-16808-1_43

16. Muhammad, G.: Multi-scale local texture descriptor for image forgery detection. In: 2013 IEEE International Conference on Industrial Technology (ICIT), pp. 1146–1151 (2013)

17. Ojala, T., Pietikäinen, M., Mäenpää, T.: Multiresolution Gray-Scale and rotation invariant texture classification with local binary patterns. IEEE Trans. Pattern Anal. Mach. Intell. **24**(7), 971–987 (2002)

18. Ojansivu, V., Rahtu, E., Heikkila, J.: Rotation invariant local phase quantization for blur insensitive texture analysis. In: 2008 19th International Conference on Pattern Recognition, pp. 1–4 (2008).https://doi.org/10.1109/ICPR.2008.4761377

19. Perronnin, F., Sánchez, J., Mensink, T.: Improving the Fisher kernel for large-scale image classification. In: Daniilidis, K., Maragos, P., Paragios, N. (eds.) Computer Vision - ECCV 2010, vol. 6314, pp. 143–156. Springer, Heidelberg (2010). https://doi.org/10.1007/978-3-642-15561-1_11

20. Schraml, R., Debiasi, L., Kauba, C., Uhl, A.: On the feasibility of classification-based product package authentication. In: IEEE Workshop on Information Forensics and Security (WIFS 2017), Rennes, FR, p. 6 (2017). https://doi.org/10.1109/WIFS.2017.8267659

21. Schraml, R., Debiasi, L., Uhl, A.: Real or fake: mobile device drug packaging authentication. In: Proceedings of the 6th ACM Workshop on Information Hiding and Multimedia Security, IH&MMSec 2018, pp. 121–126. Association for Computing Machinery, New York (2018). https://doi.org/10.1145/3206004.3206016

22. Schuiki, J., Kauba, C., Hofbauer, H., Uhl, A.: Cross-sensor micro-texture material classification and smartphone acquisition do not go well together. In: Proceedings of the 11th International Workshop on Biometrics and Forensics (IWBF 2023), Barcelona, Spain, pp. 1–6 (2023)

23. Sun, B., Tan, S.R.X., Ren, Z., Chan, M.C., Han, J.: On utilizing smartphone cameras to detect counterfeit liquid food products. In: Proceedings of the 20th

Annual International Conference on Mobile Systems, Applications and Services, MobiSys 2022, pp. 551–552. Association for Computing Machinery, New York (2022). https://doi.org/10.1145/3498361.3538779

24. Sun, Y., Liao, X., Liu, J.: An efficient paper anti-counterfeiting method based on microstructure orientation estimation. In: ICASSP 2021 - 2021 IEEE International Conference on Acoustics, Speech and Signal Processing (ICASSP), pp. 2525–2529 (2021). https://doi.org/10.1109/ICASSP39728.2021.9415114

25. Tang, X.J., Tay, Y.H., Siam, N.A., Lim, S.C.: Mywood-id: automated macroscopic wood identification system using smartphone and macro-lens. In: Proceedings of the 2018 International Conference on Computational Intelligence and Intelligent Systems, CIIS 2018, pp. 37–43. Association for Computing Machinery, New York (2018). https://doi.org/10.1145/3293475.3293493

26. Taran, O., Tutt, J., Holotyak, T., Chaban, R., Bonev, S., Voloshynovskiy, S.: Mobile authentication of copy detection patterns: how critical is to know fakes? In: 2021 IEEE International Workshop on Information Forensics and Security (WIFS), pp. 1–6 (2021). https://doi.org/10.1109/WIFS53200.2021.9648398

27. Taran, O., Tutt, J., Holotyak, T., Chaban, R., Bonev, S., Voloshynovskiy, S.: Mobile authentication of copy detection patterns (2022). https://doi.org/10.48550/ARXIV.2203.02397. https://arxiv.org/abs/2203.02397

28. Voloshynovskiy, S., Diephuis, M., Beekhof, F., Koval, O., Keel, B.: Towards reproducible results in authentication based on physical non-cloneable functions: the forensic authentication microstructure optical set (famos). In: 2012 IEEE International Workshop on Information Forensics and Security (WIFS), pp. 43–48 (2012). https://doi.org/10.1109/WIFS.2012.6412623

29. Voloshynovskiy, S., Diephuis, M., Holotyak, T.: Mobile visual object identification: from sift-bof-ransac to sketchprint. In: Media Watermarking, Security, and Forensics 2015, vol. 9409, pp. 235–249. SPIE (2015)

30. Yan, Y., Zou, Z., Xie, H., Gao, Y., Zheng, L.: An IoT-based anti-counterfeiting system using visual features on QR code. IEEE Internet Things J. 8(8), 6789–6799 (2021). https://doi.org/10.1109/JIOT.2020.3035697

From Deconstruction to Reconstruction: A Plug-In Module for Diffusion-Based Purification of Adversarial Examples

Erjin Bao[1,2]([envelope]) [ORCID], Ching-Chun Chang[2] [ORCID], Huy H. Nguyen[2] [ORCID],
and Isao Echizen[1,2,3] [ORCID]

[1] The Graduate University for Advanced Studies, SOKENDAI, Kanagawa, Japan
[2] National Institute of Informatics, Tokyo, Japan
{bao-erjin,ccchang,nhhuy,iechizen}@nii.ac.jp
[3] The University of Tokyo, Tokyo, Japan

Abstract. As the use and reliance on AI technologies continue to proliferate, there is mounting concern regarding adversarial example attacks, emphasizing the pressing necessity for robust defense strategies to protect AI systems from malicious input manipulation. In this paper, we introduce a computationally efficient plug-in module, seamlessly integrable with advanced diffusion models for purifying adversarial examples. Drawing inspiration from the concept of deconstruction and reconstruction (DR), our module decomposes an input image into foundational visual features expected to exhibit robustness against adversarial perturbations and subsequently rebuilds the image using an image-to-image transformation neural network. Through the collaborative integration of the module with an advanced diffusion model, this combination attains state-of-the-art performance in effectively purifying adversarial examples while preserving high classification accuracy on clean image samples. The model performance is evaluated on representative neural network classifiers pre-trained and fine-tuned on large-scale datasets. An ablation study analyses the impact of the proposed plug-in module on enhancing the effectiveness of diffusion-based purification. Furthermore, it is noteworthy that the module demonstrates significant computational efficiency, incurring only minimal computational overhead during the purification process.

Keywords: Adversarial Example · Adversarial Purification · Diffusion Model · Robust Vision

1 Introduction

Deep learning has demonstrated exceptional performance across diverse fields, but it remains vulnerable to adversarial examples-carefully crafted input data designed to mislead machine learning models and produce incorrect predictions [34]. Such attacks can effectively manipulate models even if the changes

---

E. Bao and C.-C. Chang—These authors contributed equally.

© The Author(s), under exclusive license to Springer Nature Singapore Pte Ltd. 2024
B. Ma et al. (Eds.): IWDW 2023, LNCS 14511, pp. 48–62, 2024.
https://doi.org/10.1007/978-981-97-2585-4_4

to the input data are subtle and imperceptible to the human eye [13]. Real-world instances of adversarial attacks span various domains, from autonomous driving systems [12] to face recognition [30,31], and extend to natural language processing [1] and voice recognition [4]. These attacks have severe consequences, underscoring the urgent need for effective countermeasures.

In the realm of computer vision, image classification remains highly vulnerable to adversarial attacks that exploit either white [13,19] or black-box [3,6,9] access to data and models. Meanwhile, various countermeasures have emerged for different operational stages and scenarios [8,26,29]. One defense approach against adversarial examples is to purify them using image transformation before inputting them into a classifier. Effective transformation should eliminate added adversarial patterns while maintaining the accuracy of the original, unperturbed data. Such a purification technique should be versatile across classifiers and adversarial modifications, functioning without prior classifier or modification knowledge. Building on previous purification methods [11,14,28,41], that have made valuable contributions to the field, Nie et al. introduced the groundbreaking DiffPure method [24]. Using a diffusion model, this method has greatly advanced the state-of-the-art in the challenging domain of adversarial sample purification.

Taking inspiration from the defense strategies introduced in DiffPure and building upon the concept of robust features introduced by Ilyas et al. [18], we incorporate the idea of visual deconstruction and reconstruction (DR). Images are deconstructed by extracting the robust color and edge features, and then these features are used to reconstruct less harmful images. The DR mechanisms systematically discard adversarial altered patterns as non-robust, resulting in reconstructed images containing less harmful information. Our experiments, encompassing L_2 and L_∞ adversarial examples, convincingly demonstrate the effectiveness of our DR-based plug-in module. When used as a forward plug-in in conjunction with the aforementioned DiffPure method, it substantially enhances DiffPure's defensive capabilities against robust adversarial perturbations. Moreover, the additional computational time required for incorporating the DR module is negligible when compared with the computational overhead of diffusion-based setups.

Our contributions can be summarized as follows:

- **De/Reconstruction Mechanisms:** We have developed de/reconstruction mechanisms based on robust color and edge features that withstand adversarial modifications. These mechanisms mitigate adversarial effects by eliminating detrimental patterns as non-robust, resulting in less harmful images.
- **Efficient Plug-In Module:** Our approach seamlessly integrates the proposed plug-in pre-purification DR module with the diffusion-based post-purification model, providing an effective defense mechanism for stable classification.
- **Superior Defense Performance:** Experimental results demonstrate that our method outperforms state-of-the-art methods in defending against strong adversarial noise while incurring minimal additional computing time.

This paper is structured as follows: In Sect. 2, we offer an overview of related work, delving into various attack scenarios, types of attacks, and existing defense strategies. Section 3 introduces our proposed defensive DR module and integrated DR-Pure method, providing a detailed exposition of the DR mechanisms and the overarching purification process. In Sect. 4, we describe the experimental setup and present evaluation results demonstrating the effectiveness of our integrated DR-Pure approach, along with its minimal computational overhead. Lastly, Sect. 5 serves as the conclusion, summarizing the key findings and offering insights into future research directions.

2 Related Work

In this section, we provide a brief overview of the fundamental concepts related to adversarial attacks. We describe adversarial attacks in various application scenarios and examine potential strategies for detecting and defending against them.

2.1 Adversarial Attacks Scenarios

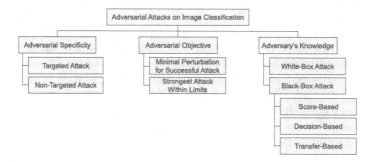

Fig. 1. Taxonomy of adversarial attacks on image classification

An adversarial example is an altered input designed to trick AI models [34]. Although the human eyes might overlook these slight changes, they can lead to substantial errors in the machine's response [13]. The crafting of adversarial examples, or "adversarial attacks," aims to mislead AI models. A relevant context for discussing adversarial attacks is image classification. Attacks on image classification can be targeted or non-targeted depending on the desired misclassification [42]. In a targeted attack, the goal is to create specific false predictions for a particular class, while a non-targeted attack aims for any incorrect predictions, showcasing a lack of specificity. An adversarial task combines achieving erroneous performance with controlled modifications. The actual expression of the adversarial objective adapts depending on the integration of these two elements. In essence, there are two prevalent types of adversarial tasks: one seeks to identify the minimal perturbation necessary to achieve a successful attack [5,23,34], while the other focuses on crafting the strongest attack feasible within the modification limits [13,19,21].

Adversarial examples stem from differing model knowledge. White-box attacks utilize abundant information, encompassing results and gradients. They use gradient-based [13,19,21,23], optimization-based [5,34], and generation-based [38] approaches. Black-box attacks lack target model access and often resort to gradient-free queries. Score-based black-box methods craft attacks using probability scores from queries, approximating gradients through gradient-free methods [1,6,17,35]. In more intricate black-box scenarios with only hard-label decisions, techniques like random walks [3] and evolution strategies [10] are used to create adversarial examples. Another black-box approach assumes transferability, crafting examples against surrogate models that can effectively deceive target models [9,34,40].

2.2   Countermeasures Against Adversarial Attacks

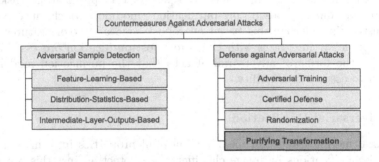

Fig. 2. Taxonomy of adversarial attack countermeasures: detection and defense

In response to adversarial attacks, a variety of approaches can be used to detect adversarial examples and prevent classification services for problematic samples. These approaches are based on feature learning [41], distribution statistics [15,16], or intermediate layer outputs [22]. In addition, a range of defense strategies customized for various stages and availability scenarios have been introduced, bolstering the robustness of classification models against adversarial attacks. Adversarial training [13] can improve a classifier's internal adversarial robustness through training using adversarial samples as augmented data [29,43]. Certified defense [26,32,36,37] establishes conditions under which classifiers can be formally proven to withstand specific adversarial perturbations. Randomization introduces randomness into inputs [25,39] or models [8] to mitigate the effect of adversarial attacks.

The defense strategy known as adversarial purification, explored in this paper, involves a defensive transformation during image preprocessing. It is primarily aimed at correcting adversarial predictions while preserving the accuracy of the original data. This universal defensive preprocessing strategy is designed to be effective across various classifiers and adversarial modifications and requires no prior knowledge of classifiers or attack strategies. Within the realm of adversarial purification studies, previous effective techniques have included methods such as JPEG compression [11], bit-depth reduction [41], total variance minimization [14], autoencoder-based denoising [20], and data distribution projection using generative models [28,33]. Nie et al. greatly advanced the state-of-the-art in this challenging domain with their introduction of a groundbreaking adversarial purification solution called DiffPure [24]. Utilizing a diffusion model, DiffPure first diffuses an adversarial image by adding a small amount of noise in a forward diffusion step and then recovers a clean image using a reverse generative process. This innovative approach not only enhances the effectiveness of adversarial purification but also provides a valuable foundation for further research aimed at advancing defense strategies.

3   Diffusion Purification with DR Mechanisms

In this section, we first introduce the adversarial purification task and then present our proposed pre-purification module, which incorporates de/reconstruction mechanisms based on robust edge and color features. This module seamlessly integrates with DiffPure's post-purification process. Finally, we explore the forms of adversarial attacks that the purification model may encounter to assess its reliability.

3.1   Adversarial Purification

Adversarial modifications can introduce harmful properties into images, leading to misclassifications by image classifiers. To protect against this, we apply defensive transformations to adversarial examples before they undergo classification, with the aim of eliminating adversarial features. Our approach focuses on developing a transformation that not only corrects classification for modified examples but also maintains the accuracy of unaltered images (Fig. 3).

Given a dataset $\mathbb{D}$ comprising N images x_i with their true labels y_i, a classification model f is evaluated on the basis of its average correctness in predicting $pred_i = f(x_i)$, referred to as classification accuracy:

$$\text{Acc}(\mathbb{D}, f) = \frac{\sum_{i=1}^{N} \delta(f(x_i), y_i)}{N}, \tag{1}$$

where

$$\delta(f(x_i), y_i) = \begin{cases} 1 & \text{if } f(x_i) = y_i, \\ 0 & \text{otherwise.} \end{cases} \tag{2}$$

Adversarial attack algorithms, denoted as $\mathcal{A}_f$, aim to generate adversarial samples $x_{adv,i} = \mathcal{A}_f(x_{ori,i})$ from clean images $x_{ori,i}$. Their primary objective is to induce incorrect predictions $pred_{adv,i} \neq y_i$, thereby reducing the classification accuracy Acc_{adv} for modified images $\mathbb{D}_{adv}$.

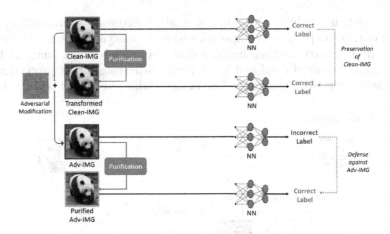

Fig. 3. Overview of adversarial purification: preserving clean performance and defending against adversarially modified images.

Figure 3 illustrates the defensive transformation and compares it with the conventional approach. In this context, adversarial (Adv) purification takes place during image (IMG) preprocessing before the images are fed into a classifier. The final prediction outcome after applying the purification transformation defense $\mathcal{D}$ is represented as $pred_{\mathcal{D}}$, calculated as $pred_{\mathcal{D}} = f(\mathcal{D}(x))$ for both adversarial and clean images. The effectiveness of the defense is evaluated using classification accuracy, where higher accuracy indicates better performance of the defense mechanism. An effective defense method should meet the following criteria:

a) Enhanced correction of adversarial examples $\mathbb{D}_{adv}$:

$$\text{Acc}_{\mathcal{D}}(\mathbb{D}_{adv}, f) > \text{Acc}(\mathbb{D}_{adv}, f). \tag{3}$$

b) Maintaining of accuracy for clean, unmodified images $\mathbb{D}_{ori}$:

$$\text{Acc}_{\mathcal{D}}(\mathbb{D}_{ori}, f) \approx \text{Acc}(\mathbb{D}_{ori}, f). \tag{4}$$

3.2 De/Reconstructive Module Based on Robust Features

Expanding upon DiffPure's exceptional performance in safeguarding unknown classifiers from unforeseen adversarial modifications, we introduce the efficient plug-in module known as the DR module. When applied prior to DiffPure and in conjunction with it, the DR module substantially enhances the effectiveness of adversarial purification, particularly when facing strong adversarial perturbations. This comprehensive approach is labeled DR-Pure, where the DR module plays a central role. Figure 4 illustrates the DR-Pure workflow. The DR-Pure workflow begins with the DR module deconstructing the input image into robust color and edge feature maps and then using these retrieved maps to reconstruct the image. In the subsequent DiffPure phase, the reconstructed image is diffused and recovered, ultimately achieving the goal of adversarial purification. The strength of this comprehensive approach resides in its synergistic use of the DR module's de/reconstruction mechanisms centered around robust image features and the diffusion model employed by DiffPure, delivering robust protection for adversarial purification, even in the face of formidable adversarial perturbations.

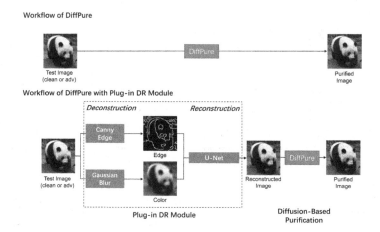

Fig. 4. Diffusion purification against adversarial examples based on deconstruction and reconstruction mechanisms.

The design of the DR module is built upon the notion of adversarial robust features, as introduced by Ilyas et al. [18]. They categorize features into adversarially robust and non-robust based on their vulnerability to adversarial noise. In light of this, features such as image structure and color exhibit higher resistance to various adversarial alterations, making them inherently more robust. Furthermore, by leveraging the structure and color feature maps deconstructed from the original image, we can reconstruct the entire image. This approach enables a reconstructed image with fewer harmful features to be obtained through a two-step process. We begin by deconstructing images that may potentially contain harmful adversarial attacks to isolate resilient color and structural features. We then use these decomposed robust features to reconstruct the images. Adversarial noise itself, introduced through adversarial modifications, lacks robustness and is vulnerable to adversarial attacks. By filtering out and discarding the harmful properties during the deconstruction process, we obtain reconstructed images that are generally less harmful.

In the initial phase of image deconstruction, we prioritize extracting robust features, specifically Canny edge maps and Gaussian blurred color maps, while excluding all other features. Canny edge maps emphasize edges and boundaries, inherently containing structural information that is less susceptible to adversarial manipulations. Gaussian blurred color maps retain essential color information while simultaneously reducing noise and fine details, which may potentially include harmful adversarial patterns. Canny edge maps are generated using

OpenCV's Canny edge detection function after applying a Gaussian kernel to reduce image noise. This results in a single-channel Canny edge feature map with the same spatial dimensions as the original image. Similarly, Gaussian blurred color maps are created using a kernel size that balances color preservation and detail reduction. The resulting Gaussian blurred color map is a three-channel RGB image with the same spatial dimensions as the original. After the robust edge and color feature map are extracted and harmful adversarial patterns are removed, the image is reconstructed. This is done using a U-Net neural network model, treating the reconstruction as an image-to-image translation. U-Net takes feature maps as input and generates an RGB color image as output. Each input and corresponding output have the same spatial size as the original image. The input feature maps contain four channels, whereas the output RGB images have three channels.

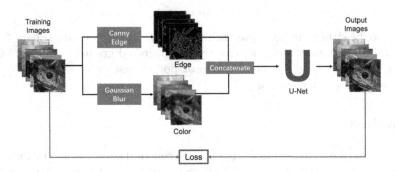

Fig. 5. Workflow of the U-Net training process, involving deconstructed features as inputs and reconstructed images as outputs.

Prior to integrating the U-Net model into our defense framework for purifying adversarial samples, we train it using feature maps generated from clean, unaltered images in the training set, as shown in Fig. 5. The model's training objective is to minimize the mean squared error (MSE) loss between the original clean image and the corresponding reconstructed version. The model's generalization ability is enhanced by randomly selecting the thresholds for Canny edge detection and varying the blur kernels for color maps during training. These variations go beyond the condition used in the actual defense. The U-Net network model is exclusively trained on feature maps extracted from clean, unaltered images to capture inherent data characteristics and generate reconstructed images closely resembling the originals. Importantly, we refrain from any adversarial-related augmentation during training to align with our method's objective of operating without prior knowledge of classification models or adversarial attack techniques.

3.3 L_p-Constrained Adversarial Attacks

The effectiveness of our purification defense is not limited to specific attack types or classification models. This means that our adversarial purification model, in addition to being trained on clean images, doesn't require prior knowledge of adversarial attacks or the specific classification models it aims to protect.

To assess our defense efficacy, we consider adversarial attacks with non-targeted specificity and white-box knowledge. In this scenario, attackers have complete access to the classification model, and an attack is considered successful if the classifier's output is any category other than the correct label. We use a widely adopted threat model based on the L_p norm [7], whose objective is to find the strongest adversarial attack within a perturbation budget denoted as ε. This budget limits the L_p norm of the vector difference between the adversarial example x_{adv} and the clean image x_{ori}, ensuring that all feasible modifications stay within this constraint. Typically, norms with p values of 2 or infinity are considered for adversarial testing. Mathematically, an adversarial example x_{adv} derived from a clean image x_{ori} adheres to the following L_p norm constraint:

$$\|x_{adv} - x_{ori}\|_p \leq \varepsilon. \tag{5}$$

where p can be either 2 or ∞, representing the choice of norm used to quantify the perturbation. For our attacks, we employ the projected gradient descent (PGD) method [21], recognized as one of the most potent first-order attacks [29], and acknowledged by Athalye et al [2].

4   Evaluation

In this section, we start by outlining our experimental setup, which includes the training and test settings, the DR module parameter settings, and the benchmarks. We then present experimental results illustrating the performance of our DR-Pure method on adversarial and clean images, along with the computation times for each processing step.

4.1   Experimental Setup

Training and Test Settings. To assess purification performance, we employ the Tiny-ImageNet dataset, a subset of ImageNet, comprising 500 training, 50 validation, and 50 test images, each initially at 64×64–pixel resolution. We upscale all images to 128×128 pixels and convert grayscale ones to RGB for consistency. We then apply our DR module, introduce adversarial modifications, and perform classification at this higher resolution. This higher resolution is used to improve edge extraction, particularly for fine details. In practice, edges often manifest as line-like structures, which require greater resolution for accurate representation compared with low-resolution images in which a single edge pixel might span a larger area. For purification using DiffPure, we utilize their official diffusion model designed for ImageNet at 224×224 resolution. Given that Tiny-ImageNet is an ImageNet subset, we directly adapt this model by adjusting the image resolution to 224×224.

As the test set comprises unlabeled images, we train our DR module on the training set and evaluate purification performance on the validation set.

We test our purification DR-Pure method against diverse adversarial attacks on three representative classification models with distinct architectures: ResNet-18, VGG-16, and ConvNeXt. These models are originally pre-trained on ImageNet, but we adapt them for our experiments by adjusting them for variations in image resolution and class count. We fine-tune them using the Tiny-ImageNet training dataset and adjust the final layer for the transition from 1000 to 200 classes. Adversarial attacks are performed using Foolbox's PGD function [27], starting from the original data points without random initialization to ensure experiment determinism and reproducibility. We maintain consistency in other experiment-influencing factors, such as relative step size and step number, throughout the study by using their default settings. The perturbation budget is the only parameter adjusted to control adversarial perturbation strength.

DR Module Parameter Settings. We use OpenCV's Canny function to extract edge structure features, with fixed thresholds of 45 (low) and 150 (high) during transformation deployment. The thresholds used for U-Net training are randomly selected between 10–80 (low) and 100–150 (high). In both cases, a 3×3 Gaussian kernel is applied for noise reduction before the Canny edge function. For color feature extraction during defense, we employ OpenCV's Gaussian blur function with a 13×13 kernel size. For U-Net training, we use various square kernel sizes, ranging from 3×3 (smallest) to 13×13 (actual size for robust color maps in defense). The U-Net training encompasses broader conditions than those encountered during defense predictions to enhance the network's generalization ability.

Benchmark. We test adversarial purification methods on both adversarial and clean samples without disclosing details of the L_2 and L_∞ adversarial attacks or the use of the ResNet-18, VGG-16, or ConvNeXt classification networks to the purification model. To assess the performance of our proposed DR-Pure method in adversarial purification, we compare it to that of the state-of-the-art method, DiffPure. Since the post-purification step of DR-Pure is built upon DiffPure's diffusion model, our comparison of the DR-Pure with the DiffPure constitutes an ablation study of pre-purification process using our designed DR module. Instances initially misclassified by the classifier, resulting in further misclassifications after adversarial attacks, involve multiple contributing factors, not simply the adversarial attack itself. To gain a better understanding of DR-Pure's effectiveness in purifying adversarial samples, it is crucial to isolate these initially misclassified samples to prevent potential confounding of experimental outcomes. We thus curate, in addition to the standard Tiny-ImageNet dataset, a selected dataset consisting exclusively of images correctly classified by the three neural network classifiers. The use of this selected dataset ensures that the adversarial attacks modified only images that are originally classified correctly. Our experiments uses 1000 images from both the standard and selected datasets.

4.2   Results

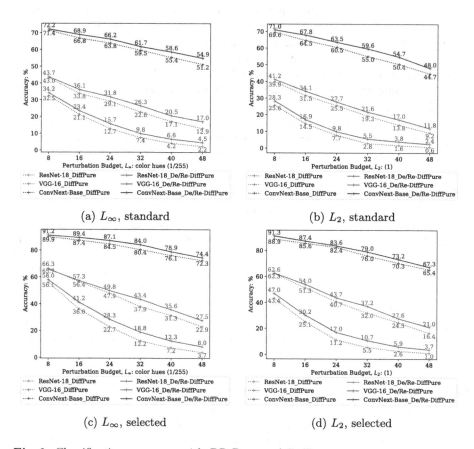

(a) L_∞, standard

(b) L_2, standard

(c) L_∞, selected

(d) L_2, selected

Fig. 6. Classification accuracy with DR-Pure and DiffPure for various L_2 and L_∞ adversarial perturbation strengths, tested on standard and selected images.

Purification Performance on Adversarial Images. Figure 6 compares the classification accuracies of the three classifiers between DR-Pure and the baseline (DiffPure) for various L_2 and L_∞ adversarial perturbation strengths. DR-Pure outperforms the baseline except in a specific scenario (L_∞ and ResNet-18 at a perturbation budget of 8, for which the performances are comparable). In other scenarios across different perturbation strengths and classifiers, DR-Pure typically improves classification accuracy by over 2%. In certain scenarios, the improvement is as much as 4.6% on the standard dataset and 6.6% on the selected dataset. These results demonstrate that our plug-in DR module substantially enhances adversarial purification.

Table 1. Classification accuracy for clean images with DiffPure and DR-Pure for three classifiers on standard and selected images.

(a) Standard images

Acc %	None	DiffPure	DR-Pure
ResNet-18	67.1	59.1	56.7
VGG-16	64.9	55.1	53.1
ConvNeXt	87.5	79.7	77.9

(b) Selected images

Acc %	None	DiffPure	DR-Pure
ResNet-18	100	86.1	85.7
VGG-16	100	85.2	82.3
ConvNeXt	100	95.5	96.5

Purification Performance on Clean Images. Table 1 summarizes the results for DR-Pure and the baseline on both standard and selected images for the three classifiers. For the standard images, the accuracy with DR-Pure is about 2% lower than that with the baseline. For the selected images, the performance varies: 2.9% lower for VGG-16 and 1% higher for ConvNeXt. Overall, the plug-in DR module has a balanced effect on clean sample accuracy.

Table 2. Computation Time (in seconds)

Plug-in DR module		DiffPure (GPU)
Deconstruction (CPU)	Reconstruction (GPU)	
0.7–1.5	1.5–1.8	$(5.3–5.4) \times 10^3$

Time Consumption. We run the experiment on a server powered by two AMD EPYC 7543P 32-Core CPUs and one NVIDIA A100 Tensor Core GPU. Table 2 shows the computation times for the two processing steps, where each data point includes 1000 images. The total execution time of the plug-in DR module with CPU-based deconstruction and GPU-based reconstruction is less than one-thousandth that of DiffPure. This means that the extra computation time introduced by our plug-in DR module is negligible.

5   Conclusion

With the spread of AI-driven automated technologies, the robustness of classification neural networks against adversarial attacks has become a persistent and urgent challenge. Our plug-in DR module substantially enhances the performance of an advanced diffusion-based purification model. The module deconstructs an image into robust features (e.g., colors and edges) and reconstructs it using an image-to-image transformation neural network (i.e., U-Net). In collaboration with a pre-trained diffusion model, we achieve state-of-the-art performance in purifying adversarial examples while preserving classification accuracy on clean samples. We test our model's purification performance on three representative classification networks (ResNet, VGG, and ConvNeXt) pre-trained on the ImageNet dataset and fine-tuned on the Tiny-ImageNet dataset. The module is computationally efficient since it adds minimal computational overhead to the purification process. While we have demonstrated the potential of fortifying deep learning classifiers with basic visual features, other robust features are worth further investigation. The development of an adaptive approach that learns the robust features automatically may also be a possible direction. This work should pave the way for a more secure and robust landscape in the field of AI research.

Acknowledgments. This work was partially supported by JSPS KAKENHI Grants JP18H04120, JP20K23355, JP21H04907, and JP21K18023, and by JST CREST Grants JPMJCR18A6 and JPMJCR20D3, Japan.

References

1. Alzantot, M., Sharma, Y., Elgohary, A., Ho, B.J., Srivastava, M.B., Chang, K.W.: Generating natural language adversarial examples. In: Proceedings of Conference on Empirical Methods Natural Language Processing (EMNLP) (2018)
2. Athalye, A., Carlini, N., Wagner, D.: Obfuscated gradients give a false sense of security: circumventing defenses to adversarial examples. In: Proceedings of International Conference on Machine Learning (ICML) (2018)
3. Brendel, W., Rauber, J., Bethge, M.: Decision-based adversarial attacks: reliable attacks against black-box machine learning models. In: Proceedings of International Conference on Learning Representations (ICLR) (2018)
4. Carlini, N., et al.: Hidden voice commands. In: Proceedings of USENIX Security Symposium (USENIX Security) (2016)
5. Carlini, N., Wagner, D.: Towards evaluating the robustness of neural networks. In: Proceedings of IEEE Symposium on Security and Privacy (SP) (2017)
6. Chen, P.Y., Zhang, H., Sharma, Y., Yi, J., Hsieh, C.J.: Zoo: zeroth order optimization based black-box attacks to deep neural networks without training substitute models. In: Proceedings of ACM Workshop Artificial Intellgient Security (AISec) (2017)
7. Croce, F., et al.: Robustbench: a standardized adversarial robustness benchmark. In: Proceedings of Advance Neural Information Processing System (NeurIPS) (2021)
8. Dhillon, G.S., et al.: Stochastic activation pruning for robust adversarial defense. In: Proceedings of International Conference on Learning Representations (ICLR) (2018)
9. Dong, Y., et al.: Boosting adversarial attacks with momentum. In: Proceedings of IEEE Conference on Computer Vision on Pattern Recognition (CVPR) (2018)
10. Dong, Y., et al.: Efficient decision-based black-box adversarial attacks on face recognition. In: Proceedings of IEEE Conference on Computer Vision on Pattern Recognition (CVPR) (2019)
11. Dziugaite, G.K., Ghahramani, Z., Roy, D.M.: A study of the effect of JPG compression on adversarial images. arXiv preprint arXiv:1608.00853 (2016)

12. Eykholt, K., et al.: Robust physical-world attacks on deep learning visual classification. In: Proceedings of IEEE Conference on Computer Vision and Pattern Recognition (CVPR) (2018)
13. Goodfellow, I.J., Shlens, J., Szegedy, C.: Explaining and harnessing adversarial examples. In: Proceedings of International Conference on Learning Representations (ICLR) (2015)
14. Guo, C., Rana, M., Cisse, M., van der Maaten, L.: Countering adversarial images using input transformations. In: Proceedings of International Conference on Learning Representations (ICLR) (2018)
15. Hendrycks, D., Gimpel, K.: A baseline for detecting misclassified and out-of-distribution examples in neural networks. In: Proceedings of International Conference on Learning Representations (ICLR) (2017)
16. Hendrycks, D., Gimpel, K.: Early methods for detecting adversarial images. In: Proceedings of International Conference on Learning Representations Workshop (ICLR) (2017)
17. Ilyas, A., Engstrom, L., Athalye, A., Lin, J.: Black-box adversarial attacks with limited queries and information. In: Proceedings of International Conference on Machine Learning (ICML) (2018)
18. Ilyas, A., Santurkar, S., Tsipras, D., Engstrom, L., Tran, B., Madry, A.: Adversarial examples are not bugs, they are features. In: Proceedings of Advance Neural Information Processing System (NeurIPS) (2019)
19. Kurakin, A., Goodfellow, I.J., Bengio, S.: Adversarial examples in the physical world. In: Proceedings of International Conference on Learning Representations Workshop (ICLR) (2017)
20. Liao, F., Liang, M., Dong, Y., Pang, T., Hu, X., Zhu, J.: Defense against adversarial attacks using high-level representation guided denoiser. In: Proceedings of IEEE Conference on Computer Vision Pattern Recognition (CVPR) (2018)
21. Madry, A., Makelov, A., Schmidt, L., Tsipras, D., Vladu, A.: Towards deep learning models resistant to adversarial attacks. arxiv:1706.06083 (2017)
22. Metzen, J.H., Genewein, T., Fischer, V., Bischoff, B.: On detecting adversarial perturbations. In: Proceedings of International Conference on Learning Representations (ICLR) (2017)
23. Moosavi-Dezfooli, S.M., Fawzi, A., Frossard, P.: Deepfool: a simple and accurate method to fool deep neural networks. In: Proceedings of IEEE Conference on Computer Vision and Pattern Recognition (CVPR) (2016)
24. Nie, W., Guo, B., Huang, Y., Xiao, C., Vahdat, A., Anandkumar, A.: Diffusion models for adversarial purification. In: Proceedings of International Conference on Machine Learning (ICML) (2022)
25. Pang, T., Xu, K., Zhu, J.: Mixup inference: better exploiting mixup to defend adversarial attacks. In: Proceedings of International Conference on Learning Representations (ICLR) (2020)
26. Raghunathan, A., Steinhardt, J., Liang, P.: Certified defenses against adversarial examples. In: Proceedings of International Conference on Learning Representations (ICLR) (2018)
27. Rauber, J., Brendel, W., Bethge, M.: Foolbox: a python toolbox to benchmark the robustness of machine learning models. In: Proceedings of International Conference on Machine Learning (ICML) (2017)
28. Samangouei, P., Kabkab, M., Chellappa, R.: Defense-GAN: protecting classifiers against adversarial attacks using generative models. In: Proceedings of International Conference on Learning Representations (ICLR) (2018)

29. Shafahi, A., et al.: Adversarial training for free! In: Proceedings of Advance Neural Information Processing System (NeurIPS) (2019)
30. Sharif, M., Bhagavatula, S., Bauer, L., Reiter, M.K.: Accessorize to a crime: real and stealthy attacks on state-of-the-art face recognition. In: Proceedings of ACM SIGSAC Conference on Computer Communication Security (CCS) (2016)
31. Sharif, M., Bhagavatula, S., Bauer, L., Reiter, M.K.: A general framework for adversarial examples with objectives. ACM Trans. Priv. Secur. (TOPS) **22**(3), 1–30 (2019)
32. Sinha, A., Namkoong, H., Duchi, J.: Certifying some distributional robustness with principled adversarial training. In: Proceedings of International Conference on Learning Representations (ICLR) (2018)
33. Song, Y., Kim, T., Nowozin, S., Ermon, S., Kushman, N.: Pixeldefend: Leveraging generative models to understand and defend against adversarial examples. In: Proceedings of International Conference on Learning Representations (ICLR) (2018)
34. Szegedy, C., et al.: Intriguing properties of neural networks. In: Proceedings of International Conference on Learning Representations (ICLR) (2014)
35. Uesato, J., O'Donoghue, B., van den Oord, A., Kohli, P.: Adversarial risk and the dangers of evaluating against weak attacks. In: Proceedings of International Conference on Machine Learning (ICML) (2018)
36. Wong, E., Kolter, Z.: Provable defenses against adversarial examples via the convex outer adversarial polytope. In: Proceedings of International Conference on Machine Learning (ICML) (2018)
37. Wong, E., Schmidt, F., Metzen, J.H., Kolter, J.Z.: Scaling provable adversarial defenses. In: Proceedings of Advance Neural Information Processing System (NeurIPS) (2018)
38. Xiao, C., Li, B., Zhu, J.Y., He, W., Liu, M., Song, D.: Generating adversarial examples with adversarial networks. In: Proceedings of International Joint Conference on Artificial Intelligence (IJCAI) (2018)
39. Xie, C., Wang, J., Zhang, Z., Ren, Z., Yuille, A.: Mitigating adversarial effects through randomization. In: Proceedings of International Conference on Learning Representations (ICLR) (2018)
40. Xie, C., et al.: Improving transferability of adversarial examples with input diversity. In: Proceedings of IEEE Conference on Computer Vision Pattern Recognition (CVPR) (2019)
41. Xu, W., Evans, D., Qi, Y.: Feature squeezing: detecting adversarial examples in deep neural networks. In: Proceedings of Network Distribution System on Security Symposium (NDSS) (2018)
42. Yuan, X., He, P., Zhu, Q., Li, X.: Adversarial examples: attacks and defenses for deep learning. IEEE Trans. Neural Netw. Learn. Syst. **30**, 2805–2824 (2019)
43. Zhang, D., Zhang, T., Lu, Y., Zhu, Z., Dong, B.: You only propagate once: accelerating adversarial training via maximal principle. In: Proceedings of Advances Neural Information Processing System (NeurIPS) (2019)

Privacy-Preserving Image Scaling Using Bicubic Interpolation and Homomorphic Encryption

Donger Mo[1][ID], Peijia Zheng[1]([✉])[ID], Yufei Zhou[1][ID], Jingyi Chen[1][ID], Shan Huang[1][ID], Weiqi Luo[1][ID], Wei Lu[1][ID], and Chunfang Yang[2][ID]

[1] School of Computer Science and Engineering, Guangdong Key Laboratory of Information Security Technology, Sun Yat-sen University, Guangzhou 510006, China
zhpj@mail.sysu.edu.cn
[2] Henan Key Laboratory of Cyberspace Situation Awareness, Zhengzhou 450001, China

Abstract. With the advancement of cloud computing, outsourced image processing has become an attractive business model, but it also poses serious privacy risks. Existing privacy-preserving image scaling techniques often use secret sharing schemes or the Paillier cryptosystem to protect privacy. These methods require the collaboration of multiple servers or only support additive operations, which increases the difficulty of data storage and complicates image processing. To address these issues, this paper focuses on cloud-based privacy-preserving image scaling in the encrypted domain. We propose an image scaling scheme based on homomorphic encryption, which allows cloud servers to perform scaling operations on encrypted images using bicubic interpolation. To avoid the high storage and communication costs of per-pixel encryption, we introduce an efficient data encoding method where a single ciphertext contains the information of an entire image. This significantly reduces the storage space and communication overhead with the cloud server, while our scheme supports computational operations in this data format. Our experimental results validate the feasibility of the proposed scheme, which outperforms existing schemes in terms of storage overhead and operational efficiency.

Keywords: Homomorphic encryption · bicubic interpolation · single instruction multiple data · cloud computing

1 Introduction

In the emerging era of big data, multimedia data, especially image data such as personal photos and satellite images, is swelling at an unprecedented rate. The local storage and processing of such a colossal amount of image data is becoming an arduous task for resource-strapped data owners. Cloud computing, with its powerful computing capabilities and expansive storage, provides a

B. Ma et al. (Eds.): IWDW 2023, LNCS 14511, pp. 63–78, 2024.
https://doi.org/10.1007/978-981-97-2585-4_5

viable solution, offering data owners the convenience of outsourcing data storage and processing to cloud servers. However, this convenience comes at the risk of potential privacy breaches and security threats, as users relinquish control over their images once they are uploaded to the cloud. Therefore, the establishment of privacy-preserving image processing techniques in cloud computing is imperative.

To address this issue, many researchers have used signal processing techniques [1] in the encrypted domain to develop numerous privacy-preserving image processing solutions, including image denoising [2], signal and image processing [3], and discrete wavelet transforms in the encrypted domain [4]. These techniques enable cloud servers to directly process and compute encrypted data without accessing the private information of the data owners, thereby protecting the privacy of users while taking advantage of the convenience of cloud servers. This is mainly achieved by Secure Multi-party Computation (SMPC) or Homomorphic Encryption (HE) techniques. For example, Mohanty et al. [5] proposed a secret image sharing scheme that allows users to scale or crop ciphertext images directly through bilinear interpolation in the encrypted domain, demonstrating the potential of using SMPC.

Homomorphic encryption also plays an important role in signal processing in the encrypted domain. HE allows cloud servers to perform additions or multiplications on encrypted data without decryption, ensuring identical results to operations performed on plaintext data. This revolutionary concept ensures the confidentiality of user data during computations on cloud servers. In this paper, we adopt the CKKS scheme [8], a type of Somewhat Homomorphic Encryption (SWHE) scheme, which is known to provide an excellent balance between efficiency and precision in practical applications that do not require absolute accuracy [9,26].

Although there have been researches using HE for image scaling, including efforts by Mohanty et al. [10] and Anurenjini et al. [11] using extended Paillier cryptosystems [7] to implement bilinear interpolation algorithms in encrypted domains for multi-user setups, the challenges remain. The Paillier system [7] only facilitates addition operations and restricts the encryption to the integer domain, requiring decimal data to be scaled to integers prior to encryption.

To address these shortcomings, this paper introduces a scheme that uses bicubic interpolation for image scaling in the encrypted domain. We use the CKKS encryption system to protect private images, which allows cloud servers to scale images using bicubic interpolation on encrypted data. We also use the Single Instruction Multiple Data (SIMD) technique to compute multiple samples simultaneously in the HE domain. Furthermore, to mitigate the high storage requirements of traditional per-pixel encryption, we propose a data encoding method that embeds an entire image into a single ciphertext. Our image scaling scheme supports computation on this data format.

Our contributions can be summarised as follows.

- Based on homomorphic encryption, we propose a scheme for image scaling in the encrypted domain using bicubic interpolation. In addition, a novel data encoding method is proposed, which greatly reduces the storage space and communication overhead.
- Experimental results show that our scheme can effectively zoom in and out the encrypted image and maintain good visual quality after zooming, which is better than the existing schemes in many aspects.

The rest of the paper is organised as follows: Sect. 2 reviews existing researches on image scaling processing in encrypted domains. Section 3 explains the preliminary knowledge related to this study. Section 4 describes the proposed scheme in detail, followed by a security analysis in Sect. 5. Section 6 presents experimental results and comparisons with existing works. Finally, Sect. 7 concludes the paper and discusses future prospects.

2   Related Works

In existing studies, Deepthi et al. [6] implemented methods for image linear interpolation and two-dimensional discrete cosine transformation in the encrypted domain using secret sharing techniques. Tanwar et al. [12] accomplished the bicubic interpolation algorithm in the encrypted domain using the same Privacy-protecting techniques. Their algorithm first generates obfuscated images through scaling and randomization, followed by modular operations to create shared images. The servers then use two-dimensional bicubic interpolation to scale the secret images within the encrypted domain, which has shown promising results. Mohanty et al. [5] proposed a secret image sharing scheme that achieves bilinear interpolation in the encrypted domain. This approach allows users to retrieve scaled or cropped versions of the secret image by directly manipulating the shadow images, thus reducing the amount of data transmitted from cloud servers to users. Later, Mohanty et al. [10] used an extended Paillier encryption scheme to facilitate image scaling and cropping for multi-user settings in the encrypted domain, allowing cloud servers to scale and crop images in the encrypted domain. To avoid the high storage overhead of the original per-pixel encryption, they introduced a space-saving tiling scheme that allows scaling and cropping of images at the tiled level, thereby encrypting an entire block of pixels instead of encrypting each pixel value separately. Anurenjini et al. [11] also used the improved Paillier encryption scheme for bilinear interpolation in the encrypted domain and integrated it with image watermarking for image authentication, which significantly improves image confidentiality.

Nevertheless, the image quality produced by bilinear interpolation schemes is suboptimal, and secret sharing schemes require collaborative computation across multiple servers, which increases data storage and complicates image processing. Although the Paillier encryption scheme has been used for privacy in some studies [10,11], it only supports additive operations, making multiplicative operations

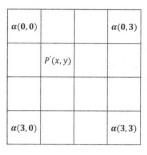

Fig. 1. Distance from the point $P'(x, y)$ 16 nearest pixels.

inconvenient. In addition, the Paillier encryption scheme does not allow for real number encryption, requiring convert decimals into integers, which compromises usability.

3   Preliminary

3.1   Bicubic Interpolation

Bicubic interpolation is a method used for image scaling, where the pixel values of the scaled image are computed using the weighted sum of 16 neighboring pixel values. The specific positional relationship of these neighboring points is illustrated in Fig. 1. In the detailed calculation process, we assume the dimensions of the original image are $m \times n$, and the dimensions of the target image are $m' \times n'$. Let $P(X, Y)$ represent a pixel in the target image, and $P'(x, y)$ represent a pixel in the original image, where (x, y) and (X, Y) denote the pixel value positions in the images. The pixel value in the target image can be obtained using the following formula:

$$P(X, Y) = \sum_{i=0}^{3} \sum_{j=0}^{3} a(i, j) \times W(i, j) \tag{1}$$

where $a(i, j)$ denotes the value of the pixel $P'(x + i - 1, y + j - 1)$ in the original image. Given known (X, Y), we can determine its corresponding position in the original image using $(x, y) = (X \times \lfloor \frac{m}{m'} \rfloor, Y \times \lfloor \frac{n}{n'} \rfloor)$. To calculate the weights $W(i, j)$ in the one-dimensional direction, we follow the method described in [14], i.e.,

$$W(l) = \begin{cases} (\alpha + 2)|l|^3 - (\alpha + 3)|l|^2 + 1, & |l| \leq 1 \\ \alpha|l|^3 - 5\alpha|l|^2 + 8\alpha|l| - 4\alpha, & 1 < |l| < 2 \\ 0, & \text{otherwise} \end{cases} \tag{2}$$

where α ($\alpha = -0.5$) is a specified parameter, l is the distance between two points $a(i, j)$ and $P'(x, y)$ on rows or columns. The two-dimensional weights $W(i, j)$ are obtained by multiplying the weights in horizontal and vertical directions.

3.2 CKKS Cryptosystem

In this work, we employ CKKS [8] or RNS-CKKS [25] based on Ring Learning with Errors problem (RLWE) [13]. We denote the cyclotomic polynomial ring of dimension N by $\mathcal{R} = \mathbb{Z}[X]/(X^N + 1)$, where N is a powers-of-two and $\mathbb{Z}[X]$ is the polynomial ring with integer coefficients. This structure delineates the plaintext and ciphertext domains as residing in $\mathcal{R}$ and $R_l = \mathcal{R}/2^l\mathcal{R}$, respectively, with l belonging to the set of positive integers. The CKKS scheme exploits the single instruction-multiple data (SIMD) strategy, facilitated through a complex canonical embedding map, $\phi : \mathbb{C}^k \to \mathcal{R}$, to batch process numerous plaintexts in the homomorphic encryption (HE) space.

Within this framework, we consider two complex vector variables, namely $\mathbf{x} = (x_0, x_1, \ldots, x_{k-1})$ and $\mathbf{y} = (y_0, y_1, \ldots, y_{k-1})$, constrained by $k \leq N/2$. By passing the detailed representation of the canonical map ϕ during encryption augments the computational efficiency. Here, we introduce the notations $[\![\cdot]\!]$ and $\text{Dec}[\cdot]$ to represent the encryption and decryption processes, respectively. Consequently, $[\![\mathbf{x}]\!]$ exemplifies the ciphertext corresponding to plaintext vector $\mathbf{x}$.

Further, we elucidate several pivotal SIMD-type homomorphic operations pertinent to the CKKS scheme as delineated below:

$$\mathbf{Dec}[[\![\mathbf{x}]\!] \oplus [\![\mathbf{y}]\!]] = (x_0 + y_0, x_1 + y_1, \ldots, x_{k-1} + y_{k-1}) \tag{3}$$

$$\mathbf{Dec}[[\![\mathbf{x}]\!] \otimes z] = (x_0 \times z, x_1 \times z, \ldots, x_{k-1} \times z), \; z \in \mathbb{C} \tag{4}$$

$$\text{rot}([\![\mathbf{x}]\!], \ell) = [\![(x_\ell, \ldots, x_{k-1}, 0, 0, \ldots, x_0, \ldots, x_{\ell-1})]\!], \; \ell \in \mathbb{Z} \tag{5}$$

where the operators $\oplus$ and $\otimes$ are designated to represent scalar and non-scalar multiplication, respectively. Additionally, we define $\text{rot}(\cdot)$ to denote the homomorphic rotation functionality, capable of repositioning the plaintext vector elements within the HE framework. Noteworthy is the flexibility in the choice of ℓ, permitting both positive and negative integers; an inverse rotation by $-\ell$ parallels a forward rotation by $N/2 - \ell$. For a comprehensive understanding of the technical intricacies and security underpinnings of the CKKS and RNS-CKKS schemes, we direct the readers to references [8,25].

3.3 Threat Model

We assume that the cloud server operates under a "honest-but-curious" model. That is, once the client sends the encrypted data to the cloud server, the server performs computational tasks according to a pre-agreed security protocol. However, it may attempt to glean additional confidential information from the encrypted data, intermediate results, and exchanged messages. During this process, without the decryption key, the server cannot decrypt the encrypted data to directly access the client's information. However, it can infer the client's private information indirectly through intermediate information. This threat model has been widely adopted in the relevant literature [15–17].

4   Overview of the Proposed Scheme

4.1   System Model

We propose a privacy-preserving image scaling scheme that delegates image data storage and computation to a cloud server. As shown in Fig. 2, the model includes two entities: the client (data owner) and the cloud server. We assume that the user has limited storage and computational resources, so the user seeks to out-source the tasks of image data storage and image scaling computation to the cloud server. To ensure image privacy security, the user encrypts the private image data before sending it to the cloud server for storage. It is important to note that all image pixels are bundled and encrypted into a single ciphertext $f(x) \in \mathcal{R} = \mathbb{Z}[X]/(X^N + 1)$, which facilitates the user to transmit a single cipher-text, thus saving transmission bandwidth, and helps the cloud server to store and manage the encrypted image data. Upon receiving a request from the data owner, the cloud server uses the proposed secure algorithm to perform bicubic interpolation calculations on the ciphertext image to obtain a scaled ciphertext image, which is then transmitted to the user. After receiving the encrypted scaled image, the user decrypts it to obtain the post-scaling plaintext image. Through-out this process, the cloud server neither decrypts the data nor interacts with the client to obtain intermediate results, thus remaining unaware of any sensitive user information.

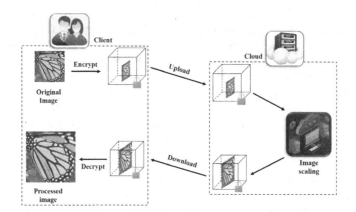

Fig. 2. System model of our scheme.

4.2   Generate Encrypted Images on the Client

This approach primarily describes operations on grayscale images. For RGB color images, similar operations can be applied independently to the three color channels. First, the user rearranges the source image of size $m \times n$ into a one-dimensional vector in a row-major order, represented as $S =$

Algorithm 1. Bicubic interpolation in the encrypted domain.

1: **Input:** $[\![\mathbf{S}]\!]$ with size $m \times n$, the output size $m' \times n'$.
2: **Output:** $[\![\mathbf{D}_{11}]\!], \cdots, [\![\mathbf{D}_{1n'}]\!], [\![\mathbf{D}_{21}]\!], \cdots, [\![\mathbf{D}_{2n'}]\!], \cdots, [\![\mathbf{D}_{m'n'}]\!]$.
3: **for** $X = 1, 2, \cdots, m'$ **do**
4:     **for** $Y = 1, 2, \cdots, n'$ **do**
5:         $x = X \times \lfloor \frac{m}{m'} \rfloor$
6:         $y = Y \times \lfloor \frac{n}{n'} \rfloor$
7:         **for** $i = 0, \cdots, 3$ **do**
8:             **for** $j = 0, \cdots, 3$ **do**
9:                 Compute $W(i, j)$ following Eq. (2).
10:                $p = n \times (x + i - 1) + (y + j - 1)$
11:                $[\![\mathbf{D}_{XY}]\!] = V \times W(i, j) \odot \text{rot}([\![\mathbf{S}]\!], p) \oplus [\![\mathbf{D}_{XY}]\!]$
12:            **end for**
13:        **end for**
14:    **end for**
15: **end for**
16: **return** $[\![\mathbf{D}_{11}]\!], \cdots, [\![\mathbf{D}_{1n'}]\!], [\![\mathbf{D}_{21}]\!], \cdots, [\![\mathbf{D}_{2n'}]\!], \cdots, [\![\mathbf{D}_{m'n'}]\!]$

$(s_{11}, \ldots, s_{1n}, s_{21}, \ldots, s_{2n}, \ldots, s_{mn})$. This vector is then encrypted into a ciphertext $[\![\mathbf{S}]\!]$ using the CKKS encryption system and sent to the cloud server along with the source image dimensions $m \times n$ and the destination image dimensions $m' \times n'$. The encrypted objects of the CKKS system are one-dimensional vectors, so the ciphertext result obtained by this encoding method contains all the information of the whole image. Compared to the traditional method of encrypting individual pixel values, this encryption method significantly reduces data storage space and communication overhead with the cloud server, while also simplifying data storage and management for the cloud server. In addition, this method fully utilizes the potential of encrypting the entire image into a single ciphertext to speed up image operation, thereby increasing efficiency for future practical applications. However, this method does not allow direct operations on individual pixels, which necessitates the design of a bicubic interpolation scheme compatible with this form of data encoding.

4.3   Image Scaling on Cloud Server

Upon receiving the encrypted image data $[\![\mathbf{S}]\!]$, the cloud server initiates the image scaling process. Since the image is encrypted into a single ciphertext, direct access to any pixel value is impossible, making the computations described in Sect. 3.1 infeasible. The crux of the problem is extracting specific pixel values from the ciphertext without decryption or user interaction, which poses a challenge to the implementation of image scaling techniques in the encrypted domain. To address this, we have developed a bicubic interpolation algorithm compatible with this data encryption scheme, which uses the homomorphic rotation operation $\text{rot}(\cdot)$ in the CKKS scheme to extract image pixels from the ciphertext.

Algorithm 2. Data Encoding

1: **Input:** The output of the scaled image$[\![D_{11}]\!], \cdots, [\![D_{1n'}]\!], [\![D_{21}]\!], \cdots, [\![D_{2n'}]\!], \cdots, [\![D_{m'n'}]\!]$.
2: **Output:** $[\![D']\!]$.
3: **for** $i = m', m' - 1, \cdots, 1$ **do**
4:     **for** $j = n', n' - 1, \cdots, 1$ **do**
5:         $[\![D']\!] = [\![D']\!] \oplus [\![D_{ij}]\!]$
6:         $[\![D']\!] = \text{rot}([\![D']\!], -1)$
7:     **end for**
8: **end for**
9: **return** $[\![D']\!]$

Specifically, we first generate all-zero initial ciphertexts $[\![D_{11}]\!], \cdots, [\![D_{1n'}]\!]$, $[\![D_{21}]\!], \cdots, [\![D_{m'n'}]\!]$ and an auxiliary vector $V = [1, 0, \ldots, 0]$. Next, we determine the corresponding positions p in the source image based on the pixel value positions required by the target image. Using the auxiliary vector V and the homomorphic rotation operation of CKKS, we extract the ciphertext of the required pixel values from the source image. We then compute the corresponding weights according to the Eq. 2, multiply them with the extracted ciphertexts of the source image, and add them to the ciphertexts $[\![D_{11}]\!], \cdots, [\![D_{1n'}]\!], [\![D_{21}]\!], \cdots, [\![D_{m'n'}]\!]$ to obtain the pixel values of the target image. The detailed process is shown in Algorithm 1.

After scaling the image, it is essential to re-encode it into a single ciphertext to reduce data communication costs, which primarily uses the CKKS homomorphic rotation operation $\text{rot}(\cdot)$ to compress the ciphertext. The specific operation is demonstrated in the Algorithm 2. An initial ciphertext $[\![D']\!]$ is created, followed by the cumulative addition of individual ciphertexts $[\![D_{11}]\!], \cdots, [\![D_{1n'}]\!], [\![D_{21}]\!], \cdots, [\![D_{2n'}]\!], \cdots, [\![D_{m'n'}]\!]$ using the homomorphic rotation operation $\text{rot}(\cdot)$, which generates the final destination image ciphertext $[\![D']\!]$.

4.4 Decrypting Images on the Client

After downloading the scaled ciphertext image $[\![D']\!]$ from the cloud server, the client uses the key from the CKKS encryption system to decrypt the ciphertext image and obtain the one-dimensional vector representation of the target image. Subsequently, the final target image is obtained through rearrangement and reordering. Throughout this process, the cloud server does not have access to the client's confidential image information because it only handles encrypted data, which ensures the client's privacy.

5 Security Analysis

In this section, we evaluate the security of our approach with respect to the threats mentioned in Sect. 3.3. In our system, the cloud server has no decryption keys, and all results obtained throughout the computation process remain encrypted. The images are encrypted by the users using the CKKS encryption system before they are transferred to the cloud server. Throughout the computation process on the cloud server, homomorphic computations can be performed on the encrypted images, although decryption is impossible due to the lack of decryption keys. Moreover, the cloud server does not interact with the clients to obtain exchangeable information or intermediate results.

Using a proof by contradiction, we argue that if the cloud server manages to extract a bit of information from the received encrypted images or the computed result, it implies the potential to extract a bit of information from any CKKS ciphertext, which contradicts the known security property of the CKKS encryption system as cited in [18]. Consequently, the security of our approach is synonymous with the security of the encryption scheme used. Since the CKKS scheme has been shown to be secure under selected plaintext attack (CPA) scenarios in [18], our system satisfies the security requirements.

Furthermore, CKKS requires an appropriate choice of parameters to ensure security. We restrict the maximum size of the ciphertext modulus q to a constant Q, with $q^L = Q$, where L represents the multiplication depth - the number of homomorphic multiplications allowed before ciphertext decryption errors occur. The amount of noise contained in the ciphertext grows exponentially with the number of multiplication instances. To prevent noise overflow, after each homomorphic multiplication a rescaling by a factor p is performed to keep the noise growth at a manageable level. If the multiplication depth exceeds L, it becomes impossible to decrypt correctly.

By fixing p, a larger Q allows a larger number of feasible HE multiplications. The multiplication depth L can be increased by using an appropriate value of Q. At the same time, the ciphertext dimension N should increase with Q to maintain a given level of security. Detailed guidance on parameter selection is available in the Homomorphic Encryption Security Standard white paper [19].

6   Experimental Results

The experimental results were generated on a 64-bit Windows PC equipped with an AMD Ryzen 5 5600U with Radeon Graphics CPU @ 2.30 GHZ and 16.0 GB RAM. Using the Tenseal cryptographic library [20] for the CKKS encryption system, we implemented our scheme in Python. For a set of cryptographic parameters chosen in this paper, we have $N = 32768$, $\log_2 Q = 880$, $\log_2 p = 40$, ensuring a security level of 128 bits.

We evaluated the effectiveness and efficiency of our proposed image scaling method on publicly available datasets, including Set5 [21], Set14 [22], B100 [23], and Urban100 [24], and we select several images from these datasets for experimentation.

6.1   Image Quality

Figure 3 shows the output of our method after cloud server scaling, with part (a) representing the original image and part (b) representing images after various

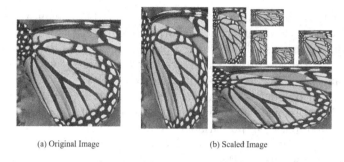

(a) Original Image   (b) Scaled Image

Fig. 3. The effect of different image sizes generated by our scheme.

degrees of scaling. As can be seen in Fig. 3, our system accommodates different degrees of scaling and maintains high quality and clarity in the scaled images without significant degradation.

To better illustrate the advantages of our approach, we compare it with existing solutions, such as those presented in [12] and [5]. We mainly use the popular metrics PSNR and SSIM to evaluate the quality of the generated images.

Peak Signal to Noise Ratio. The Peak Signal to Noise Ratio (PSNR) is commonly defined by the Mean Square Error (MSE). For two images X and Y with dimensions $H \times W$, the MSE is defined as

$$MSE = \frac{1}{H \times W} \sum_{i=0}^{H-1} \sum_{j=0}^{W-1} (X(i,j) - Y(i,j))^2 \qquad (6)$$

Then PSNR is defined as:

$$PSNR = 10 \times \log_{10} \frac{(2^b - 1)^2}{MSE} \qquad (7)$$

where b is the bit depth per pixel. A higher PSNR value indicates a smaller difference between the images, implying a similarity between the reconstructed and the original images.

Table 1. Comparison of PSNR and SSIM for different image sizes.

Number	Original Image	Target Image	PSNR			SSIM		
			Tanwar [12]	Mohanty [5]	Proposed scheme	Tanwar [12])	Mohanty [5]	Proposed scheme
1	32 × 32	64 × 64	6.73	18.61	21.38	0.3615	0.6066	0.78044
2	64 × 64	128 × 128	11.21	23.23	25.5	0.3342	0.57	0.7266
3	128 × 128	256 × 256	7.03	19.83	23.4	0.4621	0.7012	0.8424
4	256 × 256	128 × 128	11.16	35.38	32.44	0.5463	0.9718	0.9488
5	32 × 32	128 × 128	11.29	22.73	21.92	0.2540	0.4838	0.4782
6	256 × 256	64 × 64	6.88	27.36	25.87	0.6107	0.9411	0.9182
7	64 × 128	64 × 64	6.67	24.88	32.13	0.5190	0.9079	0.9802
8	128 × 256	128 × 128	10.96	28.02	36.39	0.4485	0.8701	0.9808

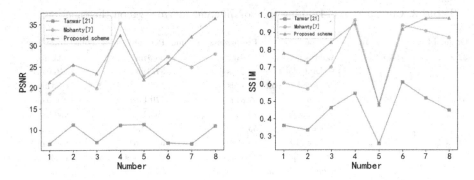

Fig. 4. Comparison of PSNR and SSIM for scaling images of different sizes using different schemes.

Structural Similarity. Structural Similarity (SSIM) is another widely used metric to measure the similarity of two images. For images X and Y, their SSIM is computed as:

$$SSIM = \frac{(2\mu_x\mu_y + c_1)(2\sigma_{xy} + c_2)}{(\mu_x^2 + \mu_y^2 + c_1)(\sigma_x^2 + \sigma_y^2 + c_2)} \tag{8}$$

where μ_x is the average value of X, μ_y is the average value of Y, σ_x^2 is the variance of X, σ_y^2 is the variance of Y, σ_{xy} is the covariance between X and Y, and c_1 and c_2 are constants that maintain stability, related to the dynamic range of pixel values. A higher SSIM value indicates greater similarity between the two images. When the value is 1, it signifies that the images are identical.

To more effectively utilize PSNR and SSIM for comparing our solution with the existing ones [5,12], we downscaled the source image by a certain factor, then used the proposed image scaling technique to restore the image size, and subsequently compared the original image with the reconstructed image to calculate the evaluation values. As shown in Table 1, we present the reconstruction effects for eight different images. Our approach significantly outperforms the scheme proposed by Tanwar et al. [12] in terms of both PSNR and SSIM. In the cases of images 4, 5, and 6, our results are slightly inferior to those achieved with the method suggested by Mohanty et al. [5], yet on the whole, our solution yields better PSNR and SSIM values. Therefore, employing our scheme can generate higher quality images. Figure 4 visually illustrates the comparison results of PSNR and SSIM between our solution and the existing ones [5,12].

6.2 Running Time

Tanwar et al. [12] and Mohanty et al. [5] use secret sharing techniques to ensure privacy and security of user data, which requires client-side pre-processing of the original images and reconstruction after downloading the scaled images from the cloud server. Given the significant resource constraints on the client side compared to the cloud server, the primary focus is on the resource consumption

Table 2. Comparison of running time of different image sizes on the client side.

Original Image	Target Image	Scheme	Pre-processing Time	Post-processing Time	Total Time
32 × 32	64 × 64	Tanwar [12]	13.3 ms	38.9 ms	52.2 ms
		Mohanty [5]	28.6 ms	87.3 ms	116.0 ms
		Proposed scheme	51.1 ms	13.2 ms	64.3 ms
64 × 64	128 × 128	Tanwar [12]	57.8 ms	179.1 ms	236.9 ms
		Mohanty [5]	107.5 ms	353.3 ms	460.9 ms
		Proposed scheme	55.8 ms	15.1 ms	71.0 ms
128 × 128	256 × 256	Tanwar [12]	195.5 ms	651.0 ms	846.5 ms
		Mohanty [5]	427.5 ms	1393.6 ms	1821.1 ms
		Proposed scheme	222.2 ms	62.5 ms	284.7 ms

on the client side. In this section, we will analyze and compare the client-side resource consumption in detail.

As shown in Table 2, we examined the client run time by enlarging three images of different sizes by a factor of two. The preprocessing time pertains to the time needed for operations performed by the client before the images are transmitted to the cloud server. In contrast, post-processing time refers to the operations performed after downloading the images from the server to generate the final target image. The total time accounts for all operations executed by the client throughout this process. In the context of enlarging a 32 × 32 image twice, Tanwar et al. [12]'s approach is faster than ours, while Mohanty et al. [5]'s approach consumes more time overall, although it is faster in the preprocessing phase. Our approach consistently outperforms in terms of post-processing time, attributing to the mere necessity of decryption. This efficiency is due to the alternative approaches relying on Chinese Remainder Theorem and Lagrange interpolation for image reconstruction, thus consuming more time. As the image size for processing increases, the time requirement for all strategies escalates, but ours maintains the minimum time consumption, a gap that widens as the image size increases. Consequently, our technique proves to be more advantageous for larger images, a merit that is vividly illustrated in Fig. 5 (a). Using bicubic interpolation on the client side, the time required to upscale 32 × 32, 64 × 64, and 128 × 128 images by a factor of two is 192.6 ms, 743.5 ms, and 2919.2 ms, respectively. In contrast, our approach requires only 64.3 ms, 71.0 ms and 284.7 ms on the client side, so our solution still has an advantage.

6.3   Data Volume

We introduce a space-saving data encoding method, which results in encrypted images being the size of a single ciphertext, considerably reducing data storage requirements and the volume of data transmitted during server communication. We then compare the storage requirements of our method with those of Tanwar et al. [12] and Mohanty et al. [5]. In particular, since these techniques use secret sharing techniques that require uploading data to different cloud servers, the subsequent data volume refers to the transmission to a single server.

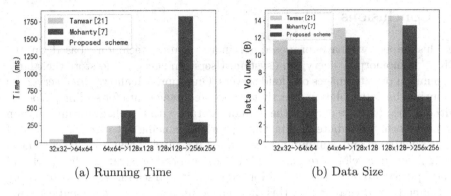

(a) Running Time   (b) Data Size

Fig. 5. Comparison of total client runnning time and total data transferred using different schemes for scaling images of different sizes.

As shown in Table 3, we evaluated the transmission data size while enlarging three different dimensional images by a factor of two. The data size includes the volume to be sent to a cloud server after client-side processing and the volume received from a server, aggregating to the total data size exchanged between the client and server during the operation. Our strategy maintains a virtually constant data size, facilitated by the use of a fixed size N that determines the plaintext data that a single ciphertext can carry. Changes in the encryption parameter N would slightly affect the data size. In contrast, the strategies of Tanwar et al. [12] and Mohanty et al. [5] entail a burgeoning data size with increasing image dimensions, culminating in a larger memory footprint. Our method significantly reduces the data storage requirements compared to the others, promoting efficient data storage and management on servers, while significantly reducing the communication overhead, thus confirming its superiority in terms of data volume. Figure 5 (b) vividly illustrates this advantage, where a logarithmic representation in bytes (B) was used for comparison due to the considerable data disparity.

Table 3. Comparison of data size for sending and receiving images of different sizes.

Original Image	Target Image	Scheme	Sent Data Volume	Received Data Volume	Total Communication Data Volume
32×32	64×64	Tanwar [12]	24.1172 KB	96.1172 KB	120.2344 KB
		Mohanty [5]	8.1172 KB	32.1172 KB	40.2344 KB
		Proposed scheme	0.08599 KB	0.0859 KB	0.1719 KB
64×64	128×128	Tanwar [12]	96.1172 KB	384.1172 KB	480.2344 KB
		Mohanty [5]	32.1172 KB	128.1172 KB	160.2344 KB
		Proposed scheme	0.08599 KB	0.0859 KB	0.1719 KB
128×128	256×256	Tanwar [12]	384.1172 KB	1536.1172 KB	1920.2344 KB
		Mohanty [5]	128.1172 KB	512.1172 KB	640.2344 KB
		Proposed scheme	0.08599 KB	0.0859 KB	0.1719 KB

7   Conclusions

In this paper, we have proposed an image scaling technique based on the CKKS homomorphic encryption system. Users can encrypt and store their large amounts of private images on cloud servers. Our approach allows cloud servers to apply bicubic interpolation directly to encrypted image data for scaling purposes, without the ability to decrypt the data or interact with the client for information exchange or intermediate results, thus greatly enhancing data security. To avoid the high storage and communication costs of per-pixel encryption, we use SIMD technology to develop a resource-efficient data encoding strategy that packs all image pixels into one ciphertext. Moreover, our image scaling solution effectively accommodates operations on this data encoding method. Experimental results show that our scheme not only yields better image results, but also stands out in terms of time and space efficiency. Future efforts could potentially focus on exploring more efficient image scaling techniques, such as super-resolution reconstruction, and extending our work to encrypted domain video processing.

Acknowledgement. This work was supported in part by the Guangdong Basic and Applied Basic Research Foundation under Grant 2022A1515011897, Grant 2023A1515030087, and Grant 2022A1515011512, and in part by the National Natural Science Foundation of China (NSFC) under Grant 62272498.

References

1. Lagendijk, L., Erkin, Z., Barni, M.: Encrypted signal processing for privacy protection: conveying the utility of homomorphic encryption and multiparty computation. IEEE Signal Process. Mag. **30**(1), 82–105 (2013)
2. Pedrouzo-Ulloa, A., Troncoso-Pastoriza, J., Pérez-González, F.: Image denoising in the encrypted domain. In: IEEE International Workshop on Information Forensics and Security (WIFS), pp. 1–6 (2016)
3. Kiya, H., Fujiyoshi, M.: Signal and image processing in the encrypted domain. ECTI Trans. Comput. Inf. Technol. (ECTI-CIT) **6**(1), 10–17 (2012)
4. Zheng, P., Huang, J.: Implementation of the discrete wavelet transform and multiresolution analysis in the encrypted domain. In: Proceedings of the 19th ACM International Conference on Multimedia, pp. 413–422 (2011)
5. Mohanty, M., Ooi, T., Atrey, K.:Scale me, crop me, knowme not: supporting scaling and cropping in secret image sharing. In: Proceedings of the 2013 IEEE International Conference on Multimedia and Expo (ICME), pp. 1–6. IEEE Computer Society, San Jose (2013)
6. Deepthi, S., Lakshmi, S., Deepthi, P.: Image processing in encrypted domain for distributed storage in cloud. In: International Conference on Wireless Communications, Signal Processing and Networking (WiSPNET), pp. 1478–1482 (2017). https://doi.org/10.1109/WiSPNET.2017.8
7. Paillier, P.: Public-key cryptosystems based on composite degree residuosity classes. In: Advances in Cryptology - EUROCRYPT '99 International Conference on the Theory and Application of Cryptographic Techniques(1999)

8. Cheon, H., Han, K., Kim, A., Kim, M., Song, Y.: A full RNS variant of approximate homomorphic encryption. In: International Conference on Selected Areas in Cryptography (2018)

9. Kim, A., Song, Y., Kim, M., Lee, K., Cheon, H.: Logistic regression model training based on the approximate homomorphic encryption. BMC Med. Genom. 11(4), 83 (2018)

10. Mohanty, M., Asghar, R., Russello, G.: 2dcrypt: image scaling and cropping in encrypted domains. IEEE Trans. Inf. Forensics Secur. 11(11), 2542–2555 (2016). https://doi.org/10.1109/TIFS.2016.2585085

11. Anurenjini, R.: Encrypted domain image scaling and cropping in cloud. In: First International Conference on Secure Cyber Computing and Communication (ICSCCC), pp. 283–288 (2018). https://doi.org/10.1109/ICSCCC.2018.8703205

12. Tanwar, K., Rajput, S., Raman, B.: Privacy preserving image scaling using 2d bicubic interpolation over the cloud. In: IEEE International Conference on Systems, Man, and Cybernetics, pp. 2073–2078 (2018). https://doi.org/10.1109/SMC.2018.00357

13. Vadim, L., Chris, P., Oded, R.: On ideal lattices and learning with errors over rings. In: Henri, G. (ed.) EUROCRYPT 2010. LNCS, vol. 6110, pp. 1–23. Springer, Heidelberg (2010)

14. Robert, K.: Cubic convolution interpolation for digital image processing. IEEE Trans. Acoust. Speech Signal Process. 29(6), 1153–1160 (1981)

15. Elmehdwi, Y., Samanthula, K., Jiang, W.: Secure k-nearest neighbor query over encrypted data in outsourced environments. IEEE (2013)

16. Araki, T., Furukawa, J., Lindell, Y., Nof, A., Ohara, K.: High-throughput semi-honest secure three-party computation with an honest majority. In: ACM SIGSAC Conference on Computer & Communications Security (2016)

17. Zhang, Y., Zheng, P., Luo, W.: Privacy-preserving outsourcing computation of QR decomposition in the encrypted domain. In: 18th IEEE International Conference on Trust, Security and Privacy in Computing and Communications/13th IEEE International Conference on Big Data Science and Engineering (Trust-Com/BigDataSE) (2019)

18. Jung, C., Andrey, K., Miran, K., Yongsoo, S.: Homomorphic encryption for arithmetic of approximate numbers. In: International Conference on the Theory and Application of Cryptology and Information Security (2017)

19. Albrecht, M., et al.: Homomorphic encryption security standard. HomomorphicEncryption.org, Toronto, Canada (2018)

20. Ayoub, B., Bilal, R., Bogdan, C., Alaa, B.:TenSEAL: a library for encrypted tensor operations using homomorphic encryption. CoRR(2021)

21. Marco, B., Aline, R., Christine, G., Marie-line, M.: Low-complexity single-image super-resolution based on nonnegative neighbor embedding. In: Proceedings of the British Machine Vision Conference, pp. 1–135 (2012)

22. Roman, Z., Michael, E., Matan, P.: On single image scale-up using sparse-representations. In: International Conference on Curves and Surfaces, pp. 711–730 (2010)

23. Radu, T., Vincent, S., Luc, G.: A+: adjusted anchored neighborhood regression for fast super-resolution. In: Asian Conference on Computer Vision, pp. 111–126 (2014)

24. Jia-Bin, H., Abhishek, S., Narendra, A.: Single image super-resolution from transformed self-exemplars. In: Proceedings of the IEEE Conference on Computer Vision and Pattern Recognition, pp. 5197–5206 (2015)

25. Jung, C., Kyoohyung, H., Andrey, K., Miran, K., Yongsoo, S.: Bootstrapping for approximate homomorphic encryption. In: Annual International Conference on the Theory and Applications of Cryptographic Techniques (2018)
26. Kim, M., Song, Y., Wang, S., Xia, Y., Jiang, X.: Secure logistic regression based on homomorphic encryption: design and evaluation. JMIR Med. Inf. (2018)

PDMTT: A Plagiarism Detection Model Towards Multi-turn Text Back-Translation

Xiaoling He[1] , Yuanding Zhou[1] , Chuan Qin[1(✉)] , Zhenxing Qian[2] ,
and Xinpeng Zhang[2]

[1] University of Shanghai for Science and Technology, Shanghai, China
{212190381,221240081}@st.usst.edu.cn, qin@usst.edu.cn
[2] Fudan University, Shanghai, China
{zxqian,zhangxinpeng}@fudan.edu.cn

Abstract. With the development of communication technologies, the practice of creating new texts by manipulating original sentence structures through multi-turn machine translation is widespread across various domains. Existing plagiarism detection models often treat different features uniformly and overlook the significance of disparities within high-dimensional features. Therefore, this paper proposes a novel plagiarism detection model towards multi-turn text back-translation (PDMTT), adopting a novel mechanism that combines local and global features and enhances them. The grouping enhancement fusion (GEF) mechanism assigns importance coefficients to sub-features, reinforcing critical aspects while diminishing less relevant ones. These enhanced features, generated by the GEF mechanism, are leveraged to extract high-quality text representations, thereby improving the precision of the model in distinguishing original content from back-translated texts. Furthermore, we improve the back-translation plagiarism detection capability of our model by optimizing the contrastive loss function and utilizing the fused translated representations as targets. To validate the effectiveness of our model, we also constructed a multi-tuple back-translation plagiarism dataset for model training and validation. Experimental results demonstrate that the proposed PDMTT outperforms previous methods in back-translation plagiarism detection, yielding superior text representations. The ablation study further confirms that the incorporation of the GEF mechanism effectively enhances the discrimination capability of our model.

Keywords: Back-Translation Plagiarism · Plagiarism Detection · Contrastive Learning · Grouping Enhancement Fusion

1 Introduction

The development of communication technology represented by online self-media platforms [1] stimulates the spread of back-translation plagiarism in different fields, and its influence extends from news reporting to literary creation. Back-translation plagiarism [2] involves the act of paraphrasing by translating a text from one language to another and subsequently retranslating it into the original language. This unethical practice infringes

B. Ma et al. (Eds.): IWDW 2023, LNCS 14511, pp. 79–94, 2024.
https://doi.org/10.1007/978-981-97-2585-4_6

on the intellectual property rights of the original authors. To solve this issue, various plagiarism detection methods have emerged within the field of natural language processing [3], including the traditional but costly approach of manual judgment, which is rendered inefficient by the volume of digital content produced daily. Furthermore, manual detection introduces subjectivity, as individual judgments may vary.

Alzahrani et al. [4] categorized plagiarism detection tasks into two formal classes: external and intrinsic. Extrinsic plagiarism detection [5] evaluates plagiarism by comparing a document to one or more source documents. Conversely, intrinsic plagiarism detection [6] focuses on discerning instances of plagiarism within individual suspicious documents by identifying similarities with their source texts. This paper specifically addresses intrinsic plagiarism detection, which encompasses the identification of similarities between suspicious documents and their source texts. Modern plagiarism detection methods can be categorized into three types: traditional manual methods, deep learning-based methods, and pre-trained model-based methods. Traditional manual methods [7] necessitate the laborious construction of feature vectors, the analysis of feature changes before and after back-translation, and the design of binary classifiers to distinguish between original and back-translated texts. These methods are impractical due to limited feature extraction capabilities and reliance on discrete hand-crafted features. Furthermore, advancements in machine translation [8], especially neural networks, can produce back-translated texts that are nearly indistinguishable from originals, posing challenges for conventional methods.

In contrast to traditional manual plagiarism detection, deep learning-based methods automatically learn features from extensive training data, eliminating the need for upfront feature design. However, certain deep learning-based methods overlook contextual dependencies within the text, relying solely on Convolutional Neural Networks (CNN) for local semantic feature extraction [9], while others employ Recurrent Neural Networks (RNN) to capture contextual information [10]. Nevertheless, RNN structures face challenges such as limited long-range context utilization and gradient-related issues. Pennington et al. [11] addressed these challenges by combining local information with global word frequency statistics to extract comprehensive semantic features.

The introduction of BERT [12] in late 2018 marked a turning point in natural language processing, ushering in the era of pre-trained models. BERT's self-attention mechanism enables the capture of both local features and global semantic relations. However, the performance of using BERT directly as an embedding model is not as good as expected [13]. Consequently, Reimers et al. proposed SBERT [14], incorporating a Siamese network structure to generate fixed-length sentence vectors with semantic information, facilitating similarity comparison. Subsequent developments, including BERT-Flow [15] and BERT-Whitening [16], addressed anisotropy issues, while Yan et al. [17] introduced a universal framework for contrastive learning in NLP semantic computation in 2021. Su et al. [18] introduced CoSENT, a novel supervised sentence vector scheme, in 2022, demonstrating enhanced convergence speed and outcomes compared to SBERT. However, these models treat all features equally, failing to account for disparities in high-dimensional features or feature quality's impact on model performance. For existing plagiarism detection methods, assessing the similarity between back-translated texts and originals poses a challenge. Enhancing feature quality is essential for improving detection model performance. Currently, there is a dearth of machine detection methods for back-translation plagiarism. Thus, this paper proposes a novel multi-turn back-translation plagiarism detection model, i.e., PDMTT, by enhancing and integrating local and global features based on pre-trained models and contrastive learning. Our PDMTT model extracts latent representations of language style, idiomatic expressions and global long-term dependencies from back-translated texts, and employs a grouping enhancement fusion (GEF) mechanism to generate importance coefficients for grouped features. This mechanism can reinforce important features and weaken less significant ones, ultimately enhancing text representation quality [19]. Our model further adopts a contrastive loss function to optimize the detection capability through text representation fusion from multiple translation rounds. The main contributions of this work are listed as follows:

- An effective plagiarism detection model, PDMTT, based on a contrastive learning architecture is proposed, which can extract richer features and obtain better detection results compared with existing models.
- A GEF mechanism is proposed, which can take into account the differences between group-wise features for coefficient assignment. The mechanism can make text representation more accurate by strengthening important features and weakening unimportant ones.
- A new quadruple dataset for the back-translation plagiarism detection task is constructed, which contains plagiarized texts in both Chinese and English as source languages after multiple rounds of translation through different translation processes.

2 Proposed Method

The overall framework of the proposed PDMTT model is given in Fig. 1, which contains three modules: text encoding, grouping enhancement fusion and loss function. Each module is described detailedly in the following.

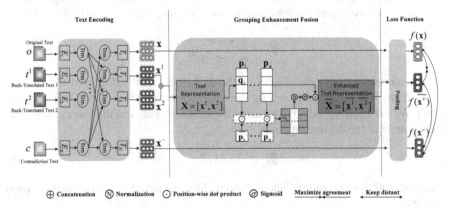

⊕ Concatenation Ⓝ Normalization ⊙ Position-wise dot product ⊘ Sigmoid   Maximize agreement   Keep distant

Fig. 1. The Overall Framework of the Proposed PDMTT Model.

2.1 Text Encoding

The text encoding module captures both local and global features in the input text through the multi-layer self-attention mechanism in the BERT model and then fuses them into the embedding matrix.

Firstly, the original text o, multi-turn back-translated texts t^1 and t^2, and contradiction text c are fed into a shared BERT model stacked by multiple transformer encoders to obtain the text embedding $\mathbf{x}$, $\mathbf{x}^1$, $\mathbf{x}^2$, and $\mathbf{x}^-$ respectively. Then the joint embedding map $\mathbf{X} = [\mathbf{x}^1, \mathbf{x}^2]$ is obtained by concatenating the text embeddings $\mathbf{x}^1$ and $\mathbf{x}^2$.

2.2 Grouping Enhancement Fusion

The purpose of the grouping enhancement fusion module is to enhance the obtained joint embedding map $\mathbf{X}$ through group-wise enhancement [19], aiming to reinforce important features while weakening less important ones, ultimately yielding more accurate text embedding.

Firstly, the embedding map $\mathbf{X}$ is divided into n groups according to vector dimensions. Then a group-wise embedding map can be obtained:

$$\mathbf{P} = \begin{bmatrix} \mathbf{p}_1, \mathbf{p}_2, \cdots, \mathbf{p}_n \end{bmatrix} \in \mathbf{R}^{m \times k}. \tag{1}$$

The embedding map of a certain group is denoted as:

$$\mathbf{p}_j = \begin{bmatrix} \mathbf{q}_1, \mathbf{q}_2, \cdots, \mathbf{q}_m \end{bmatrix} \in \mathbf{R}^{m \times h}, j = 1, 2, \cdots, n, \tag{2}$$

where $h = \frac{k}{n}$, and the sub-feature vector can be represented as:

$$\mathbf{q}_i = [a_1, a_2, \cdots, a_h]^{\mathrm{T}} \in \mathbf{R}^h, i = 1, 2, \cdots, m. \tag{3}$$

Assuming that each group carries specific semantic information during the network learning process, we can approximate the semantic feature of the group learning representation by averaging the sub-feature, and a certain group is calculated separately:

$$\overline{\mathbf{p}_j} = \frac{1}{m} \sum_{i=1}^{m} \mathbf{q}_i. \tag{4}$$

Next, the average semantic vector of the back-translated text is used to calculate the importance coefficients for each sub-feature vector through a simple positional dot product. We use the dot product to measure the similarity between the average semantic vector $\overline{\mathbf{p}_j}$ and the sub-feature vector $\mathbf{q}_i$:

$$\omega_i = \overline{\mathbf{p}_j} \cdot \mathbf{q}_i, i = 1, 2, \cdots, m. \tag{5}$$

To eliminate the bias of coefficients among different samples, data normalization is necessary:

$$\widehat{\omega_i} = \frac{\omega_i - E(\omega)}{\sqrt{D(\omega) + \varepsilon}}, \tag{6}$$

$$E(\omega) = \frac{1}{m} \sum\nolimits_{i=1}^{m} \omega_i, \tag{7}$$

$$D(\omega) = \frac{1}{m} \sum\nolimits_{i=1}^{m} (\omega_i - E(\omega))^2, \tag{8}$$

where ε is a constant added for numerical stability, $E(\omega)$ and $D(\omega)$ are the mean and variance of the coefficients for the j-th group, respectively. The normalization accelerates convergence and ensures that the correlations between features remain unaltered. Normalization of the input alters the representation of the output. To ensure that the normalization operation inserted in the network can effectively capture the transformation of text representation, we introduce a pair of parameters α and β for each coefficient $\widehat{\omega}_i$. These parameters enable scaling and shifting of the normalized values. Thus, a new normalized importance coefficient is generated using the sigmoid function $\sigma(\cdot)$:

$$\overline{\omega}_i = \sigma(\alpha\widehat{\omega}_i + \beta). \tag{9}$$

Finally, we scale the original sub-feature $\mathbf{q}_i$ using the normalized importance coefficients $\overline{\omega}_i$ to reinforce important features while weakening less important ones, and result in an enhanced sub-feature:

$$\overline{\mathbf{q}}_i = \overline{\omega}_i \cdot \mathbf{q}_i. \tag{10}$$

All the enhanced sub-features form an enhanced embedding map $\overline{\overline{\mathbf{p}}}_j = [\overline{\mathbf{q}_1}, \overline{\mathbf{q}_2}, \cdots, \overline{\mathbf{q}_m}] \in \mathbf{R}^{m \times h}$. After enhancing all the groups, we can obtain an enhanced joint representation map, denoted as $\overline{\mathbf{X}} = [\mathbf{x}^1, \mathbf{x}^2] \in \mathbf{R}^{m \times k}$.

2.3 Loss Function

The text embeddings $\mathbf{x}$, $\mathbf{x}^-$ and $\overline{\mathbf{X}}$ undergoes a pooling operation to obtain the text representations $f(\mathbf{x})$, $f(\mathbf{x}^-)$, and $f(\overline{\mathbf{x}^+})$. The pooling operation is performed to reduce the dimensionality of the features while preserving the main characteristics. We use the CLS pooling method to filter the text representation. The relationship among the three can be described as follows:

$$D(f(\mathbf{x}), f(\overline{\mathbf{x}^+})) \gg D(f(\mathbf{x}), f(\mathbf{x}^-)). \tag{11}$$

We optimize the contrastive loss function and name it enhanced positive contrastive (EPC) loss function, which adopts the fused text representation, as the objective for contrastive learning. The approach maximizes the similarity between a text representation and its back-translation plagiarism text while maintaining a sufficient distance from other text representations within the same batch. The contrastive loss function is defined as Eq. (12).

$$\ell_i = -\log \frac{e^{sim(f(\mathbf{x}_i), f(\overline{\mathbf{x}_i^+}))/\tau}}{\sum_{j=1}^{N} (e^{sim(f(\mathbf{x}_i), f(\overline{\mathbf{x}_j^+}))/\tau} + e^{sim(f(\mathbf{x}_i), f(\mathbf{x}_j^-))/\tau})}, \tag{12}$$

where the function $sim(\cdot)$ represents the cosine similarity between samples, N denotes the number of mini-batches, $f(\mathbf{x}_i)$ represents the original text representation, $f(\overline{\mathbf{x}_i^+})$ represents the fused representation, which is obtained by multiple rounds of back-translation plagiarism through group-wise enhancement, and $f(\mathbf{x}_j^-)$ represents the representation of contradiction text within the mini-batch.

2.4 Dataset Construction

We select 50,000 concise records from the LCSTS news summarization dataset [20]. This dataset is the news articles posted on Weibo collected and organized by the Harbin Institute of Technology. Among the ensemble of machine translation engines—namely Baidu Translator, Google Translator, Youdao Translator, and Bing Translator—one is arbitrarily selected as the designated translator. Furthermore, we make a random choice from a set of languages including Chinese, English, Spanish, German, Japanese and Tibetan, to serve as the pivot language.

Subsequently, we configure 2 to 4 translation iterations to obtain back-translated text 1 and back-translated text 2 as well as contradictory text without plagiarism. This process creates a quadruple back-translation plagiarism detection dataset. The test set and the validation set, each encompassing a total of 2,000 sample pairs. Each sample pair is labeled for similarity by multiplayer cross-manual labeling with the numbers 0 and 1. Here, 0 signifies the original text and its corresponding contradiction text, while 1 represents the original text and its associated translated counterpart.

3   Experiments

We evaluated the proposed PDMTT on the quadruple dataset constructed in 2.4 and the semantic text similarity (STS) dataset [21]. Besides, we chose SBERT [14], ConSERT [17], BERT-Whitening [16], and BERT [12] as the target model.

3.1 Experimental Setting

Parameter Setting. During the training process, the specific parameter settings of the model were as follows: the network had 12 layers, the hidden layer size was set to 768, the number of self-attention heads for the masked self-attention mechanism was 12, and the number of global self-attention heads was also set to 12. The entire model was implemented using the PyTorch framework, and the training process was performed using an NVIDIA GeForce RTX 2080 Ti GPU for accelerated computations.

In our experiments, we conducted training on the PDMTT for 100 epochs, and the validation was performed every 250 steps on back-translation datasets with Chinese and English serving as the source languages, respectively. This approach ensured a comprehensive assessment of the model's performance and generalization capabilities across different linguistic contexts. We saved the model checkpoint with the best performance on the test set. A grid search was conducted to find the optimal hyperparameters for batch size and learning rate. The results are listed in Table 1. It can be seen that by properly adjusting the learning rate, PDMTT is not sensitive to batch size. It can achieve impressive results even with smaller batch sizes.

Table 1. Batch Size and Learning Rates for PDMTT.

	BERT$_{base}$		BERT$_{Chinese}$	
Batch size	16	32	16	32
Learning rate	1e−5	2e−5	1e−5	3e−5

Evaluation Criteria. *Cosine Similarity.* In the plagiarism detection task, it is often necessary to calculate the similarity between different texts. Cosine similarity is a metric that quantifies the degree of similarity between two embeddings within a semantic space. Specifically, it calculates the cosine of the angle between two embeddings to measure the difference between them. The closer the cosine value is to 1, the more similar the two vectors are. Here, we used cosine similarity to calculate the distance between samples, see Eq. (13).

$$\cos\theta = \frac{\mathbf{X} \cdot \mathbf{Y}}{\|\mathbf{X}\|\|\mathbf{Y}\|}, \tag{13}$$

where $\mathbf{X}$ and $\mathbf{Y}$ denote the two text representations to be detected, respectively.

Spearman's Rank Correlation Coefficient. In the back-translation plagiarism detection task, we transformed two text sequences into vector representations, enabling the computation of their similarity. As recommended by Reimers et al. [22], Spearman's rank correlation coefficient is a non-parametric statistical method because it does not rely on the specific distribution of data but assesses the relationship between two variables by ranking the data. Firstly, the text pairs were separately input to the model to obtain the text representation pairs, and then the cosine similarity of the text representation pairs was calculated. After calculating the cosine similarity of all text representation pairs, Spearman's rank correlation coefficient was finally used to compare the correlation between model-generated cosine similarity scores and hand-labeled similarity scores. Ranging from −1 to 1, the closer the correlation coefficient is to 1 or −1, the stronger the correlation, and the closer the correlation coefficient is to 0, the weaker the correlation. The mathematical formulation of Spearman's rank correlation coefficient is given by Eq. (14).

$$\rho_s = 1 - \frac{6\sum_{i=1}^{n} d_i^2}{n(n^2 - 1)}, \tag{14}$$

where d_i represents the difference in ranks between corresponding variables, which is the difference in positions (ranks) of paired variables after sorting the two variables separately, and n denotes the number of observations.

Alignment and Uniformity. We followed the two crucial characteristics of contrastive learning: alignment and uniformity, as established by Wang et al. [23]. Given a distribution p_{pos} of positive sample pairs, the objective of alignment is to calculate the expected distance between the embedding vectors of positive samples:

$$\ell_{align} \triangleq E_{(x,x^+)\sim p_{pos}} \|f(\mathbf{x}) - f(\mathbf{x}^+)\|^2, \tag{15}$$

where $f(\mathbf{x})$ represents the embedding obtained by encoding the sample and $f(\mathbf{x}^+)$ represents the embedding of its positive sample. A smaller value of alignment indicates that the positive sample pairs are closer in the vector space. Uniformity is used to assess whether the embedding vectors have a well-unified distribution:

$$\ell_{\text{uniform}} \triangleq \log E_{x,y \sim p_{\text{data}}} e^{-2\|f(\mathbf{x})-f(\mathbf{y})\|^2}, \tag{16}$$

where p_{data} represents the data distribution, $f(\mathbf{x})$ and $f(\mathbf{y})$ represent the embeddings of two random samples. A smaller value ℓ_{uniform} of uniformity implies a larger distance between two random samples, leading to a more dispersed embedding vector for the entire dataset. Consequently, a smaller value ℓ_{uniform} is desired.

3.2 Results of Multi-turn Back-Translation Plagiarism Detection

We conducted experiments on back-translation datasets with Chinese and English as source languages by setting different pivot languages and translation rounds, respectively. This approach allowed us to systematically explore the impact of pivot languages and translation iterations on the performance of our model. The pivot languages are represented by the English initials of the respective languages, see Table 2.

Table 2. Pivot Language Abbreviations.

Language	English	Chinese	Spanish	German	Japanese	Tibetan
Capital Initial	E	C	S	G	J	T

We performed r round translations of the news texts in the LCSTS dataset via the Baidu Translator, where $r = 2, 3, 4$. During the experimental verification, the pivot language and translation rounds were selected randomly. The results of Spearman's rank correlation coefficient (ρ_s) for back-translation plagiarism detection are listed in Table 3.

Table 3. Results of Spearman's Rank Correlation Coefficient by Our PDMTT.

Round	English as the Source Language		Chinese as the Source Language	
	Translation Process	ρ_s	Translation Process	ρ_s
$r = 2$	E→C→E	0.8574	C→E→C	0.8564
	E→G→E	0.8583	C→G→C	0.8544
	E→T→E	0.8563	C→T→C	0.8611
$r = 3$	E→C→G→E	0.8561	C→E→G→C	0.8413
	E→G→T→E	0.8555	C→G→T→C	0.8547

(continued)

Table 3. (*continued*)

Round	English as the Source Language		Chinese as the Source Language	
	Translation Process	ρ_s	Translation Process	ρ_s
	E→T→J→E	0.8549	C→T→J→C	0.8551
$r = 4$	E→T→S→C→E	0.8543	C→T→S→E→C	0.8547
	E→G→C→S→E	0.8442	C→G→E→S→C	0.8473
	E→C→J→T→E	0.8511	C→E→J→T→C	0.8445

It can be seen that the experiments involving English as the source language achieved optimal performance through a two-round translation process utilizing the "E→G→E" approach. On the other hand, in the case of experiments with Chinese as the source language, the most favorable results were achieved by implementing a two-round translation procedure using the "C→T→C" process. English and German both belong to the Germanic language family, whereas Chinese and Tibetan are part of the Sino-Tibetan language family. Languages within the same language family share a significant degree of similarity in terms of their etymology and grammar. As a result, the ease of mutual translation between languages from the same family is notably higher compared with the languages from different language families.

Table 4. Back-Translated Text Pairs and Contradiction Text Pairs.

	Text Samples
Original Text 1	It is currently the peak season for tourism, and there are many cars coming to Sanya, so the demand for fuel is naturally much higher than usual. However, now we do not allow tank trucks to enter the city, and without fuel, we cannot replenish it in a timely manner. Therefore, many gas stations can only close, the staff said. Sanya's phased restrictions on the entry of engineering vehicles, large trucks, and agricultural vehicles also confirm the above statement
Back-Translated Text	"Now is the peak season for tourism. There are a lot of cars coming to Sanya, so the demand for fuel is naturally much higher than usual. However, now we do not allow Tank trucks to enter the city, and there is no fuel, so we cannot replenish it in time, so many gas stations have to close", said the staff member. The phased restrictions on construction vehicles, large trucks, and agricultural vehicles entering Sanya also confirm the above statement

(*continued*)

Table 4. (*continued*)

	Text Samples
Original Text 2	The dispute over the illegal construction of shops in Huaqiang North has lasted for more than three years, and Nandu has followed up many times. Yesterday, the Futian District Planning and Land Supervision Bureau posted a reminder letter on the first floor of the Modern Window Building on Huaqiang North Road. The value of the shops involved in illegal construction is nearly 100 million yuan, and the dispute is twists and turns. The dispute is expected to end in three years
Contradiction Text	At 5 a.m. in Bozhou, Anhui Province, 10-year-old Xiao Deqi and his grandmother were already cleaning the road. When the children in the villa drove to school, Shodditch drove the garbage truck to the transfer station and walked back and forth for three or four miles. Adults can't afford to work overtime every day, but Yuan's performance has always been in the top ten of the class

We employed our PDMTT model to detect both back-translated text pairs and contradiction text pairs. The back-translated text pairs in Table 4 consist of the original text 1 and the back-translated text, where the back-translated text was plagiarized from the original text through the "E→G→E" translation process using Baidu Translator. The contradiction text pairs in Table 4 are composed of sentences that have not been plagiarized, namely original text 2 and contradiction text.

According to the detection results of the PDMTT, the cosine similarity for back-translated text pairs is 0.9713, while the cosine similarity for contradiction text pairs is 0.4508. In our back-translation detection task, a threshold of 0.65 was set, where values exceeding 0.65 were considered back-translation plagiarism. The results are listed in Table 5.

Table 5. Cosine Similarity Results for Plagiarism Detection by Our PDMTT.

Text Pair	Results of Cosine Similarity
Back-Translated Text Pair	0.9713
Contradiction Text Pair	0.4508

3.3 Performance Comparison

Text Semantic Similarity Detection. To validate the performance of our model in the domain of the traditional text semantic similarity detection task, we first conducted a comparison between our PDMTT and state-of-the-art language models previously employed in the STS dataset. It is important to note that all experimental configurations in this paper were kept consistent.

In the natural language inference (NLI) dataset, given a premise, it is required to generate a correct sentence (entailment), a potentially correct sentence (neutral), and an incorrect sentence (contradiction). Thus, for each premise and its entailment hypothesis, there exists an accompanying contradiction hypothesis, forming a triplet (x_i, x_i^+, x_i^-), where x_i is the premise, x_i^+ and x_i^- are entailment and contradiction hypotheses [24]. We trained our model using a supervised NLI dataset [25], by predicting the relationship between two sentences as entailment, neutral, or contradiction. The results were evaluated using Spearman's rank correlation coefficient on the STS dataset in both Chinese and English, as listed in Table 6. It can be seen that our PDMTT significantly improves the results on the dataset, surpassing the previous models. Specifically, our PDMTT improves the previous best Spearman's rank correlation coefficient from 73.29% to 78.50%.

Table 6. Spearman's Rank Correlation Coefficient Results of Different Models on STS Dataset.

	PDMTT	SBERT [14]	ConSERT [17]	BERT-Whitening [16]	BERT [12]
English	**0.7641**	0.6881	0.6748	0.6755	0.5818
Chinese	**0.7850**	0.7329	0.7175	0.6722	0.5600

Back-Translation Plagiarism Detection. Furthermore, we conducted experiments on the back-translation plagiarism dataset. The results of Spearman's rank correlation coefficient for various models on both English and Chinese back-translation plagiarism datasets are shown in Fig. 2. In Fig. 2(a), we chose "E→C→E", "E→C→G→E", and "E→T→S→C→E" as representatives for the 2–4 rounds of back-translation plagiarism with English as the source language. In Fig. 2(b), we utilized the processes of "C→E→C", "C→E→G→C", and "C→T→S→E→C" with Chinese as the source language. The compared results highlight significant improvements by the proposed PDMTT model on the back-translation plagiarism detection task.

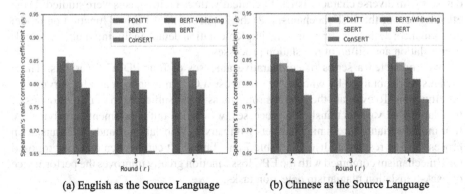

(a) English as the Source Language   (b) Chinese as the Source Language

Fig. 2. Performance of Different Models on the Back-Translation Detection Task.

Visualization Illustration. For the back-translation plagiarism detection task, obtaining accurate text representation is of utmost importance. The text representation generated by the model should be evenly distributed in the semantic space [26], with the back-translated text being closer to the original and the contradiction text further away.

To show the results intuitively, we visualized text representations generated by our PDMTT along with those generated by other compared models. Firstly, we randomly selected 20 positive pairs and 20 negative pairs from the test set and then input them into different models to obtain their corresponding text representations. Finally, we used the t-SNE algorithm to reduce the dimensionality and visualize them in the two-dimensional semantic space. The distributions of processed text representations are illustrated in Fig. 3(a)–(d), where the red numbers represent positive text pairs and the blue numbers denote negative text pairs.

Based on the aforementioned two metrics in Sect. 3.1, models with better alignment and uniformity exhibit superior performance. On one hand, we expect positive samples to be close to each other, and on the other hand, semantic vectors should be uniformly distributed on the hypersphere because a more uniform distribution maximizes information entropy, resulting in the preservation of more information. Therefore, we aim for both of these metrics to be as low as possible. As shown in Fig. 3(c) and (d), the text representations generated by the SBERT [14] and ConSERT [17] partially overlap and lack unity and distinguishability. The BERT-Whitening [16] exhibits good uniformity but lacks alignment, see Fig. 3(b). In comparison, the text representations generated by our PDMTT are more evenly distributed in the semantic space while maintaining proximity to the back-translated texts and remaining relatively distant from their contradictory counterparts, see Fig. 3(a). Therefore, the proposed PDMTT model can make the text representation after back-translation plagiarism more effective for plagiarism detection.

3.4 Ablation Study

We performed ablation studies to analyze the importance of each scheme based on nine datasets with diverse characteristics. Specifically, three main aspects were studied: 1) the effectiveness of the GEF mechanism; 2) the impact of the EPC loss function on model performance; 3) the necessity of some implementation details including multiple rounds of translation and different translation processes.

We conducted a series of comparative analyses of four PDMTT variants. These variants were obtained by whether or not to use a GEF mechanism and by varying the loss function. To evaluate the model's robustness when subjected to multiple rounds of translation and various translation processes, we carried out experiments involving the four model variants across nine datasets with varying translation rounds and translation processes. The results are listed in Table 7 and Table 8. It can be seen that the proposed GEF mechanism combined with the EPC loss function greatly improves the performance of back-translation plagiarism detection tasks.

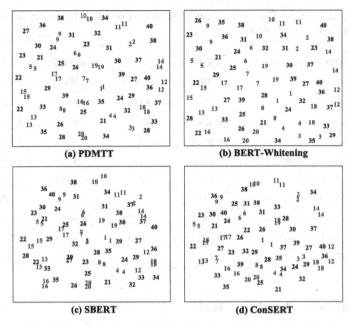

Fig. 3. Text Representations Visualization. (Color figure online)

Table 7. Spearman's Rank Correlation Coefficient Results of GEF Mechanism and EPC Loss Function with English as the Source Language.

Round	Translation Process	w/GEF		w/o GEF	
		w/EPC	w/o EPC	w/EPC	w/o EPC
$r = 2$	E→C→E	0.8574	0.8166	0.8301	0.7914
	E→G→E	0.8583	0.8285	0.8386	0.8139
	E→T→E	0.8563	0.8283	0.8385	0.7717
$r = 3$	E→C→G→E	0.8561	0.8161	0.8275	0.7869
	E→G→T→E	0.8555	0.8168	0.8035	0.7506
	E→T→J→E	0.8549	0.8122	0.7932	0.7631
$r = 4$	E→T→S→C→E	0.8543	0.8126	0.8141	0.7272
	E→G→C→S→E	0.8442	0.8243	0.8228	0.7452
	E→C→J→T→E	0.8511	0.8125	0.8139	0.7278

Table 8. Spearman's Rank Correlation Coefficient Results of GEF Mechanism and EPC Loss Function with Chinese as the Source Language.

Round	Translation Process	w/GEF		w/o GEF	
		w/EPC	w/o EPC	w/EPC	w/o EPC
$r = 2$	C→E→C	0.8564	0.8311	0.8189	0.7738
	C→G→C	0.8544	0.8261	0.8028	0.7635
	C→T→C	0.8611	0.8375	0.8267	0.7293
$r = 3$	C→E→G→C	0.8413	0.8221	0.8135	0.7454
	C→G→T→C	0.8547	0.8271	0.8083	0.7183
	C→T→J→C	0.8551	0.8328	0.8145	0.7771
$r = 4$	C→T→S→E→C	0.8547	0.8327	0.8079	0.7649
	C→G→E→S→C	0.8473	0.8223	0.8139	0.7409
	C→E→J→T→C	0.8445	0.8313	0.7870	0.7563

4   Conclusion and Future Work

This paper proposes a back-translation plagiarism detection model named PDMTT that combines the contrastive learning framework with the pre-trained BERT model. Our model adopts a GEF mechanism to make the text representation after back-translation plagiarism more effective for plagiarism detection and optimizes the plagiarism detection capability of the model by improving the contrastive loss function. Experimental results demonstrate the effectiveness of PDMTT, outperforming other models in the back-translation plagiarism detection task and the text semantic similarity detection task. Moving forward, we plan to extend the applicability of our model to longer texts, addressing its current limitations. Additionally, we'll continue refining our PDMTT for broader real-world scenarios, experimenting with diverse datasets and languages.

Acknowledgements. This work was supported in part by the National Natural Science Foundation of China under Grants U20B2051 and 62172280; in part by the Natural Science Foundation of Shanghai under Grants 21ZR1444600; and in part by the Shanghai Science and Technology Committee Capability Construction Project for Shanghai Municipal Universities under Grant 20060502300.

References

1. Lu, L., Zhou, L.: DNAP: detection of news article plagiarism. In: 2021 IEEE 6th International Conference on Cloud Computing and Big Data Analytics (ICCCBDA), Chengdu, China, pp. 337–341 (2021)
2. Jones, M.: Back-translation: the latest form of plagiarism. In: The 4th Asia Pacific Conference on Educational Integrity, Wollongong, Australia, pp. 1–7 (2009)

3. Anchal, P., Urvashi, G.: A review on diverse algorithms used in the context of plagiarism detection. In: 2023 International Conference on Advancement in Computation & Computer Technologies (InCACCT), Gharuan, India, pp. 1–6 (2023)
4. Salha, A., Naomie, S., Ajith, A.: Understanding plagiarism linguistic patterns, textual features, and detection methods. IEEE Trans. Syst. Man Cybern. Part C **42**(2), 133–149 (2012)
5. Alzahrani, S., Salim, N.: Fuzzy semantic-based string similarity for extrinsic plagiarism detection lab report for PAN at CLEF 2010. In: CLEF 2010 LABs and Workshops, Notebook Papers, Padua, Italy, 22–23 September 2010 (2010)
6. zu Eissen, S.M., Stein, B.: Intrinsic plagiarism detection. In: Lalmas, M., MacFarlane, A., Rüger, S., Tombros, A., Tsikrika, T., Yavlinsky, A. (eds.) ECIR 2006. LNCS, vol. 3936, pp. 565–569. Springer, Heidelberg (2006). https://doi.org/10.1007/11735106_66
7. El-Rashidy, M.A., Mohamed, R.G., El-Fishawy, N.A., et al.: An effective text plagiarism detection system based on feature selection and SVM techniques. Multimedia Tools Appl. **83**, 2609–2646 (2023). https://doi.org/10.1007/s11042-023-15703-4
8. Poibeau, T.: Machine Translation. MIT Press, Cambridge (2017)
9. Yoon, K.: Convolutional neural networks for sentence classification. In: Proceedings of the 2014 Conference on Empirical Methods in Natural Language Processing (EMNLP), Doha, Qatar, 25–29 October 2014, pp. 1746–1751 (2014)
10. Cho, K., Van, M.B., Gulcehre, C., et al.: Learning phrase representations using RNN encoder-decoder for statistical machine translation. In: Proceedings of the 2014 Conference on Empirical Methods in Natural Language Processing (EMNLP), Doha, Qatar, 25–29 October 2014, pp. 1724–1734 (2014)
11. Jeffrey, P., Richard, S., Christopher, D.M.: GloVe: global vectors for word representation. In: Proceedings of the 2014 Conference on Empirical Methods in Natural Language Processing (EMNLP), Doha, Qatar, 25–29 October 2014, pp. 1532–1543 (2014)
12. Jacob, D., Ming-Wei, C., Kenton, L., Kristina, T.: BERT: pre-training of deep bidirectional transformers for language understanding. In: Proceedings of NAACL-HLT 2019, pp. 4171–4186 (2019)
13. Jun, G., Di, H., Xu, T., et al.: Representation degeneration problem in training natural language generation models. In: International Conference on Learning Representations, New Orleans, America, 6–9 May 2018 (2018)
14. Nils, R., Iryna, G.: Sentence-BERT: sentence embeddings using siamese BERT-networks. In: Proceedings of the 2019 Conference on Empirical Methods in Natural Language Processing and the 9th International Joint Conference on Natural Language Processing (EMNLP-IJCNLP), Hong Kong, China, 3–7 November 2019, pp. 3982–3992 (2019)
15. Li, B., Zhou, H., He, J.X., et al.: On the sentence embeddings from pre-trained language models. In: Proceedings of the 2020 Conference on Empirical Methods in Natural Language Processing (EMNLP), 16–18 November 2020, pp. 9119–9130 (2020)
16. Su, J.L., Cao, J.R., Liu, W.J., Ouyang, Y.W.: Whitening sentence representations for better semantics and faster retrieval. CoRR abs/2103.15316 (2021)
17. Yan, Y.M., Li, R.M., Wang, S.R., et al.: ConSERT: a contrastive framework for self-supervised sentence representation transfer. In: Proceedings of the 59th Annual Meeting of the Association for Computational Linguistics and the 11th International Joint Conference on Natural Language Processing (Volume 1: Long Papers), pp. 5065–5075. Association for Computational Linguistics (2021)
18. Spaces.Ac.cn. https://spaces.ac.cn/archives/8860. Accessed 12 June 2022
19. Li, X., Hu, X.L., Yang, J.: Spatial group-wise enhance: improving semantic feature learning in convolutional networks. CoRR abs/1905.09646 (2019)
20. Hu, B.T., Chen, Q.C., Zhu, F.Z.: LCSTS: a large scale chinese short text summarization dataset. In: Proceedings of the 2015 Conference on Empirical Methods in Natural Language Processing, Lisbon, Portugal, 17–21 September 2015, pp. 1967–1972 (2015)

21. Cer, D., Diab, M., Agirre, E., et al.: SemEval-2017 task 1: semantic textual similarity multilingual and crosslingual focused evaluation. In: The 11th International Workshop on Semantic Evaluation (SemEval-2017), Vancouver, Canada, August 2017, pp. 1–14 (2017)
22. Nils, R., Philip, B., Iryna, G.: Task-oriented intrinsic evaluation of semantic textual similarity. In: Proceedings of COLING 2016, The 26th International Conference on Computational Linguistics: Technical Papers, pp. 87–96. The COLING 2016 Organizing Committee, Osaka (2016)
23. Wang, T.Z., Isola, P.: Understanding contrastive representation learning through alignment and uniformity on the hypersphere. In: Proceedings of the 37th International Conference on Machine Learning. PMLR, vol. 119, pp. 9929–9939 (2020)
24. Gao, T.Y., Yao, X.C., Chen, D.Q.: SimCSE: simple contrastive learning of sentence embeddings. In: 2021 Conference on Empirical Methods in Natural Language Processing, Virtual, Punta Cana, 7–11 November 2021, pp. 6894–6910 (2021)
25. Conneau, A., Kiela, D., Schwenk, H., et al.: Supervised learning of universal sentence representations from natural language inference data. In: Palmer, M., Hwa, R., Riedel, S. (eds) Proceedings of the 2017 Conference on Empirical Methods in Natural Language Processing. Conference on Empirical Methods in Natural Language Processing, 07–11 September 2017, pp. 670–680. Association for Computational Linguistics, Copenhagen (2017)
26. Feng, M.F., Chen, Y.S., Guo, Y.C., et al.: Learning text representations for finding similar exercises. In: 2019 IEEE International Conference on Consumer Electronics - Taiwan (ICCE-TW), Yilan, Taiwan, 20–22 May 2019, pp. 1–2 (2019)

An Image Perceptual Hashing Algorithm Based on Convolutional Neural Networks

Meihong Yang[1,2] (ID), Baolin Qi[1,2] (ID), Yongjin Xian[1,2(✉)] (ID), and Jian Li[1,2] (ID)

[1] Key Laboratory of Computing Power Network and Information Security,
Ministry of Education, Shandong Computer Science Center (National Supercomputer Center in Jinan), Qilu University of Technology (Shandong Academy of Sciences), Jinan, China
matxyj@163.com

[2] Shandong Provincial Key Laboratory of Computer Networks, Shandong Fundamental Research Center for Computer Science, Jinan, China

Abstract. The conventional perceptual hashing algorithms are constrained to a singular global feature extraction algorithm and lack efficient scalability adaptation. To address this problem, an image-perceptual hashing algorithm based on convolutional neural networks is proposed in this paper. First of all, the entire image is convolved by the backbone network to obtain a feature map. The Region Proposal Network (RPN) is employed to generate multiple-sized proposal frames at each location by using sliding windows. Considering the complexity and diversity of the object, proposal boxes of various sizes and shapes are formulated, and the local features are comprehensively exploited in an image, thereby, generating a perceptual hash code that can represent the semantic features of an image strongly. Moreover, The Mean Square Error (MSE) loss is incorporated into the optimization process to evaluate the coincidence between the proposal frame and the actual frame, generating more representative hash codes. Finally, an image perceptual hash code with high intuitive features can be formulated through iterative training of the proposed convolutional neural networks. Extensive experimental results demonstrate that the proposed image perceptual hashing algorithm based on a convolutional neural network surpasses other state-of-the-art methods.

This work was supported by National Natural Science Foundation of China (62272255, 62302248, 62302249); National key research and development program of China (2021YFC3340600, 2021YFC3340602); Taishan Scholar Program of Shandong (tsqn202306251); Shandong Provincial Natural Science Foundation (ZR2020MF054, ZR2023QF018, ZR2023QF032, ZR2022LZH011), Ability Improvement Project of Science and Technology SMES in Shandong Province (2022TSGC2485, 2023TSGC0217); Jinan "20 Universities"-Project of Jinan Research Leader Studio (2020GXRC056); Jinan "New 20 Universities"-Project of Introducing Innovation Team (202228016); Youth Innovation Team of Colleges and Universities in Shandong Province (2022KJ124); The "Chunhui Plan" Cooperative Scientific Research Project of Ministry of Education (HZKY20220482); Achievement transformation of science, education and production integration pilot project (2023CGZH-05), First Talent Research Project under Grant (2023RCKY131, 2023RCKY143), Integration Pilot Project of Science Education Industry under Grant (2023PX006, 2023PY060, 2023PX071).

B. Ma et al. (Eds.): IWDW 2023, LNCS 14511, pp. 95–108, 2024.
https://doi.org/10.1007/978-981-97-2585-4_7

Keywords: perceptual hashing · region proposal network · hash code · mean square error · image content authentication

1   Introduction

With the continuous development of intelligent terminals and digital image processing technology [1,2], the cost of image modification has been consistently reduced. Consequently, copyright verification and source detection of digital images have become a prominent research topic in the field. Image-aware hashing, as an emerging multimedia security technology [3,4], has garnered significant attention from scholars due to its ability to generate fixed-length hash codes based on visual content features of images [5]. These algorithms can map digital images with similar perceptual content into highly similar summaries while maintaining unidirectionality and collision resistance [6]. Therefore, image-aware hashing algorithms are considered an ideal solution for image copyright authentication and source detection as they strike a balance between robustness within the same image perception and vulnerability across different images. Classical image-aware hashing techniques can be broadly classified into three categories: based on statistical features [7–12], local feature point extraction [13–17], frequency domain transform [18–21], and feature dimensionality reduction [22–24]. These algorithms rely on pre-designed feature extractors to generate hash sequences, which require extensive expert knowledge and make it difficult to capture the intrinsic or abstract visual features of digital images. Moreover, the performance optimization of perceptual hashing algorithms is not strong. Qin et al. [4] proposed a multi-constraint convolutional neural network (CNN) scheme for perceptual image hashing, constructing two pairs of constraints and integrating them into an overall constraint function through a weight assignment strategy, dynamically adjusting the training set structure according to the changes in constraint values, and employ a lightweight CNN to automatically learn image features and generate perceptual hash codes, achieving a balance between robustness and discriminative nature of the perceptual hash algorithm. The proposed algorithm by Ma et al. [6] presents an unsupervised perceptual hash image content forensics approach based on a two-way generative adversarial network. This approach effectively generates perceptual hash codes with a robust representation of image semantic features through iterative adversarial interactions among the encoding network, generative network, and discriminative network.

Conventional image hashing schemes typically consist of three main stages: preprocessing, feature extraction, and hash generation. These stages aim to convert images into compact hash code representations. Perform effective comparison and retrieval of hash codes based on similarity metrics. However, the features used in traditional image hashing algorithms are often manually designed and extracted, making it difficult to achieve an optimal trade-off between robustness and recognition and failing to represent the image's semantic information better. Conventional hashing schemes for image recognition typically exhibit robustness only towards certain geometric transformations, such as translation and rotation, while being susceptible to other transformations like affine transformations, resulting in subpar performance in perceptual hashing. Deep learning-based perceptual hashing algorithms offer improved image recognition capabilities compared to traditional methods; however, deep learning-based image-aware hashing content forensics algorithms necessitate the extraction of features from the entire image, introducing irrelevant information related to non-target objects and rendering them less resilient against content retention attacks. In order to enhance the resistance of perceptual hashing algorithms against high-intensity content retention attacks requires accelerating the generation efficiency of perceptual hash codes. This paper proposes a perceptual hash image content forensic algorithm based on convolutional neural networks, which leverage the detailed feature extraction capabilities of regional proposal networks and employ the MSE loss optimization algorithm to enhance the ability of the region proposal network for generating the image intrinsic features of the perceptual hash code representation, Consequently, this approach significantly enhances the performance of image content authentication and copyright protection.

The majority of algorithms primarily focus on the global features of an image while disregarding its local features, resulting in a lack of desired balance in terms of robustness and differentiation among perceptual hashing algorithms tested in experiments. To address the issue of under-utilization of key regions within images, this paper proposes an image-aware hashing algorithm based on convolutional neural networks (CNNs), which has the following contributions compared with other image-aware hashing algorithms:

(1) The region proposal network(RPN) technique is first employed to facilitate the perceptual hash algorithm to extract image features deeply, enabling the proposed scheme to remain robust to changes in target position, size, and pose even in complex scenes.
(2) A region optimization strategy based on Mean Square Error (MSE) loss is presented to enhance the detail representation capability of an image hash code by optimizing proposal regions. Thereby, improving the sensitivity of perceptual hash codes in distinguishing images with subtle differences and representing the accuracy of image content.
(3) An alternating optimization strategy is designed in the model training process, Therefore, the excellent feature extraction capability of convolutional neural networks is further exploited, resulting in the training time and the consumption of computational resources being greatly reduced.

2   A Perceptual Hash Forensic Model Based on Convolutional Neural Networks

The Structure of Algorithm Model. The proposed CNN based image-aware hashing algorithm effectively extracts regional features with rich semantic information and captures details through fine representation of different regions in

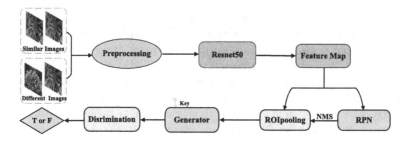

Fig. 1. Structure Diagram of Perceptual Hash Image Source Detection Algorithm Based on CNN

the preprocessed image. By quantifying local information, it generates a more intuitive and interpretable perceptual hash code, improving the algorithm's performance for authentication and copyright protection. The structure of the algorithm is shown in Fig. 1. Where Resnet50 refers to the Feature Extraction Network, RPN stands for Region Proposal Network [25], and ROI pooling represents the Regional Feature Coding Network. Similar images denote both the original images and those subjected to content retention operations, while different images refer to the original images and those with different contents. The main role of NMS (Non-Maximum Suppression) is to remove overlapping bounding boxes in the candidate box (bounding boxes) generation task, and to ensure that there is only one corresponding bounding box for each target by removing redundant detection results, thus avoiding double counting and redundant information. The image perceptual hashing algorithm based on convolutional neural network automatically identifies regions of interest in an image and extracts features related to its content through RPN. Compared to traditional image hashing algorithms, it provides a more accurate representation of detailed content such as texture, edges, color, and other features. The utilization of a multi-layered network structure enables feature extraction at various scales, facilitating effective feature fusion for improved accuracy and robustness in image perceptual hashing. The model's ability to represent image features is enhanced through the mapping of the feature coding network and optimization of the mean square error loss (MSE), ultimately achieving an effective balance between robustness and discriminative nature of the hash code.

The Design of Backbone Network. The backbone network in an CNN based image-aware hash generation network plays a role in extracting the feature representation of the input image to provide rich semantic and contextual information for subsequent networks. In this study, a Resnet50 feature extraction network is utilized. The Resnet50 network includes Conv2d, BatchNorm, Relu activation, Maxpool pooling, four residual modules, and a fully connected layer (FC). The input image is gradually transformed into a high-level feature map through the convolutional and pooling layers of resnet50. These feature representations contain semantic and structural information in the image that helps in target classification.

The Design of Discriminator. The discriminator compares hash codes generated from proposal frame features of the same or different images, producing a distribution ranging from 0 to 1. A distribution close to 1 indicates similar image content, while a distribution close to zero suggests different content. By setting an appropriate threshold T, images can be classified into two categories: those with identical perceived content and those with distinct perceived content. This method is applicable in image content authentication and copyright protection for identifying original images and detecting forged ones, ensuring intellectual property rights and content authenticity.

2.1   Design of Loss Function

The perceptual hashing algorithm for images in this paper utilizes a region proposal network (RPN) to first generate target regions of interest (ROI). These ROI are obtained by the RPN, which slides anchor frames of various sizes and aspect ratios over the image. Subsequently, these proposal target frames serve as inputs to the subsequent image perceptual hashing algorithm. The RPN network is used to initialize and train the entire image perceptual hashing network in an end-to-end manner. The process involves adding an image feature extraction layer, a hash coding layer, and a loss function on top of the RPN network. This comprehensive training approach enables the entire network to acquire more expressive feature representation and achieve efficient hash coding. Ultimately, when an image is inputted into a trained network, its features are extracted by the image perceptual hashing algorithm and encoded into compact hash codes. These generated hash codes exhibit high discriminative power while effectively preserving information about the image content. In the experiment, an error iteration method based on the Adam optimization algorithm was utilized. This method considers both first and second order moment estimates of the error loss to calculate the update step size of network parameters. To enhance convergence stability, adaptive network step size optimization and multiple iterations were employed in system design. Additionally, exponential learning rate decay was implemented during training where a larger initial learning rate facilitated rapid feature learning by the model which gradually decreased as training proceeded for more careful parameter optimization.

In this study, the inclusion of Mean Square Error (MSE) loss in the generation of proposal boxes aims to enhance the detailed representation of generated images. The optimization of proposal regions improves the accuracy of perceptual hash codes, enabling better content and detail representation. Furthermore, by optimizing the strategy, the algorithm enhances the discriminative ability of perceptual hash codes to effectively distinguish between images with subtle differences. This optimization process significantly contributes to improving image quality and realism, thereby meeting higher visual perception requirements. The loss function of image perceptual hashing algorithm based on RPN is designed:

$$L_{rpn} = \frac{1}{N_{cls}} \sum_i L_{cls}\left(p_i, p_i^*\right) + \lambda \frac{1}{N_{\text{reg}}} \sum_i L_{reg}\left(t_i, t_i^*\right) \qquad (1)$$

$$L_{cls}\left(p_i, p_i^*\right) = -\log\left[p_i p_i^* + \left(1 - p_i\right)\left(1 - p_i^*\right)\right] \qquad (2)$$

$$L_{reg}\left(t_i, t_i^*\right) = R\left(t_i - t_i^*\right) \qquad (3)$$

$$L_{mse} = \frac{1}{N_{bbox}} \sum_i \left(\frac{1}{m} \sum_i \left(x_{ij} - y_{ij}\right)^2\right) \qquad (4)$$

$$L = \frac{1}{N_{cls}} \sum_i L_{cls}\left(p_i, p_i^*\right) + \frac{1}{N_{\text{reg}}} \sum_i p_i^* L_{reg}\left(t_i, t_i^*\right) + \lambda L_{mse} \qquad (5)$$

In Eq. (1), p_i represents the predicted classification probability of Anchor[i], and p_i^* is set to 1 for positive samples and 0 for negative samples. A positive sample is defined as having an IOU (Intersection-Over-Union) overlap region greater than 0.7 with the Ground Truth Box, while a negative sample has an IOU overlap region less than 0.3. The parameterized coordinates of the Bounding Box predicted by Anchor[i] are denoted as t_i, t_i^* represents the parameterized coordinates of the Bounding Box of Ground Truth for Anchor[i]; N_{cls} denotes the number of anchors; L_{reg} denotes a binary cross-entropy loss; R is the Smooth L1 function.

The mean square error (MSE) loss is utilized to quantify the disparity between the predicted frame and the actual frame. Where x_{ij} denotes the predicted value of the j_{th} dimension of the i_{th} sample within the prediction bounding box, and y_{ij} is the target value of the j_{th} dimension of the i_{th} sample within the real bounding box. m denotes the dimensions of the bounding box which is taken as 4 in this paper, and the larger the L_{mse} indicates a greater dissimilarity between prediction and reality frames, wherea a smaller MSE implies a reduced distinction between them. Thus, adding the MSE loss to the region proposal network serves two purposes: firstly, it quantifies the level of positional consistency between the predicted frame and the real frame by calculating the error in their corresponding positions, thereby enhancing its ability to capture image details; secondly, by minimizing the MSE value, it reduces discrepancies in feature values between the real and predicted frames, thus continuously improving its capacity to represent semantic features encoded by output features.

3   Experimentation

In this experiment, a training set of 19,000 images and a test set of 12,000 images were randomly selected from the PASCAL VOC dataset. The original images encompass various subjects such as characters and landscapes. To enhance the efficiency of perceptual hash operations and reduce server costs considering the complexity of image features, all images were converted to grayscale in a uniform manner using bilinear interpolation operation resulting in uniformly sized 128×128 grayscale images. Experimental results demonstrate that although scaling down the original image may result in some loss of details, the CNN based image-aware hash generation algorithm still achieves high-precision image content forensics effectively enhancing its performance in terms of image content authentication and copyright protection.

3.1 Experimental Setup

The PASCAL VOC database is utilized in this study for conducting image content forensics research based on perceptual hashing. To accurately describe the concealed spatial feature information of each image and prevent any collision of feature information, a coding scheme employing 50-bit perceptual hash coding is employed in the experiments. This ensures that the perceptual hash code remains robust for authenticating identical content images, sensitive to distinguishing different content images, while also requiring minimal storage space. In the experiments, Pytorch framework is selected to implement the CNN based perceptual hash image content forensic algorithm and the parameters of the CNN based perceptual hash generation network are randomly initialized.

The cross-correlation value is utilized as an evaluation metric in experiments to ensure the consistency of the algorithm's performance evaluation metrics and facilitate comparison with other exceptional perceptual hash algorithms, thereby verifying the algorithm's performance.

3.2 Algorithm Performance Validation

In perceptual hashing algorithms, there exists a trade-off between recognition robustness and distinction sensitivity. An effective perceptual hash algorithm should strike a delicate balance between these two aspects, enabling accurate differentiation of images with diverse contents while maintaining robustness. Consequently, it facilitates image content authentication and copyright protection.

Algorithm Performance Validation. To assess the robustness of the CNN based image perceptual hashing content forensic network for authenticating identical content images, a random selection of 500 images from the database was used in this experiment. Employing the content preservation attack method outlined in Table 1, each test image generated 58 replica images with identical content but differing visual perception. To verify the robust performance of the perceptual hash algorithm, we computed cross-correlation values between the original image's perceptual hash codes and those of attacked images, averaging these values for comparison. A higher cross-correlation value closer to 1 indicates a stronger representation of implicit features by the perceptual hash code; conversely, a lower value closer to 0 suggests weaker representation. The results in Table 2 show that the perceptual hash codes generated by the RPN network are robust against various attacks, maintaining significant similarity

Table 1. Correlation coefficients under different content retention operations

Operation	Parameter	Parameter value
Contrast adjustment	Photoshop scale	−20, −10, 10, 20
Gamma filtering	Gamma	0.7, 0.9, 1.1, 1.3
Image scaling	Scaling ratio	0.5, 0.7, 0.9, 1.1, 1.3, 1.5
Salt and Pepper noise	Density	0.001, 0.002, ..., 0.009, 0.01
Gaussian noise	Variance	0.001, 0.002, ..., 0.009, 0.01
Affine Transformation	Rotation angle	1, 2, 3, 4, 5
JPEG compression	Quality factor	30, 40, ..., 90, 100
Random erasing	Scaling ratio	0.005, 0.01, 0.015, 0.02
Center cutting	Scaling ratio	0.8, 0.85, 0.9, 0.95

(a) Original images   (b) Tampered images.

Fig. 2. Comparison of original and tampered images

and ensuring accurate image authentication for identical content. Incorporating MSE to counteract loss slightly decreases the cross-correlation between "affine transform" and "Gaussian noise" types. This is because it enhances the network's ability to learn image details, strengthens the semantic representation of the hash code, increases discriminative requirements, and leads to greater differences between similar images. The results in Table 2 show that the average cross-correlation value of the perceptual hash codes generated solely by the RPN network is 0.9921. To achieve a balance between robustness and discriminative capability in image perceptual hash algorithms, we introduce the RPN network along with an increase in MSE loss. The cross correlation value of the generated perceptual hash code of the attacked image slightly decreases after enhancing the proposal frame detail description capability using MSE loss, but it still remains at 0.9573. By utilizing the RPN network and MSE loss, we can find an optimal balance between recognition robustness for identical content images and differentiation sensitivity for different content images, resulting in a better equilibrium between robustness and differentiation. Figure 2 is the original image and the corresponding tampered version image.

Table 2. Image cross-correlation coefficient of the same source

Operation	Mean		
	No RPN	RPN	RPN+mse
Contrast adjustment	0.9223	0.9984	0.9965
Gamma filtering	0.9154	0.9924	0.9862
Image scaling	0.9212	0.9878	0.9679
Salt and Pepper noise	0.8553	0.9965	0.9469
Gaussian noise	0.8843	0.9886	0.8474
Affine Transformation	0.9256	0.9870	0.9692
JPEG compression	0.9312	0.9821	0.9676
Random erasing	0.9174	0.9987	0.9862
Center cutting	0.8245	0.9975	0.9482
Average cross-correlation of different attacks	0.8997	0.9921	0.9573

Validation of Image Discrimination Between Different Content Sources. The efficacy of the perceptual hash algorithm in this paper was assessed in terms of its ability to discriminate between distinct content images. For this experiment, 500 images were randomly selected from the test set and subjected to nine attacks listed in Table 1. The cross-correlation value was used as an evaluation index to measure the correlation between perceptual hash codes of different content images. A total of 124750 mutual correlation values were generated between each image and other images to determine the differences in perceptual hash codes. As shown in Table 3, when only using the RPN network for differentiation verification of visually distinct images, the mean hash

Table 3. Image cross-correlation coefficient of the different source

Operation	Mean		
	No RPN	RPN	RPN+mse
Contrast adjustment	0.4258	0.3436	0.0927
Gamma filtering	0.4524	0.3590	0.0919
Image scaling	0.4601	0.2213	0.0199
Salt and Pepper noise	0.4126	0.3362	0.1168
Gaussian noise	0.4202	0.5157	0.1242
Affine Transformation	0.4413	0.2355	0.0424
JPEG compression	0.4240	0.3544	0.0644
Random erasing	0.4805	0.3451	0.0811
Center cutting	0.4480	0.2346	0.0566
Average cross-correlation of different attacks	0.4405	0.3273	0.0767

cross-correlation value for all attack methods was found to be 0.3476, with each attack method having a cross-correlation value less than 0.5027. This is a significant decrease compared to the cross-correlation value of visually identical image hashes. After adding MSE loss, which enhances image detail characterization ability and improves differentiation of visually distinct content images, the mean cross-correlation further decreases to 0.0733. The experimental results show that the cross-correlation values of different content image perceptual hash codes generated by the CNN based perceptual hash image content forensics algorithm are all distributed close to 0, indicating a weak correlation among distinct content image perceptual hash codes. Therefore, this algorithm effectively enhances the sensitivity of the perceptual hash code for discriminating between diverse content images.

3.3   Comparison of the Performance of Different Methods

In order to comprehensively evaluate the performance of perceptual hash codes generated by this algorithm, they were compared with hash codes generated by manually designed features and deep learning algorithms. However, there is limited research on using deep learning algorithms for generating image-aware hash codes and validating perceptual hash algorithms with large databases. To ensure fairness, several recent representative perceptual hash generation algorithms are selected for comparison experiments in this paper. These experiments use the PASCAL VOC training set to compare and validate different perceptual hash algorithms, the aim is to conduct a comprehensive investigation of perceptual hash image content forensics algorithms based on CNN.

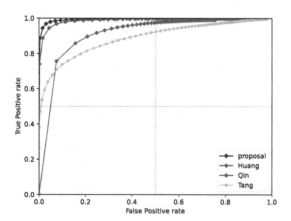

Fig. 3. ROC variation curves for different perceptual hashing algorithms

The PASCAL VOC test set randomly selects 500 images for content preservation attacks using nine different attack algorithms (listed in Table 1). Perceptual hash generation algorithms are then used to extract the perceptual hash codes of the attacked images and calculate their cross-correlation coefficients. By comparing these results, we evaluate and compare the performance of different perceptual hash generation algorithms in image content forensics. Additionally, we compare the classification performance among different algorithms by plotting ROC curves based on the joint distribution of true positive rate (TPR) and false positive rate (FPR). The experimental results presented in Fig. 3 show that the CNN based perceptual hashing algorithm for image content forensics has a higher position on the ROC curve plot's upper left corner and exhibits a larger area under curve (AUC). This demonstrates that utilizing RPN in Perceptual Hashing Algorithm possesses superior capabilities in image content forensics. The experimental findings demonstrate that the CNN based generated perceptual hash code achieves a correct recognition rate of 97.20% and an error rate of 2.60% in image recognition, exhibiting the highest area under the ROC curve (AUC). Qin employed a deep learning based perceptual hash generation network and achieved a correct recognition rate of 95.00% with an error rate of 4.24%, showing a relatively larger AUC value compared to other methods mentioned above by Huang employed a perceptual hash generation algorithm based on image texture and invariant feature vector distance, resulting in an optimal image recognition accuracy of 83.4% and an error rate of 13.27%. The proposed method exhibited commendable performance in perceptual hash based image content forensics. Tang utilized a DCT transform combined with a frequency-domain matrix compression algorithm (LLE), which led to a decline in the performance of image content forensics. The optimal image recognition accuracy achieved was 80.6%, accompanied by an error rate of 17.6%. In summary, the RPN perceptual hash image content forensics algorithm effectively combines semantic information to generate a perceptual hash code with superior representation capabilities for image semantic features, thus demonstrating enhanced capability in both image content authentication and copyright protection.

3.4 Key Dependency

The generation algorithm of a good hash code should ensure that an incorrect key produces a different hash code from the correct key, preventing forgery and ensuring uniqueness. Therefore, the experiments also investigated the dependency of keys on generated hash codes. Key dependency refers to the situation where it is difficult for another party to generate the correct hash without possessing our algorithm, even in the absence of the correct key.

To validate this, distinct encryption keys are employed to encrypt visually indistinguishable images and generate diverse hash codes. The generated hash codes are further encrypted using random perturbations during the experiments. Firstly, 1000 images are randomly selected from the PASCAL VOC dataset, and the hash code is generated using the correct key, then other parameters are kept constant, 100 different keys are used to generate the image hash code, and the mutual relation value between the correct key and the image perceptual hash code generated by other different keys is calculated and averaged, and the effect is shown in Fig. 4, with the X-axis being the index of the 100 different keys, and

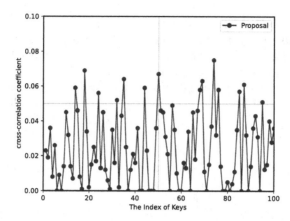

Fig. 4. The distribution of cross-correlation coefficient of hash codes generated by distinct encryption keys.

the Y-axis The number of interrelationships between the hash values generated by different keys, the results show that the number of interrelationships between the perceptual hash codes generated by using different keys are less than 0.08, indicating that it is difficult to generate the same hash without knowing the correct key. Therefore, the security of the scheme is completely dependent on the key and satisfies the security requirements in the cryptographic sense.

4   Conclusion

The present article proposes a perceptual hash algorithm based on convolutional neural networks, which integrates a region proposal network for image source detection. The comprehensive semantic information is provided by utilizing Resnet50, while the regional proposal networks are employed for detailed feature extraction in hash generation. Additionally, incorporating MSE error loss in the experiments encourages the generation of proposal frames resembling real frames closely, resulting in reduced noise in extracted local features. Moreover, a segmented alternate training strategy is applied during model training to enhance network efficiency. The performance of the hash generation algorithm is validated using the PASCAL VOC dataset, and comparative analysis with other hash algorithms demonstrates that this convolutional neural network based approach excels in image content authentication and copyright protection, effectively balancing robustness in recognizing images of identical content and sensitivity in distinguishing between images of disparate content.

References

1. Ma, B., Shi, Y.Q.: A reversible data hiding scheme based on code division multiplexing. IEEE Trans. Inf. Forensics Secur. **11**(9), 1914–1927 (2016)
2. Ma, B., Chang, L.L., Wang, C.P., et al.: Robust image watermarking using invariant accurate polar harmonic Fourier moments and chaotic mapping. Sig. Process. **172**, 107544 (2020)
3. Srivastava, M., Siddiqui, J., et al.: Local binary pattern based technique for content based image copy detection. In: 2020 International Conference on Power Electronics & IoT Applications in Renewable Energy and Its Control (PARC), pp. 374–377. IEEE, Piscataway (2020)
4. Qin, C., Liu, E.L., Feng, G.R., et al.: Perceptual image hashing for content authentication based on convolutional neural network with multiple constraints. IEEE Trans. Circ. Syst. Video Technol. **31**(11), 4523–4537 (2020)
5. Schneider, M., Chang, S.F.: A robust content based digital signature for image authentication. In: Proceedings of 3rd IEEE International Conference on Image Processing, vol. 3, pp. 227–230. IEEE, Piscataway (2002)
6. Bin, M.A., Yi-li, W.A.N.G., Jian, X.U., et al.: An image perceptual hash algorithm based on bidirectional generative adversarial network. Acta Electron. Sin. **51**(5), 1405–1412 (2023)
7. Zhao, Y., Yuan, X.R.: Perceptual image hashing based on color structure and intensity gradient. IEEE Access **8**, 26041–26053 (2020)
8. Tang, Z.J., Zhang, X.Q., et al.: Robust image hashing with ring partition and invariant vector distance. IEEE Trans. Inf. Forensics Secur. **11**(1), 200–214 (2015)
9. Srivastava, M., Siddiqui, J., Ali, M.A.: Robust image hashing based on statistical features for copy detection. In: 2016 IEEE Uttar Pradesh Section International Conference on Electrical, Computer and Electronics Engineering (UP-CON), pp. 490–495. IEEE, Piscataway (2017)
10. Yuan, X., Zhao, Y.: Perceptual image hashing based on three-dimensional global features and image energy. IEEE Access **9**, 49325–49337 (2021). https://doi.org/10.1109/ACCESS.2021.3069045
11. Yuan, X.R., Zhao, Y.: Perceptual image hashing based on three-dimensional global features and image energy. IEEE Access **9**, 49325–49337 (2021)
12. Hosny, K.M., Khedr, Y.M., Khedr, W.I., et al.: Robust image hashing using exact Gaussian-Hermite moments. IET Image Process. **12**(12), 2178–2185 (2018)
13. Ouyang, J., Liu, Y.Z., Shu, H.Z.: Robust hashing for image authentication using SIFT feature and quaternion Zernike moments. Multimedia Tools Appl. **76**(2), 2609–2626 (2017)
14. Wang, X.F., Xue, J.R., Zheng, Z.Q., et al.: Image forensic signature for content authenticity analysis. J. Vis. Commun. Image Representation **23**(5), 782–797 (2012)
15. Yu, M., Tang, Z., Zhang, X., Zhong, B., Zhang, X.: Perceptual hashing with complementary color wavelet transform and compressed sensing for reduced-reference image quality assessment. IEEE Trans. Circ. Syst. Video Technol. **32**(11), 7559–7574 (2022). https://doi.org/10.1109/TCSVT.2022.3190273
16. Tang, Z.J., Huang, L.Y., Zhang, X.Q., et al.: Robust image hashing based on color vector angle and Canny operator. AEU-Int. J. Electron. Commun. **70**(6), 833–841 (2016)
17. Vadlamudi, L.N., Vaddell, R.P.V., Devara, V.: Robust image hashing using SIFT feature points and DWT approximation coefficients. ICT Express **4**(3), 154–159 (2018)

18. Lin, C.Y., Chang, S.F.: A robust image authentication method distinguishing JPEG compression from malicious manipulation. IEEE Trans. Circ. Syst. Video Technol. **11**(2), 153–168 (2001)
19. Huang, Z.Q., Liu, S.G.: Perceptual image hashing with texture and invariant vector distance for copy detection. IEEE Trans. Multimedia **23**, 15161529 (2020)
20. Swaminathan, A., Mao, Y.N., Wu, M.: Robust and secure image hashing. IEEE Trans. Inf. Forensics Secur. **1**(2), 215–230 (2006)
21. Qin, C., Chang, C.C., Tsou, P.L.: Robust image hashing using non-uniform sampling in discrete Fourier domain. Digit. Sig. Proc. **23**(2), 578–585 (2013)
22. Huang, Z.Q., Tang, Z.J., Zhang, X.Q., et al.: Perceptual image hashing with locality preserving projection for copy detection. IEEE Trans. Dependable Secure Comput. **20**(1), 463–477 (2023)
23. Tang, Z.J., Lao, H., Zhang, X.Q., et al.: Robust image hashing via DCT and LLE. Comput. Secur. **62**, 133–148 (2016)
24. Zhu, X.F., Li, X.L., Zhang, S.C., et al.: Graph PCA hashing for similarity search. IEEE Trans. Multimedia **19**(9), 2033–2044 (2017)
25. Ren, S., et al.: Faster R-CNN: towards real-time object detection with region proposal networks. IEEE Trans. Pattern Anal. Mach. Intell. **39**(6), 1137–1149 (2017)

Finger Vein Spoof GANs: Can We Supersede the Production of Presentation Attack Artefacts?

Andreas Vorderleitner⬥, Jutta Hämmerle-Uhl, and Andreas Uhl⁽⊠⁾⬥

Visual Computing and Security Lab, University of Salzburg, J.-Haringerstr.2,
5020 Salzburg, Austria
uhl@cs.sbg.ac.at

Abstract. GAN-based I2I translation techniques for unpaired data are employed for the synthesis of biometric finger vein presentation attack instrument samples corresponding to three public presentation attack datasets. For the assessment of these synthetic samples, we analyse their behaviour when attacking finger vein recognition systems, comparing these results to such obtained from actually crafted presentation attack samples. We observe that although visual appearance and sample set correspondence are surprisingly good for some networks, respectively, the assessment of the behaviour of the data in a conducted attack is more difficult. Even if for some recognition schemes out of 11 considered we find a good accordance in terms of IAPMR (for many we don't), the attack score distributions turn out to be highly dissimilar when comparing crafted and synthetic presentation attack instrument samples. More work is needed to be able to correctly interpret corresponding diverging results with respect to the relevance in attack simulation. From the seven network architectures considered, CycleGAN provides the most useful results, but the artificially created samples do not fully mimic the behaviour of crafted ones.

Keywords: Finger veins · PAD · synthetic PAI · GANs

1 Introduction

Due to the increased usage of biometric technology in various application scenarios ranging from border control to financial services, attacks against the proper functioning of the biometric sensor in use gain increasing importance. Such attacks are usually termed "presentation attack (PA)" (aka "sensor-spoofing" – attack), performed by presenting a fraudulent biometric artefact to the systems' sensor (resulting in a so-called "presentation attack instrument (PAI) sample"). This is done by either presenting artefacts mimicking real biometric traits to the biometric sensor to be deceived or by replaying earlier captured biometric sample data on some suited device, thus also attempting to deceive the sensor ("replay attack"). In this work we consider the former attack employing PAIs mounted against finger vein recognition systems. The employment of biometric vascular patterns [1], is often attributed to increased security, caused by the demand for specialised capturing hardware and the fact that no latent prints are left behind, as it is the case with fingerprints, and yields additional levels of protection

B. Ma et al. (Eds.): IWDW 2023, LNCS 14511, pp. 109–124, 2024.
https://doi.org/10.1007/978-981-97-2585-4_8

since forgery becomes more challenging. Counter-measures to these types of attacks have of course already been developed and are termed "presentation-attack detection (PAD)" or "anti-spoofing" measures [2] (see e.g. [3,4] for PAD examples concerning finger vein recognition).

The last decade has brought forward several publications that presented multiple ways to potentially fool finger vein-based authentication systems. In first attempts, vascular PAIs are generated as easily as printing a previously captured finger vein sample image on a piece of paper or on overhead projector foil and presenting this printout (eventually manually enhanced) to the sensor [5]. There is very scarce literature on distinct approaches improving this approach: [5] also utilized a smartphone display where finger vein images were shown, [6,7] constructed silicone and wax finger artefacts into which vein printouts were sandwiched, in [8], the goal was to create a sort of artificial "wolf" master sample that exploits a weakness in a matching algorithm, and [9] used a prosthetic finger and a rubber cap with printed finger vein images glued onto it to test a hardware based liveness detection.

Obviously, the generation of PAI samples is tedious work: Generating printouts (manually enhanced) or physical models (in various materials, typically with attached printed vascular structures) and subsequent scanning with a target sensor allows to generate the forged sample data. As a consequence, available PAI sample datasets are of moderate size at best [3,4] which endangers a statistically relevant assessment of PAD techniques.

In this work, we address the issue of generating PAI samples in different manner: Based on real PAI sample data of three finger vein spoofing datasets we investigate ways to synthesize PAI samples from given real sample data, which is done by training several different I2I translation GAN structures. When pursuing this approach, a difficult question arises: As a tremendously high number of different GAN-based I2I translation techniques has been developed over the last 5 years [10], which one is best suited for this task? Are they all somewhat suited or do we have to come up with a dedicated network architecture? We will try to shed light on these pragmatic questions.

PAI samples are used for two purposes mainly: First, to evaluate the threat posed by such artefacts used in a PA against a particular recognition scheme (vulnerability assessment), and second, to train PAD techniques designed for securing the biometric system [2]. Here we focus on the first application case and evaluate synthesized PAI samples in terms of subjective quality, dataset distribution similarity to real PAI sample data, as well as in terms of behaviour when used in a PA against various finger vein recognition schemes (evaluating IAPMR and attack score distributions).

The rest of this manuscript is organised as follows: The next subsection briefly reviews related work while Sect. 2 describes the selection of suited network architectures for the task at hand. In Sect. 3 we define the experimental settings with respect to dataset and used evaluation metrics, experimental results are presented in Sect. 4. The conclusion and outlook to future work is given in Sect. 5.

1.1 Related Work

A recent survey on synthetic biometric data [11] reveals, that synthetic generation of vascular data, in particular finger vein samples, has hardly been addressed before. One of the few exceptions is [12], where it was shown that it is indeed possible to generate grayscale vascular samples (finger vein as well as hand vein data) from corresponding binary features using a learning-based approach (template inversion). An entirely different way for finger vein sample synthesis, using a model-based approach, has been demonstrated in [13]. [14] proposed a GAN-based synthesis of a finger vein sample dataset based on the prior generation of a vein pattern image, thus related to both [13] and [12], while [15] applied an end-to-end GAN-based sample generation where the samples are used to augment the training set in recognition. Similarly, also [16] applies a (Cycle)-GAN-based finger vein sample synthesis approach to improve recognition.

Finally, targeting the synthesis of PAI samples, a "SpoofGAN" has been proposed [17] for fingerprint generation, serving the same purpose as the data generated in this work. However, there are significant technical differences (besides the different modality): SpoofGAN uses synthetic imprints as input and thus, the produced PAIs cannot be used for a real-world attack (as opposed to the real samples we use for training), the actual encoder-decoder network used to create PAIs (which is not properly described in terms of reproducible research) is only refined for specific sensors (while we conduct training from scratch) and the authors do not compare their results to any publicly known network architectures (while we base our work on well known GAN variants with available implementation). A different way to increase the amount of training data for PAD network training is chosen in [18], where usual (non PAI) synthetic fingerprint samples are used for this purpose besides classical PAI samples.

Most similar to this work is [19], where hand vein PAI samples are synthesized after training two types of well-known GANs with data from the PLUS Hand Vein Spoofing database. Apart from focusing on a different modality (using NIR transillumination imaging instead of NIR reflection imaging as for hand veins), i.e. finger veins, this work differs in considering significantly more distinct GAN types and in employing data from three different PAI datasets for training.

2 Network Selection

2.1 Deep-Learning Synthesis of PAI Samples

The concept of *Generative Adversarial Networks (GANs)* was first introduced by [20] and has since become very popular and is also used in a variety of different modified versions. A GAN can be seen as a zero-sum game, where a generator G and a discriminator D compete against each other. It is also called the *minimax* game. The generator and the discriminator are implemented with two different networks, where the generator is used to generate data from a random noise variable z, trying to make it as similar as the data from the training set. The discriminator tries to identify if the data fed to it originates from the generator (fake data) or from data out of the training set (real data). The goal of this game is to find a *Nash equilibrium* between the generator and the discriminator. The loss used in optimization is called the *adversarial loss* ($\mathcal{L}_{adv}$).

GANs are trained in a special way, where the parameters of G are updated along with the ones of D. This update is done with gradient based optimization. In detail, the training is done in an alternating way between G and D. First, the discriminator is trained on the real data to perform good at classifying real data samples afterwards the discriminator is trained on the fake data from the generator. During training of the discriminator, no error is backpropagated through the generator for optimization. In the next step, the generator is trained with the discriminators feedback if the generated data from G is real or not. While training the generator, D is not optimized. Those two steps are repeated until good enough data is generated by the generator G.

Image to Image Translation (I2I) aims to translate an image from a source domain to a target domain. While during this translation the source content should be preserved, the target style should be transferred to the input image I_X. For this I2I task, GANs, in the most imaginable variations, are shown as a very good solution. I2I in conjunction with GANs can be seen as the training of a mapping $G_{X \mapsto Y}$ used to generate an image $I_{XY} \in Y$ to be indistinguishable from image $I_Y \in Y$ of the target domain Y if source image $I_X \in X$ is given. In particular, this task can be written as

$$I_{XY} \in Y : I_{XY} = G_{X \mapsto Y}(I_X).$$

First techniques involved two distinct domains in the I2I task (2-domain I2I), while more recent techniques facilitate to handle more than two domains in a single framework (multi-domain I2I). Also, one may distinguish between supervised and unsupervised I2I: The latter does not use labeled data and in context of I2I, those are unpaired images. Unpaired means that there are no corresponding pairs of images related to the background, composition, or pose of an existing object between the different domains to allow the network to learn how an image might be translated from one domain to another domain on a pixel-by-pixel mapping basis [10].

In this work, the source domain are the finger vein images from a specific database and the target domain are the manually created PAI sample images generated by the sensor, which are used to evaluate the presentation attack (PA). Thus, we clearly have an unsupervised 2-domain I2I translation task to solve, in order to supersede the physical construction of presentation attack instruments and their subsequent biometric imaging.

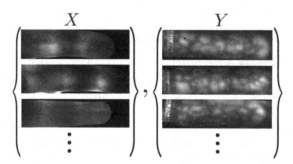

Fig. 1. Illustrating unpaired data from the PLUS Finger Vein Spoofing Data Set.

Figure 1 illustrates the "unpaired" property using data from the PLUS Finger Vein Spoofing Data Set [7] (the object displacement between the two domains is clearly visible): There is no pixel-by-pixel mapping between the image pairs of domain X and domain Y, since the PA instrument image acquisition (producing domain Y imagery) leads to a misalignment of the finger (as compared to the original X domain data), which becomes noticeable through rotation or displacement of the finger.

2.2 I2I Networks Used

Based on the observations in the last subsection (and findings in [10]), we have selected the I2I networks shown in Table 1 for a closer consideration in our task.

Table 1. An overview of the neural network architectures selected for closer inspection.

Network	Data	Domain
StarGANv2 [21]	unpaired	multi-domain
TuiGAN [22]	unpaired	two-domain
DRIT [23]	unpaired	multi-domain
CUT [24]	unpaired	two-domain
DistanceGAN [25]	unpaired	two-domain
CycleGAN [26]	unpaired	two-domain
DualGAN [27]	unpaired	two-domain

Initial results when varying standard settings led to the conclusion to discard TuiGAN, CUT, and DualGAN from further investigations as the results were far from being usable or promising. In Fig. 2, we display two samples from the PLUS Finger Vein Spoofing Data Set, "Genuine" (i.e. bona fide) and "Spoofed" (i.e. attacked) (i.e. original sample and corresponding PAI sample), and for each network the synthesized PAI sample ("Spoofed-DNN").

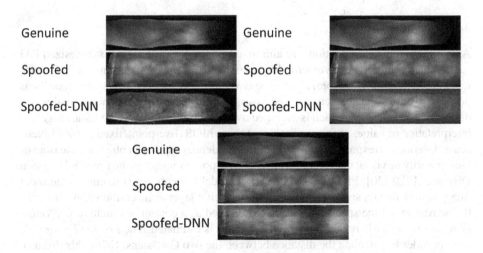

Fig. 2. Illustration of poor quality generated PAI samples: TuiGAN, CUT, DualGAN.

We clearly recognize that for TuiGAN, the generated image superimposes the original sample with high frequency speckle, while for CUT, the only effect on the original sample is an adaptation of average luminance. The DualGAN output does not exhibit any vascular structure any more. Therefore, we focus the attention to the following four I2I GAN types: CycleGAN, DistanceGAN, StarGANv2, and DRIT.

3   Experimental Settings

3.1   Datasets

The Idiap Research Institute VERA Fingervein Database (VERA) [28] consists of 220 unique fingers captured in 2 sessions from 110 subjects. Each sample has one PAI sample counterpart, which is generated by printing preprocessed samples on high quality paper using a laser printer and enhancing vein contours with a black whiteboard marker afterwards. Images come as full (250×665 pixels) or cropped (150×565 pixels) samples.

The South China University of Technology Finger Vein Database (SCUT) [29] was collected from 6 fingers of 100 subjects captured in 6 acquisition sessions. For PAI sample generation, each finger vein image is printed on two overhead projector films which are aligned and stacked. In order to reduce overexposure, additionally a strong white paper is put in-between the two overhead projector films. Images come as full (640×288 pixels) or cropped samples of variable size.

The Paris Lodron University of Salzburg Finger Vein Spoofing Data Set (PLUS) [7] uses a subset of the PLUS Vein-FV3 dataset as bona fide samples. For PAI sample generation, PC (see below) binarised vein structures from 6 fingers of 22 subjects were printed on paper and sandwiched into a top and bottom made of beeswax. Capturing is done employing two illumination variants (LED and Laser) and using two different levels of vessel thickness. Every sample is of size 192×736 pixels.

3.2   Evaluation Metrics

As outlined in the Introduction, we aim to assess the effectiveness of synthesized PAI samples in a simulated PA (vulnerability assessment), as opposed to simulate PAD training with these data. Therefore, in order to cover as many aspects with as few metrics as possible, we use (i) a subjective visual assessment of the correspondence to real-world data (opinion score OS), which is averaged over all samples and uses the same range and interpretation of values as the mean opinion score MOS (five-point, fixed-choice Likert-scale: 1 - good correspondence, 5 - bad correspondence) and (ii) an objective measure of the similarity of visual data considering entire corpora of imagery, the Fréchet Inception Distance (FID) [30]. FID uses the inception model (*inception v3*) to embed the input images into a feature space, represented by a specific layer of inception v3. Afterwards, the activations of the specific layer are represented as a continuous multivariate Gaussian distribution, where the mean and covariance is estimated. The *Fréchet Distance* is used in order to calculate the distance between the two Gaussians. Using this distance the similarity between the distributions of the real finger vein PAI data and generated

PAI data can be quantified [31]. The third metric (iii) is based on a chosen recognition scheme and determines IAPMR (as defined by the ISO/IEC 30107-3:2017 standard) for the real and synthesized PAI samples, respectively. Contrasting to ISO/IEC 30107-3:2017 we do not aim to assess a single scheme, but we need to *compare* two different datasets for the similarity of their threat potential. Thus, in addition, the intersection (in %) between the attack score distributions of real and synthesized PAIs is computed in order to get a more comprehensive view on the similarity of attack behaviour than just considering a single point of operation like IAPMR (or other comparable points of operation). We conducted the latter experiments by using the PLUS OpenVein Finger- and Hand-Vein Toolkit (http://www.wavelab.at/sources/OpenVein-Toolkit/ [32]), which is a fully open source non-commercial vein recognition software framework implemented in MATLAB.

The actual recognition schemes from this toolkit used in the evaluation of metric (iii) are of different types as follows: First, this study uses seven vein pattern based feature extraction methods, six of which generate a binary feature image as a result of extracting vein structures from the background. *Maximum Curvature (MC)* and *Repeated Line Tracking (RLT)* achieve this by looking at the cross sectional profile of the finger vein image, *Wide Line Detector (WLD)*, *Gabor Filter (GF)* and *Isotropic Undecimated Wavelet Transform (IUWT)* also consider local neighbourhood regions via filter convolution and *Principal Curvature (PC)* first computes the normalized gradient field and then looks at the eigenvalues of the Hessian matrix at each pixel. The resulting binary images are compared using a correlation measure. One more advanced vein pattern based feature extraction and matching strategy is given by *Anatomy Structure Analysis-Based Vein Extraction (ASAVE)*. This technique extracts two binary vessel structures, differing by the extent of used context in creating these.

Second, three keypoint based schemes are used, two of which being filtered versions of the general purpose keypoint detection and matching schemes *SIFT* and *SURF*. The third keypoint based method is *Deformation Tolerant Feature Point Matching (DTFPM)* which was tailored especially for vein pattern by also looking at curvature and vein directions.

Finally, as a generic texture-based technique we employ a *Local Binary Pattern* descriptor that uses histogram intersection as comparison method. Note that, as the PLUS OpenVein Toolkit supports a common evaluation protocol, we have refrained from using recognition schemes apart from the already 11 considered not included there-in (i.e. deep-learning based ones), in order to avoid incompatibilities (also, as far as we know, current commercial systems rely on traditional feature extraction so far).

3.3 Evaluation Protocol

We use the network implementations made available by the authors of the original papers, samples are fed into the networks in full size and slightly resized according to the networks need using bicubic (e.g. CycleGAN) or bilinear (e.g. DistanceGAN) interpolation. Augmentations are done within the network as supported, without any additional external augmentation.

Experiments are conducted in a five-fold cross validation, i.e. for each configuration, five different network instances have been trained from scratch to generate their share of the final data. Fold construction prevents to have distinct samples of a single subject in both the training and evaluation sets, respectively (thus we separate subjects in training and evaluation data). Note that we present the best results from a large collection of experiments, the detailed description of which parameters have been used can be retrieved from https://wavelab.at/sources/Vorderleitner23b.

4   Experimental Results

For each of the three datasets, we present visual examples of synthetic PAI samples for qualitative analysis, one for high correspondence to real PAI samples and one for low correspondence, respectively. In Fig. 3 we provide these results for the VERA database having used DistanceGAN and DRIT for the synthesis process. While for the Distance-GAN the synthesis result is rather satisfactory, DRIT generated PAI samples lack in clear vascular structure (which is much stronger present in the real PAI samples) and the luminance is too high.

In Table 2 we present the quantitative results of our assessment for data generated from the VERA dataset. In terms of FID and OS, the numerical values clearly favour DistanceGAN over DRIT, thus confirming the visual impression. StarGANv2 shows even better values in that respect, while CycleGAN exhibits the worst FID value and OS close to DRIT.

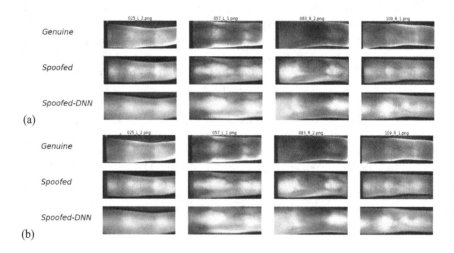

Fig. 3. DistanceGAN (a) and DRIT (b) on VERA data: Visual comparison of the synthesized PAI samples (third line) against the real PAI samples (second line), generated from the original samples.

In the subsequent lines of Table 2, we report results of applying a set of distinct vein recognition systems to the original and synthesized PAI data, respectively. In increasing darkness of gray levels, we indicate increasing correspondence of original and synthesized PAI data - i.e., the darker, the better. For IAPMR, we compare the respective value for the original PAI sample data separated by "/" from that of the synthetic data. Additionally, we provide the intersection ("int.") of the attack matching score distributions using original and synthetic PAI samples, respectively. We notice that the extent of correspondence of IAPMR depends on the recognition system in use – e.g. for CycleGAN on the tables' left side: While we notice a good match for MC, WLD and IUWT, we also observe fairly distinct IAPMR values, e.g. for RLT, SURF and SIFT. Thus, the synthetic data obviously does not allow for a good real-world "approximation" for all types of recognition system types.

Table 2. VERA - Quantitative results: FID, OS, IAPMR for the real and synthesized PAI samples, respectively, and attack score distribution intersection.

	CycleGAN		DistanceGAN		StarGANv2		DRIT	
	FID	OS	FID	OS	FID	OS	FID	OS
	189	3.0	28	2.5	24	2.0	106	3.5
Method	IAPMR [%]	int. [%]	IAPMR [%]	int. [%]	IAPMR [%]	int. [%]	IAPMR [%]	int. [%]
PC	89.90 / 97.54	43.36	89.90 / 30.35	27.35	89.90 / 49.30	34.42	89.90 / 3.68	13.69
MC	92.26 / 97.89	46.38	92.26 / 49.82	38.57	92.26 / 69.82	51.77	92.26 / 12.63	17.36
WLD	93.10 / 92.63	78.09	93.10 / 39.82	30.42	93.10 / 61.93	42.14	93.10 / 1.40	8.04
RLT	36.87 / 59.82	74.01	36.87 / 0.18	12.66	36.87 / 0.70	37.02	36.87 / 1.75	48.38
GF	85.19 / 94.39	54.90	85.19 / 41.23	38.65	85.19 / 48.95	51.44	85.19 / 23.16	30.90
IUWT	92.76 / 96.67	51.07	92.76 / 53.33	34.34	92.76 / 70.53	45.36	92.76 / 24.04	18.48
ASAVE	70.54 / 78.77	73.05	70.54 / 53.16	82.60	70.54 / 66.32	85.51	70.54 / 32.28	61.44
DTFPM	23.23 / 74.21	48.97	23.23 / 2.63	56.48	23.23 / 18.60	88.80	23.23 / 0.18	37.66
SURF	3.03 / 34.74	71.61	3.03 / 0.18	74.75	3.03 / 1.40	83.34	3.03 / 0.53	69.65
SIFT	11.45 / 55.09	54.72	11.45 / 0.35	52.47	11.45 / 1.58	58.27	11.45 / 0.53	48.26
LBP	10.77 / 38.07	73.40	10.77 / 0.18	61.64	10.77 / 6.32	84.55	10.77 / 0.18	39.82

When looking into the results with a little more depth, one might ask if a good IAPMR alignment is sufficient to claim that in this setting, we have a good model for the real world situation. Taking again the CycleGAN result as example, we note that for IUWT recognition we observe rather good IAPMR correspondence, while the intersection of attack match score distributions overlap is only at 51%.

Figure 4 visualises the mated and non-mated score distributions (including both attack score distributions) of the LBP- and PC-based recognition system, respectively, when applied to StarGANv2 generated data (for LBP a good IAPMR accordance has been reported before, while for PC accordance is poor). For LBP the IAPMR values are very close (light blue crosses), and it is clearly visible that the real attack score distribution (yellow) is highly similar to the pink score distribution of the synthetic PAI samples. However, as indicated by the low IAPMR, the attack does neither work with real PAI samples nor with synthetic ones, as also clearly indicated by the high overlap with the impostor score distribution in violet.

For PC-based recognition, the poor overlap between both attack score distributions is clearly visible. However, the attack with real PAI samples works significantly better

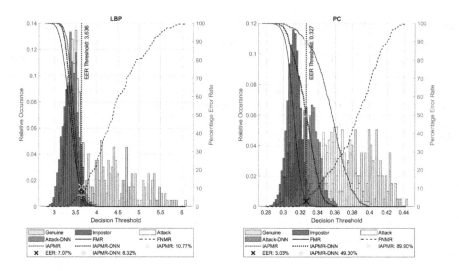

Fig. 4. Score distributions of LBP- and PC-based recognition on StarGANv2 PAI samples: high extent vs. low extent of attack score distribution intersection, respectively. (Color figure online)

as the overlap to the bona fide score distribution is much higher as compared to the synthetic data.

Overall, the poor performance of DRIT on the VERA data as observed in the visual examples is confirmed by all quantitative measures. DistanceGAN results, although visually rather good, do not convince quantitatively except for FID and OS - IAPMR correspondence and attack score distribution overlap is weaker as compared to Cycle-GAN and StyleGANv2. StarGANv2 excels in terms of FID, OS, and IAPMR correspondence in case IAPMR is low. It seems that the quality of the generated PAI samples does not allow to conduct effective attacks, as IAPMR using synthetic PAI samples is always clearly lower (this is also true for DistanceGAN and DRIT, while the opposite is true for CycleGAN). Finally, CycleGAN, while being rather poor in FID and OS, exhibits decent IAPMR correspondence for a specific set of recognition schemes (i.e. WLD, MC, IUWT), where also reasonable attack score distribution intersection can be stated. However, for other recognition systems, quantitative results are poor.

Subsequently, we discuss results obtained on the SCUT dataset. In Fig. 5 we provide visual examples for PAI samples generated with the CycleGAN and DistanceGAN networks, respectively. While the CycleGAN is able to produce synthetic PAI samples being visually close to the real ones, the DistanceGAN PAI data lack in detail of the vascular structure and their luminance is too low (note that DistanceGAN worked well on VERA data).

Table 3 displays the quantitative results for the SCUT dataset. Comparing Cycle-GAN and DistanceGAN in terms of FID and OS, the visual impression is numerically confirmed. StarGANv2 is in between the two, while FID and OS of the DRIT data is at a similar level as the DistanceGAN data.

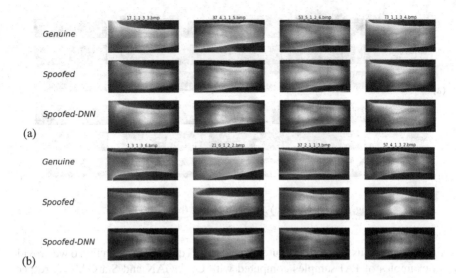

Fig. 5. CycleGAN (a) and DistanceGAN (b) on SCUT data.

Table 3. SCUT - Quantitative results.

	CycleGAN		DistanceGAN		StarGANv2		DRIT	
	FID	**OS**	**FID**	**OS**	**FID**	**OS**	**FID**	**OS**
	41	2.0	110	4.5	74	4.0	122	4.0
Method	**IAPMR** [%]	**int.** [%]	**IAPMR** [%]	**int.** [%]	**IAPMR** [%]	**int.** [%]	**IAPMR** [%]	**int.** [%]
PC	83.46 / 96.06	45.66	83.46 / 22.27	32.80	83.46 / 15.56	27.13	83.46 / 6.25	21.78
MC	87.88 / 95.76	44.97	87.88 / 28.03	34.32	87.88 / 10.83	21.79	87.88 / 25.00	29.80
WLD	70.96 / 94.09	45.05	70.96 / 21.67	50.25	70.96 / 14.03	41.59	70.96 / 8.19	34.24
RLT	33.84 / 73.03	42.83	33.84 / 0.15	25.53	33.84 / 0.14	16.31	33.84 / 7.36	71.84
GF	50.51 / 86.52	51.09	50.51 / 23.48	72.73	50.51 / 2.78	45.38	50.51 / 11.53	59.92
IUWT	71.72 / 93.64	51.84	71.72 / 12.12	40.53	71.72 / 0.69	24.51	71.72 / 8.33	36.53
ASAVE	78.66 / 93.48	63.71	78.66 / 75.15	72.53	78.66 / 43.61	56.86	78.66 / 21.94	41.28
DTFPM	71.09 / 89.39	70.08	71.09 / 11.36	40.28	71.09 / 8.61	38.16	71.09 / 38.89	61.93
SURF	6.44 / 25.30	76.34	6.44 / 0.30	56.06	6.44 / 0.83	55.18	6.44 / 1.53	80.72
SIFT	22.60 / 45.15	76.77	22.60 / 0.15	24.57	22.60 / 2.22	44.51	22.60 / 4.03	59.79
LBP	39.65 / 84.09	47.45	39.65 / 1.67	53.99	39.65 / 1.25	44.95	39.65 / 0.42	46.89

In terms of IAPMR correspondence and attack score distribution intersection, none of the four networks does a convincing job. The best results are seen for low IAPMR values, for keypoint-based recognition schemes mostly.

Again, there is an interesting observation to be made: For CycleGAN, the IAPMR values are always higher for the synthetic data (thus, attacks can be conducted more effectively), while for the other three networks the opposite is true. This indicates the option of using CycleGAN generated data for more effective attacks as compared to real data (if this is desirable from an application viewpoint).

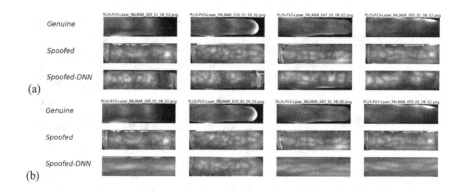

Fig. 6. CycleGAN (a) and StarGANv2 (b) on PLUS data.

Finally, we discuss results computed on the PLUS dataset. First, in Fig. 6 we provide visual examples of PAI samples computed with CycleGAN and StarGANv2, respectively. While the CycleGAN is able to create the honeycomb-like structure of the original PAI data reasonably well, this is not at all the case for the StarGANv2 data. Subjectively, real and synthetic PAI samples as created by the StarGANv2 do not have much in common.

When comparing this qualitative result to the quantitative findings in Table 4, we observe clearly inferior FID and OS values for StarGANv2 as compared to CycleGAN. DistanceGAN and DRIT are also clearly inferior to CycleGAN FID and OS values, respectively.

Table 4. PLUS - Quantitative results.

	CycleGAN		DistanceGAN		StarGANv2		DRIT	
	FID	**OS**	**FID**	**OS**	**FID**	**OS**	**FID**	**OS**
	57	2.0	111	4.5	128	4.5	163	3.5
Method	**IAPMR** [%]	**int.** [%]	**IAPMR** [%]	**int.** [%]	**IAPMR** [%]	**int.** [%]	**IAPMR** [%]	**int.** [%]
PC	58.62 / 35.28	73.80	58.62 / 4.17	43.89	58.62 / 0.83	37.84	58.62 / 0.83	36.50
MC	60.03 / 55.69	84.46	60.03 / 15.42	54.87	60.03 / 14.17	52.78	60.03 / 4.86	41.00
WLD	58.49 / 66.67	87.34	58.49 / 25.14	63.94	58.49 / 10.97	52.03	58.49 / 0.69	30.02
RLT	22.09 / 31.81	84.52	22.09 / 2.50	54.54	22.09 / 1.39	18.53	22.09 / 0.42	32.03
GF	32.44 / 29.86	90.83	32.44 / 9.17	74.32	32.44 / 6.39	64.96	32.44 / 1.81	53.61
IUWT	81.10 / 74.72	84.28	81.10 / 22.64	34.30	81.10 / 18.75	35.28	81.10 / 11.81	28.80
ASAVE	9.45 / 44.31	52.12	9.45 / 10.83	86.67	9.45 / 19.72	84.61	9.45 / 2.50	79.57
DTFPM	4.21 / 7.64	85.02	4.21 / 0.14	60.88	4.21 / 2.22	52.90	4.21 / 0.14	63.94
SURF	5.49 / 0.83	92.78	5.49 / 0.42	95.00	5.49 / 2.22	88.19	5.49 / 4.86	100.00
SIFT	0.13 / 0.28	88.45	0.13 / 0.14	94.56	0.13 / 1.25	88.59	0.13 / 0.14	92.53
LBP	0.13 / 17.36	38.11	0.13 / 0.42	56.46	0.13 / 0.14	81.86	0.13 / 0.14	87.35

When it comes to IAPMR correspondence and attack score distribution intersection, we notice that the three networks with poor FID and OS values exhibit good results for settings in which IAPMR values are low, i.e. keypoint-based recognition schemes and LBP. In these cases, also attack score distribution intersection can be rather high. It is interesting to note that IAPMR values for PAI samples generated by these networks are always lower (except for the SAVE recognition scheme) than the values of the real data. These facts taken together, it seems that these networks generate poor quality PAI sample data for the PLUS dataset, and can be only successful in mimicking real data when attacks do not work in both cases.

The situation is different when considering CycleGAN results. Also for medium and high IAPMR values, good IAPMR correspondence and high attack score distribution intersection can be observed.

Finally, in Fig. 7, we take a look into the attack score distributions of CycleGAN generated PAI samples, employing GF and ASAVE recognition. We visualize the score distributions of real and synthetic PAI samples as boxplots, showing results of the computed 5 folds and the overall dataset. Both, for each single fold and the overall data, the small extent of the intersection for ASAVE data is clearly visible, while the large intersection is somewhat impressive for GF.

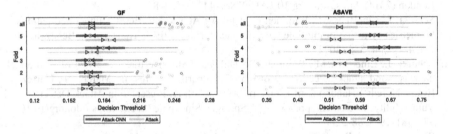

Fig. 7. Boxplots of attack score distributions of GF- and ASAVE-based recognition on CycleGAN PAI samples: high and low extent of attack score distribution intersection.

5   Conclusion

Among the huge variety of existing I2I translation GANs for unpaired training data, it is not easy to identify suited networks to synthesize finger vein PAI sample data. While the subjective visual quality of such samples and computed FID/OS values do correspond well in most cases, we have identified difficulties in the selection of a proper GAN as follows: First, and most importantly, we find distinct networks to be suited to generate data from/for different PAD datasets. Second, we find different properties of the synthetic data depending on which actual finger vein recognition scheme is being used. And third, it turns out that even good correspondence in IAPMR values does not guarantee similar behaviour in attacks, as the overall shape of the attack score distributions needs to be considered.

The overall best suited network identified is CycleGAN, not exhibiting spectacular failures for any of the considered datasets and providing good results for at least two (i.e. VERA and PLUS) out of three datasets. It is questionable if it is indeed possible to come up with a dedicated GAN network structure clearly outperforming CycleGAN in this obviously delicate task. As we have not yet been able to synthesize PAI sample data with sufficiently identical properties as the real data exhibits, the current answer to the question posed in the title is: "No". However, eventually, it needs to be reconsidered if the ultimate goal is to mimic the behaviour of real data as closely as possible.

Future work will apply diffusion models for the task pursued, and will investigate the effect of the synthetic PAI samples when training PA detectors. Also, deep-learning based recognition schemes will complete the assessment.

Acknowledgements. This work has been partially supported by the Austrian Science Fund, project no. I4272.

References

1. Uhl, A., Busch, C., Marcel, S., Veldhuis, R.: Handbook of Vascular Biometrics. Advances in Computer Vision and Pattern Recognition, Springer, Cham (2019). https://doi.org/10.1007/978-3-030-27731-4
2. Marcel, S., Nixon, M.S., Li, S.Z. (eds.): Handbook of Biometric Anti-Spoofing. Springer, London (2014). https://doi.org/10.1007/978-1-4471-6524-8
3. Schuiki, J., Linortner, M., Wimmer, G., Uhl, A.: Attack detection for finger and palm vein biometrics by fusion of multiple recognition algorithms. IEEE Trans. Biom. Behav. Identity Sci. 4(4), 544–555 (2022)
4. Schuiki, J., Linortner, M., Wimmer, G., Uhl, A.: Extensive threat analysis of vein attack databases and attack detection by fusion of comparison scores. In: Marcel, S., Fierrez, J., Evans, N. (eds.) Handbook of Biometric Anti-Spoofing: Presentation Attack Detection and Vulnerability Assessment, pp. 467–487. Springer, Singapore (2023). https://doi.org/10.1007/978-981-19-5288-3_17
5. Raghavendra, R., Busch, C.: Presentation attack detection algorithms for finger vein biometrics: a comprehensive study. In: 11th International Conference on Signal-Image Technology Internet-Based Systems (SITIS 2015), pp. 628–632 (2015)
6. Debiasi, L., Kauba, C., Hofbauer, H., Prommegger, B., Uhl, A.: Presentation attacks and detection in finger- and hand-vein recognition. In: Proceedings of the Joint Austrian Computer Vision and Robotics Workshop (ACVRW 2020), Graz, Austria, pp. 65–70 (2020)
7. Schuiki, J., Prommegger, B., Uhl, A.: Confronting a variety of finger vein recognition algorithms with wax presentation attack artefacts. In: Proceedings of the 9th IEEE International Workshop on Biometrics and Forensics (IWBF 2021), Rome, Italy (moved to virtual), pp. 1–6 (2021)
8. Otsuka, A., Ohki, T., Morita, R., Inuma, M., Imai, H.: Security evaluation of a finger vein authentication algorithm against wolf attack. In: 37th IEEE Symposium on Security and Privacy, San Jose, CA (2016)
9. Krishnan, A., Thomas, T., Nayar, G.R., Sasilekha Mohan, S.: Liveness detection in finger vein imaging device using plethysmographic signals. In: Tiwary, U.S. (ed.) IHCI 2018. LNCS, vol. 11278, pp. 251–260. Springer, Cham (2018). https://doi.org/10.1007/978-3-030-04021-5_23
10. Hoyez, H., Schockaert, C., Rambach, J., Mirbach, B., Stricker, D.: Unsupervised image-to-image translation: a review. Sensors 22(21), 8540 (2022)
11. Makrushin, A., Uhl, A., Dittmann, J.: A survey on synthetic biometrics: fingerprint, face, iris and vascular patterns. IEEE Access 11, 33887–33899 (2023)

12. Kauba, C., Kirchgasser, S., Mirjalili, V., Uhl, A., Ross, A.: Inverse biometrics: generating vascular images from binary templates. IEEE Trans. Biom. Behav. Identity Sci. **3**(4), 464–478 (2021)

13. Hillerström, F., Kumar, A., Veldhuis, R.: Generating and analyzing synthetic finger vein images. In: Proceedings of the International Conference of the Biometrics Special Interest Group (BIOSIG 2014), pp. 121–132 (2014)

14. Yang, H., Fang, P., Hao, Z.: A GAN-based method for generating finger vein dataset. In: Proceedings of the 2020 3rd International Conference on Algorithms, Computing and Artificial Intelligence, ACAI 2020. Association for Computing Machinery, New York (2021)

15. Zhang, J., Lu, Z., Li, M., Wu, H.: Gan-based image augmentation for finger-vein biometric recognition. IEEE Access **7**, 183118–183132 (2019)

16. Yang, W., Hui, C., Chen, Z., Xue, J.H., Liao, Q.: FV-GAN: finger vein representation using generative adversarial networks. IEEE Trans. Inf. Forensics Secur. **14**(9), 2512–2524 (2019)

17. Grosz, S.A., Jain, A.K.: SpoofGAN: synthetic fingerprint spoof images. IEEE Trans. Inf. Forensics Secur. **18**, 730–743 (2023)

18. Purnapatra, S., et al.: Presentation attack detection with advanced CNN models for noncontact-based fingerprint systems. In: Proceedings of the 11th International Workshop on Biometrics and Forensics (IWBF 2023), Barcelona, Spain, pp. 1–6 (2023)

19. Vorderleitner, A., Hämmerle-Uhl, J., Uhl, A.: Hand vein spoof GANs: Pitfalls in the assessment of synthetic presentation attack artefacts. In: Proceedings of the 2023 ACM Workshop on Information Hiding and Multimedia Security, IH&MMSec 2023, pp. 133–138. Association for Computing Machinery, New York (2023)

20. Goodfellow, I., et al.: Generative adversarial networks. Commun. ACM **63**(11), 139–144 (2020)

21. Choi, Y., Uh, Y., Yoo, J., Ha, J.W.: StarGAN v2: Diverse image synthesis for multiple domains. In: Proceedings of the 2020 IEEE/CVF Conference on Computer Vision and Pattern Recognition, CVPR 2020, pp. 8188–8197 (2020)

22. Lin, J., Pang, Y., Xia, Y., Chen, Z., Luo, J.: TuiGAN: Learning versatile image-to-image translation with two unpaired images. In: Vedaldi, A., Bischof, H., Brox, T., Frahm, J.-M. (eds.) ECCV 2020. LNCS, vol. 12349, pp. 18–35. Springer, Cham (2020). https://doi.org/10.1007/978-3-030-58548-8_2

23. Lee, H.-Y., Tseng, H.-Y., Huang, J.-B., Singh, M., Yang, M.-H.: Diverse image-to-image translation via disentangled representations. In: Ferrari, V., Hebert, M., Sminchisescu, C., Weiss, Y. (eds.) ECCV 2018. LNCS, vol. 11205, pp. 36–52. Springer, Cham (2018). https://doi.org/10.1007/978-3-030-01246-5_3

24. Park, T., Efros, A.A., Zhang, R., Zhu, J.-Y.: Contrastive learning for unpaired image-to-image translation. In: Vedaldi, A., Bischof, H., Brox, T., Frahm, J.-M. (eds.) ECCV 2020. LNCS, vol. 12354, pp. 319–345. Springer, Cham (2020). https://doi.org/10.1007/978-3-030-58545-7_19

25. Benaim, S., Wolf, L.: One-sided unsupervised domain mapping. In: Proceedings of the 31st International Conference on Neural Information Processing Systems (NIPS 2017), Red Hook, NY, USA, pp. 752–762. Curran Associates Inc. (2017)

26. Zhu, J.Y., Park, T., Isola, P., Efros, A.A.: Unpaired image-to-image translation using cycle-consistent adversarial networks. In: 2017 IEEE International Conference on Computer Vision (ICCV), pp. 2242–2251 (2017). ISSN: 2380-7504

27. Yi, Z., Zhang, H., Tan, P., Gong, M.: DualGAN: Unsupervised dual learning for image-to-image translation. In: Proceedings of the IEEE International Conference on Computer Vision (ICCV 2017), pp. 2849–2857 (2017)

28. Tome, P., et al.: The 1st competition on counter measures to finger vein spoofing attacks. In: International Conference on Biometrics (ICB 2015), pp. 513–518 (2015)

29. Qiu, X., Kang, W., Tian, S., Jia, W., Huang, Z.: Finger vein presentation attack detection using total variation decomposition. IEEE Trans. Inf. Forensics Secur. **13**(2), 465–477 (2018)
30. Heusel, M., Ramsauer, H., Unterthiner, T., Nessler, B., Hochreiter, S.: GANs trained by a two time-scale update rule converge to a local Nash equilibrium. In: Proceedings of the 31st International Conference on Neural Information Processing Systems (NIPS 2017), Red Hook, NY, USA, pp. 6629–6640. Curran Associates Inc. (2017)
31. Lucic, M., Kurach, K., Michalski, M., Bousquet, O., Gelly, S.: Are GANs created equal? A large-scale study. In: Proceedings of the 32nd International Conference on Neural Information Processing Systems (NIPS 2018), Red Hook, NY, USA, pp. 698–707. Curran Associates Inc. (2018)
32. Kauba, C., Uhl, A.: An available open-source vein recognition framework. In: Uhl, A., Busch, C., Marcel, S., Veldhuis, R. (eds.) Handbook of Vascular Biometrics. ACVPR, pp. 113–142. Springer, Cham (2020). https://doi.org/10.1007/978-3-030-27731-4_4

Generalizable Deep Video Inpainting Detection Based on Constrained Convolutional Neural Networks

Jinchuan Li[1,2], Xianfeng Zhao[1,2(✉)], and Yun Cao[1,2]

[1] Institute of Information Engineering, Chinese Academy of Sciences,
Beijing 100085, China
{lijinchuan,zhaoxianfeng,caoyun}@iie.ac.cn
[2] School of Cyber Security, University of Chinese Academy of Sciences,
Beijing 100085, China

Abstract. Deep video inpainting can automatically fill in missing content both in spatial and temporal domain. Unfortunately, malicious video inpainting operations can distort media content, making it challenging for viewers to detect inpainting traces due to their realistic visual effects. As a result, the detection of video inpainting has emerged as a crucial research area in video forensics. Several detection models that have been proposed are trained and tested on datasets made by three kinds of inpainting models, but not tested against the latest and better deep inpainting models. To address this, we introduce a novel end-to-end video inpainting detection network, comprising a feature extraction module and a feature learning module. The Feature extraction module is a Bayar layer and the feature learning module is an encoder-decoder module. The proposed approach is evaluated with inpainted videos created by several state-of-the-art deep video inpainting networks. Extensive experiments has proven that our approach achieved better inpainting localization performance than other methods.

Keywords: Video inpainting detection · Bayar layar · Constrained convolution

1 Introduction

Video inpainting [1–3] is to complete missing region in video spatio-temporal domain. In recent years, data-driven deep learning technology has achieved successful visual effects in video inpainting, leading to its wide adoption in various applications, including video completion, content creation, and virtual reality. However, this extensive application of video inpainting also gives rise to visual safety concerns, particularly with regards to video content forgery. In terms of video content forgery, video inpainting has great potential. Fixed spatio-temporal inpainting could remove visible copyright watermarks and subtitles. Flexible spatio-temporal video inpainting enables removal of semantic objects.

B. Ma et al. (Eds.): IWDW 2023, LNCS 14511, pp. 125–138, 2024.
https://doi.org/10.1007/978-981-97-2585-4_9

Video inpainting technology can tempt video publishers to tamper with video content, release fake news and mislead the direction of public opinion. Therefore, it is necessary to research into forensics approach against illegal video inpainting forgery.

Video inpainting forensics research is similar to the image manipulation forgery forensics. As shown in Fig. 1, given a group of frames, video inpainting is to repair the missing area labeled by a group of masks according known area textures. Video inpainting forensics is generate pixel-level masks against inpainted video frames.

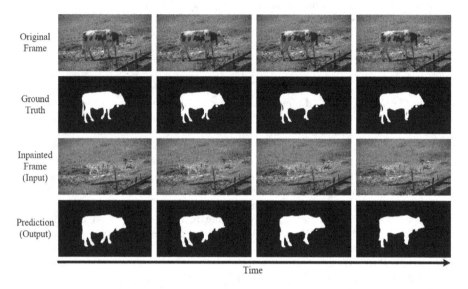

Fig. 1. Given the original frame sequence and the object marked by the mask. The inpainting model [3] generate the inpainted frames. The inpainted frames are our input. The output is the predicted inpainted masks.

In terms of deep video inpainting, researchers have proposed a variety of data-driven deep learning models. VIDNet [4] and FAST [5] models are proposed for detection against three deep video inpainting models. Two methods achieve the goal of inpainting detection by capturing error level analysis noise and frequency domain inconsistency, respectively. Both two detection models generalize poorly detection performance against state-of-the-art video inpainting models. Recently, a new detection model for advanced video inpainting models has appeared which use spatiotemporal traces to catch video inpainting traces [6]. Two advanced inpainting methods [7,8] were used to create inpainting datasets. The trained model achieved a better performance in same inpainting approach, but the generalization detection performance against other inpainting models was not discussed. Therefore, generalizable detection methods against many different deep video inpainting models are lacking in current research.

We propose an end-to-end model for generalizable video inpainting detection, comprising a feature extraction module and a feature learning module. The feature extraction module employs a Bayar convolution to enhance video inpainting traces and effectively counter fake videos generated by various video inpainting methods. Remarkably, the output feature map of the Bayar convolution layer effectively highlights the inpainting traces. The feature learning module adopts a convolutional encoder-decoder structure. The encoder is a pre-trained VGG network, while the decoder consists of multi-layer transposed convolution layers and upsampling layers. In the symmetrical structure of the encoder and decoder, skip-connections are used to enhance the feature learning performance of inpainting traces. We employ various different video inpainting methods to generate a video object removal forgery dataset. After dividing the training and test set, the model is separately trained on the datasets generated by different methods. Precise localization of inpainted regions can be achieved on the test set. A single data-driven trained model is still able to detect fake videos produced by other inpainting methods. It shows that the model is able to generalize to out-of-domain inpainted videos, which were not seen during training.

Our contributions can be summarized as follows: (1) We propose a simple yet effective end-to-end video inpainting detection model to detecting video inpainting forgeries. (2) We find that the Bayar single convolutional layer module can highlighting forgeries against inpainted videos generated by various video inpainting models. (3) We produced object removal datasets for several state-of-the-art models for deep video inpainting, and our model has achieved acceptable performance in both in-domain and cross-domain video inpainting detection.

2 Related Work

2.1 Video Inpainting

Video inpainting can be classified into traditional and deep inpainting. The traditional inpainting algorithm is an extension of image inpainting, generally based on pixel diffusion [11,12] and patch match [13,14]. [15] pointed out that the video inpainting problem involves patch-based global optimization. Specifically, the missing content is synthesizesed by sampling similar spatio-temporal domain patches from known regions based on global optimization. [16] formulate the problem as a non-parametric patch-based optimization. Video inpainting is addressed by jointly estimating optical flow and color in the missing regions. Using pixel-wise forward/backward flow fields is enable to synthesize temporally coherent colors. However, traditional inpainting algorithms encounter difficulties to truly fill texture accurately globally when facing various complex non-rigid scene movements. Deep learning-based inpainting methods have been proven to be more effective in addressing these challenges and achieving superior visual results efficiently.

Building upon the development of deep learning, great progress has been made in video inpainting. According to model structure and training strategy, it can be divided into four main categories: 2D or 3D CNN, flow guide, vision transformer, and internal learning. Early deep video inpainting models were usually composed of 2D or 3D convolutional network [1–3,17–19]. These methods employed an end-to-end 2D convolution or 3D convolution codec structure, interspersing specially designed optical flow guidance or feature matching attention modules between the encoder and the decoder to achieve reference multiframe to single frame inpainting.In recent years, flow-based methods guided by optical flow and vision transformer-based approaches have gained popularity. Flow-based method firstly completed the optical flow feature map. Then utilizes the completed optical flow to capture the correspondence between valid regions and corrupted regions in a chain manner through all video frames [8,10,20]. Benefiting from the rapid development of vision transformer, researchers also applied transformer to achieve video completion [7,9,21]. Multi-layer transformer encoder was used for feature map encoded by CNN to achieve patch matching. These approach also combined with PatchGAN to enhance the visual effect. Internal learning relies on the premise that neural networks embody implicit prior information for visual images. These methods train a model with inter-frame continuity constraints on a single video sample, facilitating the implicit propagation of spatio-temporal information [24,25]. Our designed model is trained and evaluated on nine diverse video inpainting datasets.

2.2   Video Inpainting Detection

Video inpainting can not only complete video content, but also modify video semantics, such as object deletion. Therefore, several video inpainting detection models have been proposed in recent years. VIDNet [4] was the first detection model for video deep inpainting, which composed of VGG encoder and ConvLSTM decoder. In particular, VIDNet processed RGB frames and ELA frames which highlight inpainting traces simultaneously. Furthermore, VIDNet incorporated a quad-directional local attention module between the encoder and decoder to capture inpainting feature transfer traces. FAST [5] employed pre-trained vision transformer model to extract long-range relations in temporal and spatial domain, and utilized convolutional encoder to extract high-pass, mid-pass, and low-pass features separated by DCT transform. Two features were integrated through decoder to generate pixel classification map. While VIDNet and FAST were trained and tested on three video inpainting models. Model generalizability for advanced video inpainting models was not widely discussed. Our model is discussed for the generalization against more diverse video inpainting models.

Recent model [6] fused frame difference feature warped by optical flow and high-pass filter feature into a deep feature space. Then Bi-ConvLSTM constrained the continuity between frames. These two detection models have been proposed and subsequently trained and evaluated using datasets generated by same inpainting models. Another algorithm [28] applied a two-stage refinement model. Firstly, special temporal and spatial convolutions were used applied suppress content expression and highlight inpainting traces. Then a two-stage convolutional network was employed to localize video inpainted regions. However, these models have not been subjected to testing against more advanced and contemporary deep inpainting models. The generalization of detection model has been not discussed in details.

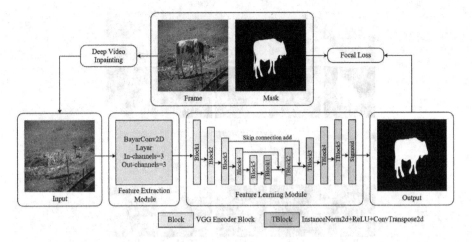

Fig. 2. Network overview. The generalizable deep video inpainting detection network includes a feature extraction module and a feature learning module. The feature extraction module consists of a BayarConv2D layer. Feature learning module is a convolutional encoder-decoder, including a VGG encoder and a deconvolutional decoder. Skip connections (add operation) are applied in the third and fourth block symmetrical positions of the deep features in this module.

3 Proposed Method

In this section, we firstly provide an overview of the proposed network structure for deep video inpainting detection. Subsequently, we introduce the feature enhancement module which comprises the Bayar convolutional layer. Following that, we present the structure of the feature learning module. Finally, we introduce the loss function employed in the model during training stage.

3.1 Overview

Our approach is shown in Fig. 2. Let $X_1^T := \{X_1, X_2, ..., X_T\}$ be a corrupted video quence of height H, width W and frames length T. $M_1^T := \{M_1, M_2, ..., M_T\}$ denotes the corresponding frame-wise masks. For each mask M_i, value "0" indicates known pixels, and value "1" indicates missing regions. We formulate deep video inpainting as vision task that create (X_1^T, M_1^T) pairs as input and reconstruct the inpainted video frames $Y_1^T := \{Y_1, Y_2, ..., Y_T\}$. We formulate deep video inpainting detection as vision task that input the reconstructed inpainted frames $Y_1^T := \{Y_1, Y_2, ..., Y_T\}$ and output the frame-wise masks $\widehat{M}_1^T := \{\widehat{M}_1, \widehat{M}_2, ..., \widehat{M}_T\}$. Our proposed network comprises two essential parts: the feature extraction module and the feature learning module. The feature extraction module utilizes Bayar convolutions to extract traces of inpainted frames. Meanwhile, the feature learning module employs a convolutional encoder-decoder to decode deep features into pixel-level localization.

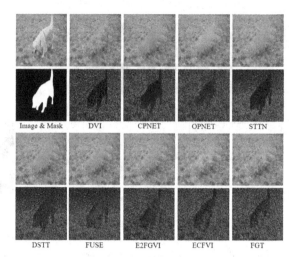

Fig. 3. Constrained convolution output visualization feature map. The frame is 51th of the "dog" sample in the DAVIS dataset. The first column is the frame and the corresponding object mask. The first row is object deletion inpainting frames generated by nine video inpainting models including DVI [1], CPNET [2], OPNET [3], STTN [7], DSTT [21], FUSE [9], E2FGVI [20], ECFVI [22], FGT [23]. The second row is a visualization of the constrained convolution output feature map for the corresponding inpainted frame.

3.2 Feature Extraction Module

As RGB channels are not sufficient to tackle all the different cases of deep video inpainting, we propose to enhance the inpainting traces through adding pre-designed input layers, Bayar layer [26]. Instead of relying on pre-determined kernels, we incorporate the Bayar layer to adaptively learn low-level prediction residual features for detecting inpainting traces. It is implemented by adding specific constraints to the standard convolutional kernels. For simplicity, we use W_b^i to represent the ith ($i = 1, 2, 3$) channel of the weights W_b in the Bayar layer, and the central values of each channel W_b^i are denoted by a spatial index $(0, 0)$. Then the following constraints are enforced on each channel of W_b before each training iteration:

$$\begin{cases} W_b^i(0, 0) = -1 \\ \sum_{m,n \neq 0} W_b^i(m, n) = 1 \end{cases} \quad for \;\; i = 1, 2, 3. \tag{1}$$

For a 3-channel input Y, Bayar layer extracts the constrained features Φ_b by using a 3×3 kernel W_b, namely,

$$\Phi_b(Y) = W_b \otimes Y, \tag{2}$$

and $\otimes$ represents the convolutional operation.

We designed a convolution layer with a convolution kernel of size 3×3, and randomly initialized parameters. Then take formula (1) to change the parameters from general convolution to constraints convolution. As shown in the Fig. 3, for the video sample that namely "dog" in the DAVIS dataset, we employ nine deep video inpainiting models to generate object deletion frames. The object deletion frames (first row) generated by different kinds of video inpainting models are processed with constraints convolutional layers to obtain visual inpainting trace feature maps $\Phi_b(Y)$ (second row). Remarkably, our observations indicate that the constraints convolution feature map exhibits effective feature extraction capabilities for inpainted frames produced by diverse video inpainting models.

3.3 Feature Learning Module

The feature learning module includes convolutional encoders and decoders. To build the encoder, we utilize the pre-trained VGG16_BN network [27]. We make modifications to replace the BatchNorm2d layer with an InstanceNorm2d layer to enhance generalization capabilities. The BatchNorm2d layer integrates multi-channel restricted convolution features for normalization, so this is detrimental to supervised learning of Bayar convolution feature maps. InstanceNorm2d performs separate normalization processing in the channel feature dimension, which retains the feature of limiting the channel independence of convolutional features and is conducive to better convergence of the supervised learning network. A more specific explanation is that, for the depth encoding feature $x \in [B, C, H, W]$ that restricts convolution, where B represents the batch size, C represents the number of channels, and H and W represent the height and width of the depth feature. The deep inpainting traces of Bayar convolution features are visually most obvious in dimensions H and W. Batch normalization calculates the mean and variance from the batch dimension of the tensor. The Batch normalization calculation process is as follows

$$\begin{cases} \mu_i = \dfrac{1}{THW} \sum_{t=1}^{T} \sum_{h=1}^{H} \sum_{w=1}^{W} x_{thw} \\[2mm] \sigma_i^2 = \dfrac{1}{THW} \sum_{t=1}^{T} \sum_{h=1}^{H} \sum_{w=1}^{W} (x_{thw} - \mu_i)^2 \end{cases}, \tag{3}$$

Instance normalization calculates the mean and variance from the channel dimension of the tensor. The Instance normalization calculation process is as follows

$$\begin{cases} \mu_{ti} = \dfrac{1}{HW} \sum_{h=1}^{H} \sum_{w=1}^{W} x_{hw} \\[2mm] \sigma_{ti}^2 = \dfrac{1}{HW} \sum_{h=1}^{H} \sum_{w=1}^{W} (x_{hw} - \mu_{ti})^2 \end{cases}, \tag{4}$$

batch normalization destroys the channel style characteristics in the depth features, but instance normalization can maintain the style characteristics of Bayar convolution.

Furthermore, we extract the outputs from the 24th and 34th layers of the encoder, corresponding to the third and fourth blocks respectively, to establish skip connections. We found that level-by-level feature skip connections like UNet cannot effectively improve network performance, but will reduce network performance. Because building skip connections stage by stage will cause feature learning to be passed through the skip links instead of through the deeper convolutional encoder. In the decoder, we adopt various combinations of "ReLU+ConvTranspose2d+InstanceNorm2d" to decode deep features into pixel local representations. Add operations are applied to deep features in skip connections. The number of output channels of the decoder is 1, and finally the Sigmoid function is used to constrain features to $(0, 1)$.

3.4 Loss Function

The models of the same task were converged by various loss constraints during training, such as dice loss [6], IoU loss [4] or the weighted sum of focal loss and IoU loss [5]. We also tried dice loss, iou loss, and focal loss during training. In the end, we choosed focal loss because of more stable convergence. The model converges better under this loss function during training. In most cases, the inpainted area in the input video frames tends to be relatively smaller than other area. The focal loss is a kind of general cross entropy loss which assigns an extra factor to the original cross entropy term, so the loss solve the problem of unbalanced distribution of positive and negative samples in the training data, thereby constrains the model to pay more attention to smaller inpainted areas. We use focal loss which is formulated as:

$$
\begin{aligned}
L_{Focal}(M, \widehat{M}) = &- \sum \alpha(1 - \widehat{M})^{\gamma} M \log(\widehat{M}) \\
&- \sum (1 - \alpha)\widehat{M}^{\gamma}(1 - M) \log(1 - \widehat{M})
\end{aligned}
\tag{5}
$$

4 Experiment

We firstly introduce the dataset and evaluation metrics in Sect. 4.1. Section 4.2 introduces some settings and hyperparameters we used during model training. Section 4.3 presents ablation experiments of our model under different settings to demonstrate the effectiveness of the model modules. Section 4.4 presents generalization detection experiments against different inpainting models. In Sect. 4.5, we compare the proposed model with other detection models. Section 4.6 presents the robustness experiments of our proposed algorithm against video compression and noise. In Sect. 4.7 we show the qualitative results of the proposed method and current models in the same task.

4.1   Dataset and Evaluation

Since DAVIS [29] dataset is the most common benchmark for video inpainting, which consists of 90 videos with frame-wise object masks, we evaluate our approach on it for inpainting detection. The DAVIS 2016 version includes 50 single-target videos, 30 for training and 20 for testing. The DAVIS 2017 version has been expanded from the 2016 version with 40 multi-target videos. We generate inpainted videos using nine video inpainting approaches including DVI [1], CPNET [2], OPNET [3], STTN [7], DSTT [21], FUSE [9], E2FGVI [20], ECFVI [22], FGT [23], with the ground truth object mask as reference. We adopted the F1 score and mean Intersection of Union (IoU) between the prediction masks the ground truth as evaluation metrics.

4.2   Implementation Details

We use PyTorch for implementation. Our model is trained on a NVIDIA Tesla A40. The network input is resized to 512×512. Our feature extraction backbone is VGG16_BN for RGB frames. The encoder is initialized from VGG16_BN model pretrained on ImageNet and the decoder is initialized by Xavier initialization. The Adam optimizer is applied during the model training. Learning rate is initialized to $1e-4$. And the batchsize is set to 4. The hyperparameters α and γ in the loss function are set to 0.25 and 2 respectively.

Table 1. IoU/F1 score ablation comparison on inpainted DAVIS 2016. The model is trained on DVI and CPNET inpainting (denoted as '*')

Methods	DVI*	CPNET*	OPNET
BayarVGGLSTM	0.48/0.60	0.48/0.62	0.12/0.20
VGG_BN	0.57/0.69	0.52/0.64	0.04/0.07
BayarVGG_IN	**0.63/0.75**	**0.56/0.68**	**0.26/0.37**

Table 2. IoU/F1 score generalization study on inpainted DAVIS 2017. The first column represents the training dataset and the first row represents the testing dataset.

	STTN	DSTT	FUSE	ECFVI	E2FGVC	FGT
STTN	**0.75/0.85**	0.66/0.78	0.62/0.75	0.17/0.25	0.50/0.63	0.14/0.22
DSTT	0.28/0.40	**0.85/0.91**	0.49/0.62	0.30/0.42	0.73/0.83	0.05/0.09
FUSE	0.58/0.72	0.72/0.81	**0.75/0.84**	0.25/0.35	0.55/0.67	0.11/0.17
ECFVI	0.22/0.33	0.34/0.45	0.26/0.37	**0.70/0.81**	0.35/0.47	0.36/0.49
E2FGVC	0.31/0.43	0.84/0.91	0.46/0.58	0.45/0.58	**0.83/0.90**	0.15/0.23
FGT	0.42/0.55	0.42/0.56	0.45/0.59	0.56/0.69	0.45/0.59	**0.52/0.66**

4.3   Ablation Analysis

We evaluate ablation performance on inpainted videos generated from the DAVIS 2016 dataset. 30 videos are used for training the model and 20 videos are used for evaluation. We compared three schemes. The first is to replace the decoder with the ConvLSTM. The second is to remove the Bayar convolutional layer. The last is our proposed model. As shown in Table 1, we choose three inpainted datasets to compare localization performance. The ablation experiment shows that our proposed method has the better performance. The comparison between VGG_BN and BayarVGG_IN models proves that Bayar layer can improve the localization performance. And the instance normalization layer can enhance the generalization of Bayar convolution.

4.4   Generalization Analysis

We evaluate the generalization performance of the model on the DAVIS 2017 dataset, which divided into 70 training videos and 20 testing videos. Table 2 presents the generalization evaluation scores of our models trained on inpainted datasets produced by different models. Our model has better generalization performance among transformer inpainting models such as STTN, DSTT, and FUSE inpainting models.

Table 3. IoU/F1 score comparison on inpainted DAVIS 2016. The model is trained on DVI and CPNET inpainting (denoted as '*')

Methods	DVI*	CPNET*	OPNET
HP-FCN	0.55/0.67	0.69/0.80	0.19/0.29
GSR-Net	0.59/0.70	0.70/0.77	0.22/0.33
VIDNet	0.59/0.71	0.76.0.85	0.25/0.34
FAST	0.57/0.68	**0.76/0.83**	0.22/0.34
BayarVGG	**0.63/0.75**	0.56/0.68	**0.26/0.37**

Table 4. IoU/F1 score on inpainted DAVIS 2017 against compression by different quality factors.

Compression	DVI	CPNET	STTN	DSTT	FUSE
95	0.70/0.80	0.61/0.72	0.75/0.85	0.85/0.91	0.75/0.84
85	0.70/0.78	0.60/0.71	0.69/0.78	0.84/0.91	0.73/0.82
75	0.68/0.77	0.59/0.69	0.61/0.69	0.82/0.89	0.75/0.84
65	0.70/0.80	0.60/0.70	0.54/0.61	0.84/0.90	0.74/0.83

Table 5. IoU/F1 score on inpainted DAVIS 2017 against adding noise by different variance.

Noise	DVI	CPNET	STTN	DSTT	FUSE
Original	0.70/0.80	0.61/0.72	0.75/0.85	0.85/0.91	0.75/0.84
Val:50	0.58/0.65	0.52/0.59	0.63/0.69	0.74/0.83	0.64/0.69
Val:100	0.52/0.61	0.47/0.53	0.54/0.62	0.68/0.76	0.53/0.61

4.5   Comparison Analysis

We compare the proposed model with five models for deep video inpainting detection including HP-FCN [30], GSR-Net [31], VIDNet [4], FAST [5]. The experiment is evaluated on the DAVIS 2016 dataset. Table 3 shows the comparative localization performance between our model and other models. Our model is most competitive against DVI model, but not competitive against CPNET model. However, our proposed model is tested for generalization on a variety of inpainting models.

4.6   Robustness Analysis

To test the robustness of our approach under noise and JPEG perturbation, we conduct experiments listed in Table 4 and Table 5 on five inpainting datasets which includes DVI, CPNET, STTN, DSTT, FUSE. We add gaussian noise (mean is 0, variance are 50 and 100) or recompress test frame with JPEG quality 95, 85, 75 and 65 for perturbation to the input frame. Our method is tested on the 2017 version of the DAVIS inpainting dataset. Table 4 shows the test results of our model for compressed frames. The experimental results show that our method is robust against compression. The reason is that the compression operation can not affect the feature extraction of Bayar convolutional layer. However, in the anti-noise experiments in Table 5, the localization performance of our model decreases with the increase of noise fluctuations. It indicating that the model is less robust against noise.

4.7   Qualitative Results

Figure 4 indicates the visualization results of our proposed model compared with other approaches under the same setting. Compared with other methods, our model can locate more inpainted areas in detail, such as arm inpainting. Because the input of the model is a single frame, the localization results are not very good in continuity between frames. But in general, our method is closer to the ground truth in qualitative comparison of localizing inpainted regions.

Fig. 4. Qualitative visualization results on DVI dataset. From the first row, we present the inpainted video frames. From the second to the sixth row, these images show the final prediction results of different methods and we utilize the green mask to highlight the results. The seventh row is the ground truth. (Color figure online)

5   Conclusions

This paper proposes a localization model for generalizable deep video inpainting detection. The model includes a feature extraction module and a feature learning module. The feature extraction module adopt Bayar convolution for extracting inpainting traces. The feature learning module learns inpainting traces to obtain pixel level localization. We perform generalization analysis on datasets made with various inpainting models by DAVIS. Extensive experiments demonstrate that our model is competitive for localizing inpainting regions. However, the performance on the generalization of inpainting detection is not enough for applications. Future work will continue to focus on the direction of high generalization detection against various deep video inpainting models.

Acknowledgements. This work was supported by National Key Technology Research and Development Program under 2020AAA0140000.

References

1. Kim, D., Woo, S., Lee, J.Y., Kweon, I.S.: Deep video inpainting. In: 2019 IEEE/CVF Conference on Computer Vision and Pattern Recognition (CVPR), pp. 5792–5801 (2019). https://doi.org/10.1109/CVPR.2019.00594
2. Lee, S., Oh, S.W., Won, D., Kim, S.J.: Copy-and-paste networks for deep video inpainting. In: 2019 IEEE/CVF International Conference on Computer Vision (ICCV), pp. 4413–4421 (2019). https://doi.org/10.1109/ICCV.2019.00451

3. Oh, S.W., Lee, S., Lee, J.Y., Kim, S.J.: Onion-peel networks for deep video completion. In: 2019 IEEE/CVF International Conference on Computer Vision (CVPR), pp. 4403–4412 (2019). https://doi.org/10.1109/ICCV.2019.00450

4. Zhou, P., Yu, N., Wu, Z., Davis, L.S., Shrivastava, A., Lim, S.N.: Deep video inpainting detection. arXiv preprint. arXiv:2101.11080 (2021)

5. Yu, B., Li, W., Li, X., Lu, J., Zhou, J.: Frequency-aware spatiotemporal transformers for video inpainting detection. In: 2021 IEEE/CVF International Conference on Computer Vision (ICCV), pp. 8188–8197 (2021). https://doi.org/10.1109/ICCV48922.2021.00808

6. Wei, S., Li, H., Huang, J.: Deep video inpainting localization using spatial and temporal traces. In: 2022 IEEE International Conference on Acoustics, Speech and Signal Processing (ICASSP), pp. 8957–8961 (2022). https://doi.org/10.1109/ICASSP43922.2022.9746190

7. Zeng, Y., Fu, J., Chao, H.: Learning joint spatial-temporal transformations for video inpainting. In: Vedaldi, A., Bischof, H., Brox, T., Frahm, J.M. (eds.) ECCV 2020. LNCS, vol. 12361, pp. 528–543. Springer, Cham (2020). https://doi.org/10.1007/978-3-030-58517-4_31

8. Gao, C., Saraf, A., Huang, J.B., Kopf, J.: Flow-edge guided video completion. In: Vedaldi, A., Bischof, H., Brox, T., Frahm, J.M. (eds.) ECCV 2020. LNCS, vol. 12357, pp. 713–729. Springer, Cham (2020). https://doi.org/10.1007/978-3-030-58610-2_42

9. Liu, R., et al.: FuseFormer: fusing fine-grained information in transformers for video inpainting. In: 2021 IEEE/CVF International Conference on Computer Vision (ICCV), pp. 14040–14049 (2021). https://doi.org/10.1109/ICCV48922.2021.01378

10. Xu, R., Li, X., Zhou, B., Loy, C.C.: Deep flow-guided video inpainting. In: 2019 IEEE/CVF Conference on Computer Vision and Pattern Recognition (CVPR), pp. 3723–3732 (2019). https://doi.org/10.1109/CVPR.2019.00384

11. Bertalmio, M., Sapiro, G., Caselles, V., Ballester, C.: Image inpainting. In: 2000 27th Annual Conference on Computer graphics and Interactive Techniques, pp. 417–424 (2000). https://doi.org/10.1145/344779.344972

12. Bertalmio, M., Bertozzi, A.L., Sapiro, G.: Navier-stokes, fluid dynamics, and image and video inpainting. In: 2001 IEEE Computer Society Conference on Computer Vision and Pattern Recognition (CVPR), p. I (2001). https://doi.org/10.1109/CVPR.2001.990497

13. Barnes, C., Shechtman, E., Finkelstein, A., Goldman, D.B.: PatchMatch: a randomized correspondence algorithm for structural image editing. ACM Trans. Graph. 28(3) (2009). https://doi.org/10.1145/1531326.1531330

14. Criminisi, A., Pérez, P., Toyama, K.: Region filling and object removal by exemplar-based image inpainting. IEEE Trans. Image Process. 13(9), 1200–1212 (2004). https://doi.org/10.1109/TIP.2004.833105

15. Newson, A., Almansa, A., Fradet, M., Gousseau, Y., Pérez, P.: Video inpainting of complex scenes. SIAM J. Imaging Sci. 7(4), 1993–2019 (2014). https://doi.org/10.1137/140954933

16. Huang, J.B., Kang, S.B., Ahuja, N., Kopf, J.: Temporally coherent completion of dynamic video. ACM Trans. Graph. (ToG) 35(6), 1–11 (2016). https://doi.org/10.1145/2980179.2982398

17. Chang, Y.L., Liu, Z.Y., Lee, K.Y., Hsu, W.: Free-form video inpainting with 3D gated convolution and temporal PatchGAN. In: 2019 IEEE/CVF International Conference on Computer Vision (CVPR), pp. 9066–9075 (2019). https://doi.org/10.1109/ICCV.2019.00916

18. Chang, Y.L., Yu Liu, Z., Hsu, W.: VORNet: spatio-temporally consistent video inpainting for object removal. In: 2019 IEEE/CVF Conference on Computer Vision and Pattern Recognition Workshops (CVPR) (2019). https://doi.org/10.1109/CVPRW.2019.00229

19. Wang, C., Huang, H., Han, X., Wang, J.: Video inpainting by jointly learning temporal structure and spatial details. In: 2019 AAAI Conference on Artificial Intelligence, pp. 5232–5239 (2019). https://doi.org/10.1609/aaai.v33i01.33015232

20. Zhang, K., Fu, J., Liu, D.: Inertia-guided flow completion and style fusion for video inpainting. In: 2022 IEEE/CVF Conference on Computer Vision and Pattern Recognition (CVPR), pp. 5982–5991 (2022). https://doi.org/10.1109/CVPR52688.2022.00589

21. Liu, R., et al.: Decoupled spatial-temporal transformer for video inpainting. arXiv preprint. https://arxiv.org/abs/2104.06637

22. Kang, J., Oh, S.W., Kim, S.J.: Error compensation framework for flow-guided video inpainting. In: Avidan, S., Brostow, G., Cissé, M., Farinella, G.M., Hassner, T. (eds.) ECCV 2022. LNCS, vol. 13675, pp. 375–390. Springer, Cham (2022). https://doi.org/10.1007/978-3-031-19784-0_22

23. Zhang, K., Fu, J., Liu, D.: Flow-guided transformer for video inpainting. In: Avidan, S., Brostow, G., Cissé, M., Farinella, G.M., Hassner, T. (eds.) ECCV 2022. LNCS, vol. 13678, pp. 74–90. Springer, Cham (2022). https://doi.org/10.1007/978-3-031-19797-0_5

24. Zhang, H., Mai, L., Xu, N., Wang, Z., Collomosse, J., Jin, H.: An internal learning approach to video inpainting. In: 2019 IEEE/CVF International Conference on Computer Vision (ICCV), pp. 2720–2729 (2019). https://doi.org/10.1109/ICCV.2019.00281

25. Ouyang, H., Wang, T., Chen, Q.: Internal video inpainting by implicit long-range propagation. In: 2021 IEEE/CVF International Conference on Computer Vision (ICCV), pp. 14579–14588 (2021). https://doi.org/10.1109/ICCV48922.2021.01431

26. Bayar, B., Stamm, M.C.: Constrained convolutional neural networks: a new approach towards general purpose image manipulation detection. IEEE Trans. Inf. Forensics Secur. **13**(11), 2691–2706 (2018). https://doi.org/10.1109/TIFS.2018.2825953

27. Simonyan, K., Zisserman, A.: Very deep convolutional networks for large-scale image recognition. arXiv preprint. http://arxiv.org/abs/1409.1556

28. Ding, X., Pan, Y., Luo, K., Huang, Y., Ouyang, J., Yang, G.: Localization of deep video inpainting based on spatiotemporal convolution and refinement network. In: 2021 IEEE International Symposium on Circuits and Systems (ISCAS), pp. 1–5 (2021). https://doi.org/10.1109/ISCAS51556.2021.9401675

29. Perazzi, F., Pont-Tuset, J., McWilliams, B., Van Gool, L., Gross, M., Sorkine-Hornung, A.: A benchmark dataset and evaluation methodology for video object segmentation. In: 2016 IEEE Conference on Computer Vision and Pattern Recognition (CVPR), pp. 724–732 (2016). https://doi.org/10.1109/CVPR.2016.85

30. Li, H., Huang, J.: Localization of deep inpainting using high-pass fully convolutional network. In: 2019 IEEE/CVF International Conference on Computer Vision (ICCV), pp. 8301–8310 (2019). https://doi.org/10.1109/ICCV.2019.00839

31. Zhou, P., et al.: Generate, segment, and refine: Towards generic manipulation segmentation. In: 2020 AAAI Conference on Artificial Intelligence, pp. 13058–13065 (2020). https://doi.org/10.1609/aaai.v34i07.7007

3DPS: 3D Printing Signature for Authentication Based on Equipment Distortion Model

Fei Peng[1]([✉])[iD] and Min Long[2][iD]

[1] Institute of Artificial Intelligence, Guangzhou University, Guangzhou 510006, Guangdong, China
eepengf@gmail.com
[2] School of Electronics and Communication Engineering, Guangzhou University, Guangzhou 510006, Guangdong, China
caslongm@aliyun.com

Abstract. To counter the counterfeit of 3D printed products, 3DPS, an improved printing signature for 3D printed objects based on our previous work is proposed for authentication. A specific hole is added to the non-critical flat portion of the 3D printed object, and the 3DPS is constructed from the contour of the hole with a hand-held microscope. Compared to the previous work, the equipment distortion model and the 3DPS's construction are improved, and the threshold can be calculated directly rather than determined by experience. Experimental results show that the 3DPS not only can effectively authenticate 3D printed objects with high accuracy but is also robust and secure.

Keywords: 3D printing signature · 3D printed object authentication · copyright protection · equipment distortion model · 3D printer

1 Introduction

Additive Manufacturing (AM) has become an important alternative paradigm in manufacturing. Due to its huge advantage in terms of cost reduction and product personalization, it is considered to be an important tool for the fourth industrial revolution [9]. However, 3D printed products could be relatively easily reverse-engineered. As a result, manufacturing companies are worried about the potential violation of the intellectual property of their 3D printed products. Therefore, effective authentication that can tell genuine 3D printed objects apart from counterfeit has started receiving attention from both the industry and academia. Recently, various methods have been proposed to authenticate different mediums, such as printed documents [1,3,4,7,16,18], digital images [12], and 2D context [16]. However, as 3D printed objects are different from them in materials, geometric appearance and utility characteristics, these methods cannot be directly applied to 3D printed objects. Currently, some specific methods had been proposed for 3D printed objects, and the typical methods include 3D printing watermarking [5,6,14,15,17], chemical fingerprint [11] and quantum materials [2,8]. Nevertheless, they all need extra expensive imaging devices [5,6,14,15,17], special materials [11] or spectrometers [8,11], which significantly increase the costs. Thus, a secure authentication method with low cost is needed. Recently, motivated by the print signature scheme for document authentication [18], we proposed a 3D printed object authentication based on printing noise and digital signature [10]. It authenticates the printed 3D objects by the digital signature that extracted from the contour of the added hole in the non-critical flat portion of the 3D printed object. Even though it can effectively distinguish the fake products from genuine ones, it can only determine the threshold by experience, which indicates that its theoretic foundation can be further developed. Here, an improved printing signature for 3D printed objects, 3DPS, is proposed based on equipment distortion model. The main contributions of this paper include:

(1) An improved equipment distortion model is developed. The distortions in the observed 3D printed object are classified into controllable equipment distortion, uncontrollable random distortion, and observation distortion.
(2) The accuracy is improved by employing the principal component analysis (PCA) rather than the discrete cosine transform (DCT). Based on the new method, a chi-square-distribution-based threshold can be calculated with a given false-negative rate.
(3) The effectiveness of the proposed method is validated by 6 combinations of 3D printers and materials. Furthermore, a basic framework for authenticating 3D printed objects based on 3DPS has been established.

2   Equipment Distortion Model for 3D Printed Objects and Construction of 3DPS

2.1   Equipment Distortion Model

Stereolithography (SLA) is one of the most widely used rapid prototyping technologies. Because of the slight imprecise of mechanical motion, the distortions are unavoidably introduced during SLA 3D printing process [10]. After analyzing hundreds of holes printed by different 3D printers, we found that the distortion introduced by 3D printers can be further classified into two categories: controllable equipment distortion and uncontrollable random distortion, and their relations are illustrated in Fig. 1.

2.2   Controllable Equipment Distortion

Controllable equipment distortion, which is represented as C, depends on a specific 3D printer. As a 3D model needs to be divided into many slices by the CAM software before printing, the 3D model is often slightly modified to adapt

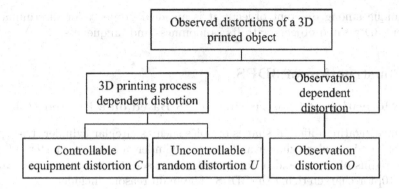

Fig. 1. Composition of the observed distortions of a 3D printed object.

to the characteristics of a specific 3D printer. This type of distortion repeatedly occurs in the printed objects manufactured by the same 3D printer, and it can be regarded as a fixed property of a specific 3D printer.

2.3   Uncontrollable Distortion

Uncontrollable random distortion, which is represented as U, is determined by the combination of the physical and chemical properties of printing materials, the precision of laser beam, etc. They vary in every 3D printed object, even though they are produced by the same 3D printer.

2.4   Observation Distortion

Observation distortion, which is represented as O, is introduced by different observation conditions, such as lighting, position, settings of the hand-held microscope, image processing, and etc. From the above definitions and analyses, three types of distortions are independent of each other. The final printed object can be regarded as the result of the impact of these three types of distortions on an ideal design model S. The equipment distortion model of 3D printed objects can be represented as

$$W' = S + C + U + O, \tag{1}$$

where $W' = \{w_1', w_2', ..., w_k'\}$ represents a feature sequence extracted from the captured images of the hole, $S = \{s_1, s_2, ..., s_k,\}$ represents the feature sequence of the ideal hole, and k represents the feature dimension. As mentioned above, C, U, and O respectively represent the feature sequence of controllable equipment distortion, uncontrollable equipment distortion, and observation distortion, and their sizes are the same as W.

For some given almost identical printed objects that are printed by a same 3D printer, S and C are identical. Since U can be regarded to be independent and unique among different objects, it is an ideal property for discriminating different 3D printed objects with its randomness and uniqueness.

3   Construction of 3DPS

3.1   Motivation of Constructing 3DPS Based on a Printed Hole

After investigating different shapes of holes such as circular cylinder, triangular cylinder, semi-circular cylinder and quarter cylinder, a concave quarter cylinder with a radius of 1 mm and a depth of 3 mm is selected as the authentication mark [10] for characterizing the 3DPS. The main reasons include:

(1) as the top view of the quarter cylinder is not centrosymmetric, the extracted features can be easily synchronized without additional marks.
(2) Both arc and lines exist in the quarter cylinder, which can better capture the characteristics of the uncontrollable random distortion.
(3) It can reduce the possibility of being worn and the influence of the light condition.

3.2   3DPS Generation

The $3DPS$ generation process is composed of three phases: signature extraction, uncontrollable random distortion extraction, and dimensionality reduction.

(1) Signature Extraction
   i) Acquire the printed hole's contour, which is represented by a binarized image.
   ii) Obtain the signature by calculating the distance from the contour centroid to the contour with a step of $1°$.
   iii) Synchronize the signatures observed from different angles by setting the distance from a specific corner to the centroid as the starting point.
   iv) Standardize the signatures as

$$w_k{}'' = \frac{R}{w'^A} w_k{}''^A, \tag{2}$$

   where $w_k{}''$ represents the kth element in the signature $w_k{}''$ with a scaling factor A, w'^A represents the average of all $w_k{}''$, and R represents the average distance from the centroid to the boundary in the ideal quarter circle, respectively.
   v) For W'' with 360 dimensions, the $1, t+1, 2t+1, ..., \lfloor 360/t \rfloor t$ dimensions are sampled as the final signature W' to ensure the independence of different sampling points, where t represents a sampling size, and $\lfloor \cdot \rfloor$ represents a floor function.

(2) Uncontrollable random distortion extraction To exactly extract the uncontrollable random distortion, the observation distortion O needs to be eliminated. To remove O, the average of the signatures from some images acquired from the same hole with a different location, rotation angle, and scaling factor is calculated, according to the generation mechanism of O. The signature of the ith printed hole W_i can be represented as

$$w_i = S + C + U_i \approx \frac{1}{m} \sum_{j=1}^{m} w'_{i,j}, \tag{3}$$

where U_i represents the uncontrollable random distortion of the ith 3D printed hole, $w'_{i,j}$ represents the signature extracted from the ith hole and its jth captured image, and S is represented by the signature extracted from the top view of the designed hole.

Since the ideal feature sequence S and the controllable equipment distortion C are assumed to be constant across the holes printed with the same printer, they can be obtained from the average of W_i, which is formulated as

$$S + C = \frac{1}{mn} \sum_{i=1}^{n} \sum_{j=1}^{m} w'_{i,j}, \tag{4}$$

where n represents the number of 3D printed holes manufactured by the same printer.

The uncontrollable random distortion U_i of the ith 3D printed hole can be extracted from (3), (4), which is represented by

$$U_i = W_i - (S + C) \approx \frac{1}{m} \sum_{j=1}^{m} w'_{i,j} - \frac{1}{mn} \sum_{i=1}^{n} \sum_{j=1}^{m} w'_{i,j} \tag{5}$$

(3) Dimensionality reduction Even though the uncontrollable random distortion is a feature sequence with a total of $\lfloor 360/t \rfloor$ dimensions, there still exists considerable redundancy in it. As the principal component analysis (PAC) is often used to reduce the dimensionality of data [13], it is applied to the feature sequence U_i.

Firstly, calculate $\lfloor 360/t \rfloor \times \lfloor 360/t \rfloor$ principal component coefficients $coeff'$ from all uncontrollable random distortion. Secondly, apply the linear conversion to every U_i with the first N vectors of the principal component coefficients $coeff$, and the principal components of U_i is represented as $F_i = \{f_{i,1}, f_{i,2}, ..., f_{i,N}\}$. The reason is that the sum of the explained variance of the first N dimensions of F_i is large enough, which indicates that the first N dimensions of F_i can represent most of the feature of the original variable. The feature sequence of different dimensions is normalized by

$$r_{i,k} = \frac{f_{i,k} - \overline{f_k}}{\delta_k} (k = 1, ..., N), \tag{6}$$

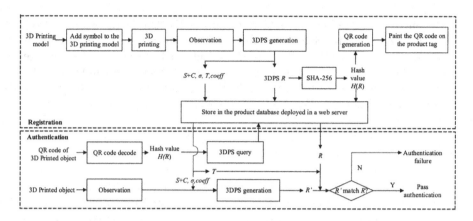

Fig. 2. Framework of 3D printed object authentication.

where $\overline{f_k}$ and δ_k represent the mean and standard deviation of the kth dimension among all F_i, respectively. As U_i is the difference between W_i and its mean value $S + C$, all $\overline{f_k}$ equal 0.

After the processing, the 3DPS R_i of the ith 3D printed object is constructed.

4 The Authentication Framework

The authentication framework based on 3DPS is illustrated in Fig. 2.

i) Registration: For a 3D model, a concave quarter cylinder is first added to the non-critical flat portion as an authentication mark. After 3D printing, the object with the hole is obtained. Secondly, the 3DPS R of the 3D printed object is obtained according to the above method. Meanwhile, SHA-256 is used to calculate the message digest $H(R)$ of the 3DPS. $H(R)$ is stored in the product database deployed in a web server, and it is used as a key value of the product database. Meanwhile, $H(R)$ is utilized to generate a QR code and is painted on the product tag of the 3D printed object. The parameters R, $S + C$, $\delta = \{\delta_1, \delta_2, ..., \delta_N\}$, $N \times \lfloor 360/t \rfloor$, matrix $coeff$ and a predefined threshold T are stored in the product database.

ii) Authentication: For a 3D printed object, its corresponding registered 3DPS R and some parameters are first obtained from the product database deployed in the web server via the QR code. After that, the hole's images are captured by a hand-held microscope, and the 3DPS is generated from them with parameters $S + C$, δ and $coeff$. If the Euclidean Square distance between the extracted 3DPS R' and the registered 3DPS R is less than the threshold T, the 3D printed object is regarded to pass the authentication; otherwise, the authentication is failed. The Euclidean Square distance is

$$D(R, R') = \sum_{k=1}^{N} N\sqrt{(r_k - r'_k)}, \qquad (7)$$

where r_k and r'_k represent the kth dimension of R and R', respectively.

5   Experimental Results and Analysis

5.1   Experimental Results

To evaluate the reliability, experiments with large scale of samples are done with the hand-held microscope Anyty 3R-MSUSB401 and six 3D printers-material combinations, and their corresponding information is listed in Table 1.

Table 1. The information of the 3D printers.

Combination	3D Pinter model	Printing materia	Sample number	δ_R
Combo A	ZRapid SL660	Resin Future8000	120	0.1344
Combo B	UnionTech Lite 600	Resin Future8000	120	0.0852
Combo C	UnionTech SLA600	Resin YGC8000	120	0.1976
Combo D	Union 3D RS6000	Resin C-UV 940	105	0.1792
Combo E	Zrapid SL660	Resin Rs14120	105	0.1106
Combo F	Zrapid SL500	Resin zr550	105	0.3401

To facilitate the experiments and save costs, the holes are printed on 4 boards to observe the contour of different printed holes for the same 3D printer. The size of the board is $40 \times 40 \times 3$ mm. For each hole, 5 images are captured to eliminate the observation distortion. For a specific hole, the others from the same 3D printer are regarded as the counterfeits. When $N = 14$, the results are shown in Fig. 3, where the presented data is in logarithmic. The minimum squared distance between the registered 3DPS and the one of counterfeit product from 6 combinations are all larger than the threshold and the maximum squared distance between the 3DPS of genuine product and the registered one. It indicates that the proposed authentication scheme can discriminate between the genuine products and counterfeit ones with 100% accuracy.

Meanwhile, cross-validation is also made between the samples from different 3D printers. Take a sample from one combination as a genuine one, while the rest are acted as counterfeits. Experimental results show that all squared distance between two 3DPSs of 3D printers are larger than those from the same 3D printer and none of them passes the authentication. It indicate its good authentication ability.

5.2   Analysis of Security

Since the purpose of the authentication is to determine whether the 3DPS extracted from a 3D printed object matches the registered one, the only way for the counterfeiter is to forge the same printed hole with high precision. An imitator may counterfeit the object with a high precision scanner and 3D printer. Nevertheless, the scanning, model re-creation, and printing process inevitably

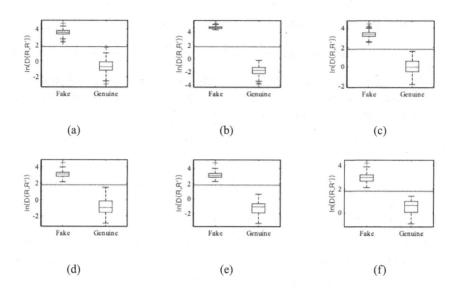

Fig. 3. Boxplot for authentication results from Combination A to F. The minimum logarithmic distance between the fake ones and the genuine ones are: (a) 0.7662; (b) 4.1119; (c) 0.7017; (d) 1.4425 (e) 2.0500; (f) 1.4328.

introduce extra distortions. For simplicity, the uncontrollable distortion of the counterfeit product can be represented as:

$$U_f = (S + C_0 + U_0) + (C_1 + U_1) - (S + C_0) = U_0 + C_1 + U_1, \quad (8)$$

where $C_0 + U_0$ represents the distortion of the original genuine product, $C_1 + U_1$ represents the distortion introduced during forging, and S represents the ideal contour.

To perform the attack, the adversary needs to make sure that $C_1 + U_1$ is smaller than the precision of distinction. However, the 3DPS is constructed in the PCA domain, and it is hard to evaluate the precision in the spatial domain from. To acquire the precision of the distinction in the spatial domain, a simulation experiment is done by adding noise to every position of the standardized signature. And the noise can be assumed to follow the uniformly distributed in $[-p/2, p/2]$ and the average is 0. Here, p represents the precision of distinction, i.e., the actual distance between the noisy position and the origin position.

The Euclidean Square distances between the 3DPSs of modified signature and that of the original one of Combo A is shown in Fig. 4. As shown in the figure, the square distance is increased with the increase of p. Among 6 combinations, if p is greater than 10 μm, the probability of failed authentication is very high. Since the current resolution of 3D printing is about 50 to 100 μm, the imitator cannot intentionally forge the 3DPS of the product from this aspect. The analysis indicates the high security of the proposed scheme.

5.3   Analysis of Robustness

As the feature sequence of the hole's contour is used for the construction of 3DPS, the robustness of the feature sequence is very important. Some experiments and analysis are done to evaluate the robustness of the proposed scheme.

Analysis of the Robustness Against Rotation, Scaling and Translation (RST). According to the above, the signature is independent of the contour's location, rotation angle or scaling factors. In order to verify the robustness against RST, all images are captured with different angles, amplifications, and locations. The experimental results show that the products still can be correctly authenticated, which indicates its good robustness against RST.

Analysis of Robustness Against Scrubbing Attack. As 3D printer solidifies the photosensitive resin layer by layer, the specific position of the contour is determined by the most outstanding layer. Assuming that each feature position is only determined by a layer of resin, and the distance between the most prominent and the sub-prominent layer of the same position is t mm in the real world. The scrubbing attack can be regarded as reducing p feature positions of the signature with a strength of t. Here, simulation experiments are done to determine the relationship between t and p. In the experiments, p positions are modified with a strength of t mm, where p is from 1 to 360 with a step of 1, and t is from 0 to 0.025 mm with a step of 0.0003. The Euclidean square distances between the 3DPSs of modified feature sequence and the original one from Combo A are shown in Fig. 4. Meanwhile, the results from other combinations are similar to Combo A.

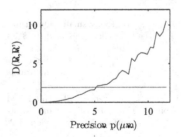

Fig. 4. The Euclidean square distances between the counterfeit feature sequence and the original one from Combo A.

Assuming that only one layer was scrubbed out. As the layer thickness is around 0.1 mm [15], it can be deduced that the printed hole is composed of 3 mm/0.1 mm = 30 layers of resin, and every layer has feature positions. From Fig. 5, when $p = 12$ and $t < 11.22$, the Euclidean square distances are still smaller than the threshold. It indicates that when the amplitude of the scrubbing is less than 10 µm, the proposed method can resist the scrubbing attack. Moreover, even if $p = 360$ and $t < 0.8$ µm, it can still resist the attack.

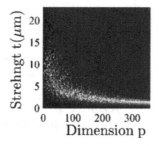

Fig. 5. The Euclidean Square distances between the modified feature sequence and the original one of Combo A.

Table 2. The comparison of different methods.

Methods	Complexity	Robustness	Impact on the 3D printing Object	Cost	Security	Accuracy
Hou [5]	Reconstruct 3D model	Signal processing; Cropping	Global distortion	3D scanner: $18,000	Cloneable	87% 100%
Hou [6]	As above	Scrubbing	As above	As above	As above	93.2% 100%
Silapasuph. [14, 15]	Measure the temperature change from videos or images	As above	Form cavities inside object	Therm.: $4,000, Near infrared Sensor: $900	As above	100%
InfraTrac [11]	Detect with a spectrometer and a handheld special light	As above	Add chemical particles inside object	Spectro.: Lab versions: $50,000; Hand-held version: $3,700	As above	100%
QMC [2,8]	Detect with fluorescent microscope	As above	Add very small particles inside object	Quantum dots: $150/ 20mg. Fluorescent microscope: $ 5,000	Unclonable	100%
Proposed and our previous work [10]	Need a hand-held microscope	RST; Scrubbing	Add very small hole in local surface	Hand-held microscope: about $ 200	As above	100%

Comparison of Different Methods. Here, an extensive comparison is listed in Table 2. Since only very small particles are added inside the 3D printed objects in [5,6], the impact on the appearance 3D printed objects is the least and the robustness is the best. While for the proposed scheme, only a hand-held microscope is needed, and the authentication procedure only needs to scan the QR code and take photos. Moreover, there is no need to encode specific information for authenticating every 3D printed object, and only one same authentication mark is added. Compared to [10], it improves the distortion model, and can achieve better authentication with convenient threshold. Thus, it can strike a better balance among complexity, cost, robustness and authentication reliability.

6 Conclusions

In this paper, a novel 3D printed object authentication framework is proposed. Experimental results and analysis show that the proposed method has good authentication ability, and the 3DPS is difficult to be forged by an adversary. Furthermore, it is robust against RST and surface scrubbing. In addition to SLA-based 3D printing, the proposed method can also be applied to fused deposition modeling (FDM) and other rapid prototyping techniques. Our future work will be towards the refining of the equipment distortion model as well as on more effective and robust features from the 3D printed object itself for authentication.

Acknowledgement. This work was supported in part by projects supported by Natural Science Foundation of Guangdong Province (Grant No. 2023A1515011575), Science and Technology Program of Guangzhou under Grant No. 2024A03J0092, National Natural Science Foundation of China (Grant No. 62072055, U1936115, 92067104).

References

1. Ahmad, F., Cheng, L.M.: Authenticity and copyright verification of printed images. Sig. Process. **148**, 322–335 (2018)
2. Elliott, A.M.: The effects of quantum dot nanoparticles on the polyjet direct 3D printing process. Ph.D. thesis, Virginia Polytechnic Institute and State University (2014)
3. Espejel-Trujillo, A., Castillo-Camacho, I., Nakano-Miyatake, M., Perez-Meana, H.: Identity document authentication based on VSS and QR codes. Procedia Technol. **3**, 241–250 (2012)
4. Ho, A.T.P., Mai, B.A.H., Sawaya, W., Bas, P.: Document authentication using graphical codes: impacts of the channel model. In: ACM Workshop on Information Hiding and Multimedia Security, p. 978. ACM (2013)
5. Hou, J.U., Kim, D.G., Choi, S., Lee, H.K.: 3D print-scan resilient watermarking using a histogram-based circular shift coding structure. In: Proceedings of the 3rd ACM Workshop on Information Hiding and Multimedia Security, pp. 115–121 (2015)
6. Hou, J.U., Kim, D.G., Lee, H.K.: Blind 3D mesh watermarking for 3D printed model by analyzing layering artifact. IEEE Trans. Inf. Forensics Secur. **12**(11), 2712–2725 (2017)
7. Huang, S., Wu, J.K.: Optical watermarking for printed document authentication. IEEE Trans. Inf. Forensics Secur. **2**(2), 164–173 (2007)
8. Ivanova, O., Elliott, A., Campbell, T., Williams, C.: Unclonable security features for additive manufacturing. Addit. Manuf. **1**, 24–31 (2014)
9. Macq, B., Alface, P.R., Montanola, M.: Applicability of watermarking for intellectual property rights protection in a 3D printing scenario. In: Proceedings of the 20th International Conference on 3D Web Technology, pp. 89–95 (2015)

10. Peng, F., Yang, J., Long, M.: 3-D printed object authentication based on printing noise and digital signature. IEEE Trans. Reliab. **68**(1), 342–353 (2018)
11. Scott, C.: InfraTrac successfully applies anti-counterfeit technology to 3D printed metal parts (2017)
12. Shehab, A., et al.: Secure and robust fragile watermarking scheme for medical images. IEEE Access **6**, 10269–10278 (2018)
13. Shlens, J.: A tutorial on principal component analysis. arXiv preprint arXiv:1404.1100 (2014)
14. Silapasuphakornwong, P., Suzuki, M., Unno, H., Torii, H., Uehira, K., Takashima, Y.: Nondestructive readout of copyright information embedded in objects fabricated with 3-D printers. In: Shi, Y.Q., Kim, H., Pérez-González, F., Echizen, I. (eds.) IWDW 2015. LNCS, vol. 9569, pp. 232–238. Springer, Cham (2016). https://doi.org/10.1007/978-3-319-31960-5_19
15. Suzuki, M., Silapasuphakornwong, P., Uehira, K., Unno, H., Takashima, Y.: Copyright protection for 3D printing by embedding information inside real fabricated objects. In: International Conference on Computer Vision Theory and Applications, vol. 2, pp. 180–185. SCITEPRESS (2015)
16. Toreini, E., Shahandashti, S.F., Hao, F.: Texture to the rescue: practical paper fingerprinting based on texture patterns. ACM Trans. Privacy Secur. (TOPS) **20**(3), 1–29 (2017)
17. Uehira, K., Suzuki, M., Silapasuphakornwong, P., Torii, H., Takashima, Y.: Copyright protection for 3D printing by embedding information inside 3D-printed objects. In: Shi, Y., Kim, H., Perez-Gonzalez, F., Liu, F. (eds.) IWDW 2016. LNCS, vol. 10082, pp. 370–378. Springer, Cham (2017). https://doi.org/10.1007/978-3-319-53465-7_27
18. Zhu, B., Wu, J., Kankanhalli, M.S.: Print signatures for document authentication. In: Proceedings of the 10th ACM Conference on Computer and Communications Security, pp. 145–154 (2003)

Multi-Scale Enhanced Dual-Stream Network for Facial Attribute Editing Localization

Jinkun Huang[1], Weiqi Luo[1(✉)], Wenmin Huang[1], Ziyi Xi[1],
Kangkang Wei[1], and Jiwu Huang[2]

[1] School of Computer Science and Engineering and Guangdong Key Laboratory of
Information Security Technology, Sun Yat-sen University, Guangzhou 510006, China
huangjk27@mail2.sysu.edu.cn, luoweiqi@mail.sysu.edu.cn
[2] Guangdong Key Laboratory of Intelligent Information Processing and Shenzhen
Key Laboratory of Media Security, Shenzhen University, Shenzhen 518060, China

Abstract. The advancement of Facial Attribute Editing (FAE) technology allows individuals to effortlessly alter facial attributes in images without discernible visual artifacts. Given the pivotal role facial features play in identity recognition, the misuse of these manipulated images raises significant security concerns, particularly around identity forgery. While existing image forensics algorithms primarily concentrate on traditional tampering methods like splicing and copy-move and are often tailored to detect tampering in natural landscape images, they fall short in pinpointing FAE manipulations effectively. In this paper, we introduce two FAE datasets and propose the Multi-Scale Enhanced Dual-Stream Network (MSDS-Net) specifically for FAE Localization. Our analysis reveals that FAE artifacts are present in both the spatial and DCT frequency domains. Uniquely, in contrast to traditional tampering methods where modifications are localized, facial attribute alterations often span the entire image. The transitions between edited and unedited regions appear seamless, devoid of any conspicuous local tampering signs. Thus, our proposed method adopts a dual-stream structure, targeting the extraction of tampering signs from both the spatial and DCT frequency domains. Within each stream, multi-scale units are employed to discern editing artifacts across varying receptive field sizes. Comprehensive comparative results indicate that our approach outperforms existing methods in the field of FAE localization, setting a new benchmark in performance. Additionally, when applied to the task of pinpointing facial image inpainting, our method demonstrated commendable results.

Keywords: Facial Attribute Editing Localization · Image Forensics ·
Dual-Stream Structure · Multi-Scale Feature · Inpainting

1 Introduction

In recent years, rapid advancements in technologies like GANs [1–3] and VAEs [4, 5], have led to the emergence of various face manipulation techniques, including

B. Ma et al. (Eds.): IWDW 2023, LNCS 14511, pp. 151–165, 2024.
https://doi.org/10.1007/978-981-97-2585-4_11

(a) From blonde hair to black hair.   (b) From young to old.

Fig. 1. Real face images and images after attribute editing using STGAN [9].

facial attribute editing (FAE) [6–12]. These techniques make it easy to edit facial attributes without leaving noticeable artifacts. In the field of FAE, IcGAN [6] stands as a groundbreaking contribution, as it achieved a novel milestone by enabling multi-attribute editing for the first time. Subsequently, methods like StarGAN [7], AttGAN [8], STGAN [9], CooGAN [10], HQ-GAN [11], and HifaFace [12] have been introduced for FAE. These methods have achieved higher precision and more realistic FAE. In Fig. 1 we present real facial images and images after attribute editing using STGAN [9]. It's evident that the edited face images, while retaining the original identity, have undergone specific attribute alterations that are difficult for the human eye to detect. The authenticity of these manipulated images can potentially give rise to significant security issues if maliciously used. Therefore, an urgent necessity exists to introduce methods for the localization of FAE, aiming to address these security concerns.

Existing forensic methods for image manipulation localization primarily targeting on traditional tampering techniques (e.g. splicing, copy-move, and region removal) on natural images [13–20]. For example, ManTra-Net [13] utilizes Bayar's constraint convolution [21] and SRM filters [22] to extract noise information. DFCN [15] combines DenseBlocks [23] with dilated convolutions to extract features at larger scales in images. SATFL [16] employs a coarse-to-fine network that decomposes mask generation into two stages, and employs a self-adversarial training strategy to enhance the model's robustness and performance during training. CAT-Net [17] employs a dual-stream structure to locate tampered regions by capturing the secondary compression artifacts of quantized DCT coefficients in the tampered regions. However, due to significant statistical differences between facial images and natural images, as well as the unique nature of FAE in contrast to traditional tampering techniques, the aforementioned methods may not be well-suited for handling FAE. In response to these challenges, some recent studies have started to focus on face manipulation localization [24–27]. Fakelocator [27] is the pioneering method to incorporate a semantic segmentation network for this purpose. Given that tampered regions in images often consist of non-semantic features, these methods face difficulties in achieving satisfactory performance.

In this paper, we propose MSDS-Net for FAE localization. MSDS-Net employs a dual-stream structure to extract features from both the RGB domain

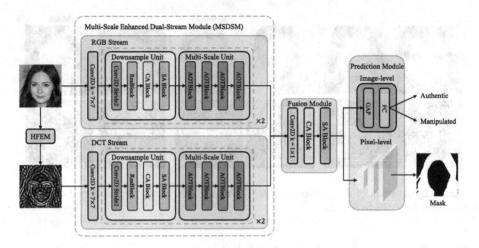

Fig. 2. The structure of the proposed MSDS-Net.

and the DCT domain. Within each stream, multi-scale units are employed to extract features across varying receptive field sizes. In order to compare the effectiveness of different methods in the task of FAE localization, we construct two datasets by utilizing two well-performing FAE methods. In addition, we have applied MSDS-Net to the task of locating facial image inpainting, and construct a dataset using an inpainting method. The experimental results on FAE localization and facial image inpainting localization demonstrate that MSDS-Net achieves the best performance. In summary, our work makes the following key contributions:

- We propose MSDS-Net, which employs a multi-scale enhanced dual-stream structure, extracting multi-scale features from both the RGB domain and the DCT domain to identify tampering artifacts of facial images.
- We provide two datasets for FAE localization, as well as one dataset for facial image inpainting localization. These datasets are generated using two mainstream FAE methods and an image inpainting method.
- Extensive experiments demonstrate that MSDS-Net consistently offers superior localization performance. Additionally, ablation experiments demonstrate the validity of our model.

2   Proposed Method

Figure 2 illustrates the structure of the MSDS-Net framework, which is composed of four primary components: the high-frequency extraction module (HFEM), the multi-scale enhanced dual-stream module (MSDSM), the feature fusion module, and the prediction module. Detailed explanations of these components are provided in the following subsections. Additionally, we will discuss the loss functions implemented in our model.

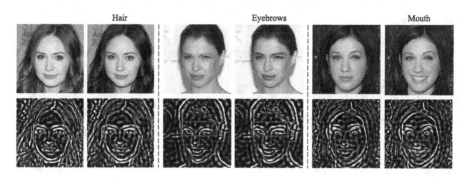

Fig. 3. Illustration of disparities in spatial and high-frequency component before and after facial attribute editing.

2.1   High-Frequency Extraction Module

Our experimental findings indicate that facial editing artifacts are present not only in the spatial domain but also in the high-frequency components of the DCT domain. To visualize these artifacts, we initially convert the images from the spatial domain to the DCT domain. We then selectively retain only the high-frequency components before converting them back to the spatial domain, resulting in the final images. As depicted in Fig. 3, the highlighted red boxes reveal discrepancies between the original and edited images in both the spatial and high-frequency domains of the DCT. While modifications to attributes may not be readily apparent in the spatial domain images, the corresponding regions in the high-frequency images exhibit more pronounced anomalies. This observation suggests that FAE not only influences color and shape within specific image regions but also leaves artifacts in the high-frequency information of these regions. This insight proves valuable for effectively localizing FAE effects.

Based on above analysis, we introduce a HFEM to extract high-frequency information from facial images. To elaborate, we apply the DCT separately to each of the RGB channels of input image, obtaining their respective DCT coefficients. We then selectively preserve only the high-frequency coefficients, setting the others to zero. Finally, we apply the inverse DCT to these coefficients to generate the high-frequency image. As shown in Fig. 2, the original image and the high-frequency image serve as inputs for the RGB stream and DCT stream in the MSDSM, respectively.

2.2   Multi-Scale Enhanced Dual-Stream Module

To leverage the complementary characteristics of artifacts in both the RGB and DCT domains, we introduce the MSDSM with a dual-stream architecture designed for efficient multi-scale feature extraction. Given the often significant disparities in the features extracted from the RGB and DCT streams, we ensure the independence of parameters for each stream to avoid any potential mutual interference. As illustrated in Fig. 2, within each stream, we incorporate two

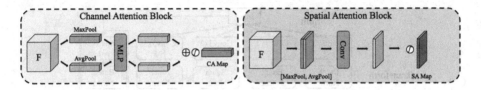

Fig. 4. The structure of CABlock and SABlock.

downsampling units (DSU) and two multi-scale units (MSU), both of which will be elaborated upon in the subsequent subsections.

Downsampling Unit. To reduce the computational complexity and memory usage of the model, existing deep learning methods often employ downsampling operations. However, downsampling inevitably leads to a loss of features. Taking inspiration from CBAM [28], we introduce the DSU in our model, which optimizes the downsampling process to retain more meaningful features during downsampling. A DSU employs a downsampling layer, a ResBlock [29], and a combination of channel attention block (CABlock) and spatial attention block (SABlock) shown in Fig. 4. The CABlock and SABlock are used to generate attention maps and then calculate the output features using *Eq.* 1:

$$F' = F \times Map + F, \tag{1}$$

where F represents the input feature map, F' represents the output feature map, and Map represents the attention weight map output by the CABlock/SABlock. By employing channel and spatial attention, the model can weigh the importance of features at both channel and spatial scales, aiming to preserve more meaningful features in the feature maps after downsampling.

Multi-Scale Unit. Unlike traditional tampering methods e.g. splicing and copy-move, facial attribute edited images, being generated by deep learning models, undergo edits not only in the target attribute regions (like eyebrows, hair, etc.) but also involve adjustments in other non-target regions to achieve better visual results. As a result, the artifacts of image editing are often distributed across the entire image rather than being confined to local regions. The transition between the edited target attribute region and the non-target region appears more natural, without clear indications of local modifications. Therefore, the model needs a larger receptive field to extract editing artifacts over a broader range. At the same time, it also needs to retain the capability to extract features from local editing regions.

In order to achieve the above goals, we introduce MSU for multi-scale feature extraction. Within the MSU, we employed the AOTBlock from AOTGAN [30], the structure of which is shown in Fig. 5. AOTBlock adopts the split-transform-merge strategy by three steps. (i) Splitting: an AOTBlock splits the kernel of a standard convolution into multiple sub-kernels, each of which has fewer output

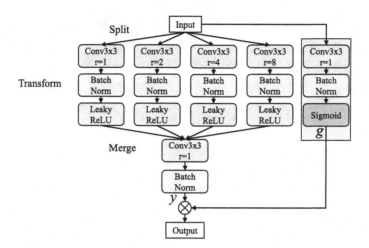

Fig. 5. The architecture of the AOTBlock, where $\otimes$ symbolizes the gated residual connection.

channels. The sum of the output channels of sub-kernels is equal to the standard convolution kernel. (ii) Transforming: each sub-kernel performs a different transformation of the input feature by using a different dilation rate. Different dilation rates correspond to different receptive field sizes, enabling multi-scale feature extraction. (iii) Merging/Aggregating: the features transformed through various receptive fields are aggregated by a concatenation followed by a standard convolution for feature fusion. The AOTBlock also incorporates a gated residual connection. As depicted by the gray box in Fig. 8, the gated convolution generates a trainable weight matrix, which is utilized to compute the ultimate output feature maps based on *Eq.* 2:

$$F' = F \times g + y \times (1 - g), \tag{2}$$

where F represents the input feature map, F' represents the output feature map, g represents the weight matrix outputted by gated convolution, y represents the output of convolutional layer.

2.3   Feature Fusion Module

Since the features extracted by MSDSM include both RGB and DCT domain features, we propose a feature fusion module to fuse the features from both streams for subsequent detection and localization tasks. As shown in Fig. 2, the feature fusion module is similar to the DSU. We employ a 1×1 convolution to reduce the dimension of the features. Then, we use CABlock and SABlock shown in Fig. 4 to enhance the features. The fused features will serve as inputs to the prediction module.

2.4   Prediction Module

After completing the feature fusion process, the extracted features are utilized for subsequent FAE detection and localization tasks. As shown in Fig. 2, the prediction module consists of two parts: the classifier and the decoder. The classifier comprises a global average pooling (GAP) layer followed by a fully connected (FC) layer, which yields a image-level classification result for FAE detection. The decoder is composed of multiple upsampling units (USU) and a 1×1 convolutional layer serving as the output layer, which yields a pixel-level classification result for FAE localization. Each USU has a similar structure to the DSU, consisting of an upsampling layer, a ResBlock, a CABlock, and a SABlock.

2.5   Loss Function

The proposed MSDS-Net primarily addresses two issues: FAE detection and localization. Based on these objectives, the loss function of this method is composed of two parts: the loss function for FAE detection L_{det} and the loss function for FAE localization L_{loc}. The specific loss function are presented in $Eq.$ 3, both L_{det} and L_{loc} are binary cross-entropy loss.

$$L = \lambda L_{det}(y, \hat{y}) + L_{loc}(M, \hat{M}), \tag{3}$$

where y and M are the image label and mask image, respectively, while $\hat{y}$ and $\hat{M}$ are the corresponding model outputs. λ is a balancing hyperparameter.

3   Experiments

3.1   Experimental Setup

Due to the absence of a standardized dataset for FAE localization, we turned to two reputable methods, STGAN [9] and AttGAN [8], to create two FAE datasets: CelebA-STGAN and CelebA-AttGAN. These datasets derive from the CelebA-HQ [31], which includes 30,000 facial images. Through the utilization of STGAN and AttGAN, we crafted four variations of facial attribute edited images focusing on hair, eyebrows, mouth, and age attributes. This yielded a collective count of 150,000 images for each dataset. Within each set, there are 140,000 images designated for training, 2,500 for validation, and 7,500 for testing. For our analysis, all images underwent a resizing to 256×256 dimensions and were subjected to JPEG compression at a quality factor of 75. Beyond the field of GANs for FAE, image inpainting emerges as an alternative technique for facial image alterations. To this end, we harnessed a state-of-the-art inpainting method, MAT [32], to curate a dataset rooted in both Places2 [33] and the aforementioned CelebA-HQ. In this configuration, 36,500 images from Places2 served the training process, while the 2,993 images from CelebA-HQ were allocated, with 993 set aside for validation and 2,000 for testing.

In our comparative analysis, we incorporated four traditional image manipulation localization methods, namely ManTra-Net [13], DFCN [15], SATFL [16],

Table 1. Comparisons of FAE localization (%). Values accompanied by an asterisk (*) indicate the best results when compared to related works in this paper.

Method	CelebA-STGAN				CelebA-AttGAN			
	AUC	F1	mIoU	MCC	AUC	F1	mIoU	MCC
ManTra-Net [13]	97.58	81.61	85.04	80.74	94.79	58.54	71.12	57.88
DFCN [15]	97.88	84.73	87.16	83.54	97.79	82.62	85.47	81.63
SATFL [16]	95.36	87.33	89.10	86.62	93.63	87.17	88.64	86.2
CAT-Net [17]	96.12	88.41	89.53	87.63	95.92	88.03	89.09	87.19
FakeLocator [27]	99.60	83.55	85.87	83.15	99.58	83.54	85.49	82.99
Ours	**99.74***	**89.32***	**90.39***	**88.58***	**99.64***	**88.90***	**89.89***	**88.02***

and CAT-Net [17]. Additionally, we included the modern face manipulation localization method, FakeLocator [27]. It's essential to note that, compared with tampering localization tasks (pixel-level), tampering detection (image-level) is relatively simple. In our experiments, the proposed method consistently achieves near-perfect detection accuracy, often around 100% for tampering detection. Consequently, our results primarily focus on a performance comparison for tampering localization due to page limitation. In line with related studies, e.g. [15–17], we utilized pixel-level AUC, F1, mIoU, and MCC scores for evaluation.

In terms of model implementation, we use Adam for optimization with a learning rate of 0.0001. We train the model for 60 epochs on 2 FAE datasets with a batch size of 32, and 100 epochs on image inpainting dataset with a same batch size. The referenced datasets and our model's source code is available at: https://github.com/kerryhhh/MSDS-Net.

3.2   Comparative Study

FAE Localization. The experimental results are shown in Table 1. It is evident that MSDS-Net outperforms all other methods across all cases. When compared with the FakeLocator [27], MSDS-Net achieves superior results with margins of 0.14%, 5.77%, 4.52%, and 5.43% on the CelebA-STGAN dataset and 0.06%, 5.36%, 4.40%, and 5.03% on the CelebA-AttGAN dataset for the four metrics, respectively. Furthermore, in comparison to traditional image manipulation localization methods, MSDS-Net surpasses CAT-Net [17] by margins of 3.62%, 0.91%, 0.86%, and 0.95% on the CelebA-STGAN dataset and 3.72%, 0.87%, 0.80%, and 0.83% on the CelebA-AttGAN dataset for the same four metrics. These improvements are significant in FAE localization.

In Fig. 6, we present the localization masks generated by various methods. The results clearly illustrate that MSDS-Net consistently delivers the most precise localization masks. Conversely, other methods tend to struggle with accurately delineating the boundaries of edited regions. For instance, in the first row, MSDS-Net effectively identifies the hair region, whereas other methods encounter difficulties in distinguishing the boundary between the hair and the

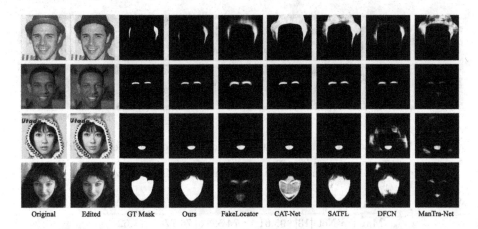

Original   Edited   GT Mask   Ours   FakeLocator   CAT-Net   SATFL   DFCN   ManTra-Net

Fig. 6. Visualization of the localization masks predicted by different methods.

Table 2. Comparisons of cross-dataset localization (%).

Method	STGAN - AttGAN				AttGAN - STGAN			
	AUC	F1	mIoU	MCC	AUC	F1	mIoU	MCC
ManTra-Net [13]	94.97	68.80	76.44	68.00	93.19	54.25	68.88	53.46
DFCN [15]	96.80	76.28	81.53	75.35	96.39	75.66	81.60	74.84
SATFL [16]	91.44	80.18	84.25	79.39	88.91	77.18	82.36	76.47
CAT-Net [17]	94.04	85.18	87.02	84.17	93.41	84.36	86.99	83.53
FakeLocator [27]	99.33	84.04	86.29	83.30	99.39	83.81	86.37	83.07
Ours	**99.47***	**86.79***	**88.37***	**85.85***	**99.52***	**87.57***	**89.12***	**86.73***

hat. In the fourth row, both FakeLocator [27] and ManTra-Net [13] exhibit errors in attribute recognition. This analysis highlights that MSDS-Net surpasses alternative methods in terms of localization accuracy, resulting in fewer artifacts and reduced instances of edits being misplaced in unrelated regions. It consistently produces superior localization outcomes, especially on high-quality facial images with precise edits and superior overall quality.

Cross-Dataset Localization. Because of the variations in editing artifacts introduced by different editing methods, there is a noticeable drop in performance when attempting cross-method localization directly. To address this limitation, similar to related methods [16,18,20], we employ a pre-training and fine-tuning strategy to evaluate the cross-dataset localization performance of different methods. Specifically, we initiate pre-training using CelebA-STGAN /CelebA-AttGAN dataset and subsequently fine-tune on a smaller subset of CelebA-AttGAN/CelebA-STGAN dataset (25,000 images, which is less than 1/5 of the training set). We then evaluate the models on the corresponding testing sets. The experimental results, as shown in Table 2, clearly demonstrate that

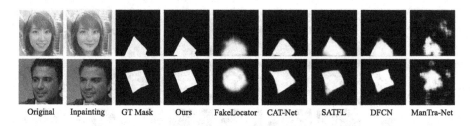

Original   Inpainting   GT Mask   Ours   FakeLocator   CAT-Net   SATFL   DFCN   ManTra-Net

Fig. 7. Visualization of the localization masks predicted by different methods.

Table 3. Comparisons of inpainting localization (%).

Method	AUC	F1	mIoU	MCC
ManTra-Net [13]	95.61	74.52	76.77	71.62
DFCN [15]	99.93	96.54	96.06	95.94
SATFL [16]	96.37	93.50	93.15	92.50
CAT-Net [17]	94.73	90.22	89.69	88.61
FakeLocator [27]	98.09	85.44	85.58	83.14
Ours	**99.99***	**99.06***	**98.91***	**98.90***

MSDS-Net achieves the best performance in cross-dataset localization. In comparison to FakeLocator [27], MSDS-Net outperforms by margins of 0.14%, 2.75%, 2.08%, and 2.55% on the STGAN-AttGAN dataset and by 0.13%, 3.76%, 2.75%, and 3.66% on the AttGAN-STGAN dataset across the four metrics, respectively. When compared with CAT-Net [17], MSDS-Net exhibits superior performance with improvements of 5.43%, 1.61%, 1.35%, and 1.68% on the STGAN-AttGAN dataset and 6.11%, 3.21%, 2.13%, and 3.20% on the AttGAN-STGAN dataset across the four metrics, respectively.

Facial Image Inpainting Localization. In this section, we will demonstrate the application of MSDS-Net in the context of facial image inpainting. The comparative results are presented in Table 3. It is evident that MSDS-Net consistently achieves the highest performance across all four metrics. To be specific, MSDS-Net outperforms FakeLocator [27] by 1.90%, 13.62%, 13.33%, and 15.76% in the respective metrics. While DFCN shows relatively strong performance in traditional methods, when compared with DFCN [15], MSDS-Net still exhibits superior performance with improvements of 0.06%, 2.52%, 2.85%, and 2.96% in the four metrics, respectively.

In Fig. 7, we present the localization masks generated by various methods. The results demonstrate that MSDS-Net consistently delivers the most precise localization masks in the facial image inpainting localization task. Conversely, existing methods face challenges in accurately distinguishing the boundaries of inpainted regions during localization, leading to a notable presence of artifacts. This observation aligns with findings in the FAE localization task.

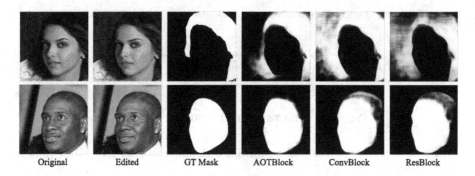

| Original | Edited | GT Mask | AOTBlock | ConvBlock | ResBlock |

Fig. 8. Illustration of localization masks predicted by some variants of our model.

Table 4. Ablation results on CelebA-STGAN dataset (%).

Variants	AUC	F1	mIoU	MCC
w/o DCT stream	99.68	88.66	89.85	87.86
w/o RGB stream	99.66	88.02	89.33	87.20
w/ ConvBlock	99.60	86.41	88.24	85.60
w/ ResBlock	99.60	87.26	88.87	86.49
w/o Attention	94.67	88.68	89.91	87.88
Ours	**99.74***	**89.32***	**90.39***	**88.58***

3.3   Ablation Study

To assess the effectiveness and rationale of MSDS-Net, we evaluate the localization performance of various model variants on the CelebA-STGAN dataset. The ablation experiments encompass the following six components.

Effectiveness of Dual-Stream Structure. We conduct two sets of ablation experiments on the dual-stream structure, where we remove the RGB stream and the DCT stream respectively. As shown in the first two rows of Table 4, the results indicate that removing one stream from the model leads to a decrease in performance compared with the complete model.

Effectiveness of Multi-Scale Feature Extraction. We evaluate the effectiveness of the MSU. Specifically, we replace the AOTBlock in the MSU with ConvBlock and ResBlock for comparison. The experimental results are shown in rows three and four of Table 4. It can be observed that AOTBlock achieves the best performance, indicating the effectiveness of multi-scale feature extraction. As shown in Fig. 8, compared with the results generated by using ConvBlock and ResBlock instead of AOTBlock, the model's generated masks exhibit noticeable artifacts and limited boundary localization capability. This observation indicates the meaningfulness of AOTBlock's multi-scale feature extraction.

Table 5. Quantitative analysis of parameter sharing in MSDSM (%).

Variant	AUC	F1	mIoU	MCC
Partial # 1	99.69	88.76	89.98	87.98
Partial # 2	99.56	85.86	87.94	85.11
Fully	97.87	56.40	71.04	56.82
Ours	**99.74***	**89.32***	**90.39***	**88.58***

Table 6. Quantitative analysis of MSDSM with different numbers of MSU (%).

Variant	AUC	F1	mIoU	MCC
1 MSU	99.54	86.67	88.39	85.80
2 MSU	**99.74***	**89.32***	**90.39***	**88.58***
3 MSU	99.68	89.15	90.16	88.36

Effectiveness of Attention Modules. We use attention modules in multiple modules of the model. To evaluate the effectiveness of the attention modules, we remove the attention modules from the model and conducted experiments. The results are shown in the fifth row of Table 4. It can be observed that removing the attention modules led to a decrease in model performance, indicating the significance of introducing attention modules.

Parameter Sharing in MSDSM. Since there are distinctions in features between the RGB domain and the DCT domain, the parameters of the two streams in MSDSM are kept independent. To confirm the effectiveness of parameter independence, we conduct a series of experiments involving parameter sharing between the two streams. The quantitative results are shown in Table 5, where Partial #1 represents sharing the parameters of input layer, Partial #2 represents sharing the parameters of the input layer, the first DSU and the first MSU, Fully represents sharing all the parameters. The result shown that when parameters are shared, the model's performance experiences a noticeable decline, and this decline becomes more pronounced as the sharing ratio increases. This indicates that the design with independent parameters is effective.

Number of MSU in MSDSM. To evaluate the impact of different numbers of MSU in MSDSM on model performance, we conduct experiments with varying numbers of MSU. It's worth noting that before each MSU, a DSU is added. Based on the results in Table 6, we use two MSUs in each stream.

Number of AOTBlock in MSU. To evaluate the impact of different numbers of AOTBlock in MSU on model performance, we conduct experiments with varying numbers of MSU. As shown in Table 7, with an increase in the number of

Table 7. Quantitative analysis of MSU with different numbers of AOTBlock (%).

Variant	AUC	F1	mIoU	MCC
1 AOTBlock	99.63	87.39	88.86	86.57
2 AOTBlocks	99.67	88.58	89.77	87.74
3 AOTBlocks	99.70	88.66	89.89	87.88
4 AOTBlocks	**99.74***	**89.32***	**90.39***	**88.58***
5 AOTBlocks	99.70	89.13	90.27	88.39

AOTBlocks, the model's performance gradually improves. However, when using five AOTBlocks, the performance declines instead. Based on this experimental result, we opt to employ four AOTBlocks in each MSU.

4   Conclusion

To tackle the security challenges posed by FAE, we introduce the innovative DSMS-Net designed specifically for FAE detection and localization. This network leverages multi-scale features from both the RGB and DCT domains to proficiently localize FAE. Recognizing the limited availability of appropriate experimental datasets, we employed two prevalent FAE methods alongside an image inpainting technique to create three datasets. Our findings suggest that the MSDS-Net stands out in delivering superior localization results for FAE and inpainting compared with some modern methods.

Despite the notable achievements of our approach in FAE localization, several facets merit deeper exploration. Moving forward, our research will concentrate on: (1) Augmenting our model's universality, allowing it to simultaneously detect and localize diverse image tampering techniques, such as copy-move and splicing. (2) Bolstering the model's resilience against common image post-processing such as adding noise, resizing and lossy compression. (3) Transitioning our current framework from a facial image localization task to a more encompassing facial video localization task.

Acknowledgments. This work is supported by the National Natural Science Foundation of China (Grant Nos. U19B2022, 61972430).

References

1. Goodfellow, I.: Generative adversarial networks. Commun. ACM **63**(11), 139–144 (2020)
2. Mirza, M., Osindero, S.: Conditional generative adversarial nets. arXiv preprint arXiv:1411.1784 (2014)
3. Zhu, J.Y., Park, T., Isola, P., Efros, A.A.: Unpaired image-to-image translation using cycle-consistent adversarial networks. In: Proceedings of the IEEE International Conference on Computer Vision, pp. 2223–2232 (2017)

4. Kingma, D.P., Welling, M.: Auto-encoding variational bayes. In: Bengio, Y., LeCun, Y. (eds.) 2nd International Conference on Learning Representations, ICLR 2014, Banff, AB, Canada, 14–16 April 2014, Conference Track Proceedings (2014). http://arxiv.org/abs/1312.6114

5. Razavi, A., van den Oord, A., Vinyals, O.: Generating diverse high-resolution images with VQ-VAE. In: Deep Generative Models for Highly Structured Data, ICLR 2019 Workshop, New Orleans, Louisiana, United States, 6 May 2019. OpenReview.net (2019). https://openreview.net/forum?id=ryeBN88Ku4

6. Perarnau, G., Van De Weijer, J., Raducanu, B., Álvarez, J.M.: Invertible conditional GANs for image editing. arXiv preprint arXiv:1611.06355 (2016)

7. Choi, Y., Choi, M., Kim, M., Ha, J.W., Kim, S., Choo, J.: Stargan: unified generative adversarial networks for multi-domain image-to-image translation. In: Proceedings of the IEEE Conference on Computer Vision and Pattern Recognition, pp. 8789–8797 (2018)

8. He, Z., Zuo, W., Kan, M., Shan, S., Chen, X.: Attgan: facial attribute editing by only changing what you want. IEEE Trans. Image Process. **28**(11), 5464–5478 (2019)

9. Liu, M., et al.: Stgan: a unified selective transfer network for arbitrary image attribute editing. In: Proceedings of the IEEE/CVF Conference on Computer Vision and Pattern Recognition, pp. 3673–3682 (2019)

10. Chen, X., et al.: CooGAN: a memory-efficient framework for high-resolution facial attribute editing. In: Vedaldi, A., Bischof, H., Brox, T., Frahm, J.-M. (eds.) ECCV 2020. LNCS, vol. 12356, pp. 670–686. Springer, Cham (2020). https://doi.org/10.1007/978-3-030-58621-8_39

11. Deng, Q., Li, Q., Cao, J., Liu, Y., Sun, Z.: Controllable multi-attribute editing of high-resolution face images. IEEE Trans. Inf. Forensics Secur. **16**, 1410–1423 (2020)

12. Gao, Y., et al.: High-fidelity and arbitrary face editing. In: Proceedings of the IEEE/CVF Conference on Computer Vision and Pattern Recognition, pp. 16115–16124 (2021)

13. Wu, Y., AbdAlmageed, W., Natarajan, P.: Mantra-net: manipulation tracing network for detection and localization of image forgeries with anomalous features. In: Proceedings of the IEEE/CVF Conference on Computer Vision and Pattern Recognition, pp. 9543–9552 (2019)

14. Hu, X., Zhang, Z., Jiang, Z., Chaudhuri, S., Yang, Z., Nevatia, R.: SPAN: spatial pyramid attention network for image manipulation localization. In: Vedaldi, A., Bischof, H., Brox, T., Frahm, J.-M. (eds.) ECCV 2020. LNCS, vol. 12366, pp. 312–328. Springer, Cham (2020). https://doi.org/10.1007/978-3-030-58589-1_19

15. Zhuang, P., Li, H., Tan, S., Li, B., Huang, J.: Image tampering localization using a dense fully convolutional network. IEEE Trans. Inf. Forensics Secur. **16**, 2986–2999 (2021)

16. Zhuo, L., Tan, S., Li, B., Huang, J.: Self-adversarial training incorporating forgery attention for image forgery localization. IEEE Trans. Inf. Forensics Secur. **17**, 819–834 (2022)

17. Kwon, M.J., Nam, S.H., Yu, I.J., Lee, H.K., Kim, C.: Learning jpeg compression artifacts for image manipulation detection and localization. Int. J. Comput. Vision **130**(8), 1875–1895 (2022)

18. Liu, X., Liu, Y., Chen, J., Liu, X.: Pscc-net: Progressive spatio-channel correlation network for image manipulation detection and localization. IEEE Trans. Circ. Syst. Video Technol. **32**(11), 7505–7517 (2022)

19. Dong, C., Chen, X., Hu, R., Cao, J., Li, X.: Mvss-net: multi-view multi-scale supervised networks for image manipulation detection. IEEE Trans. Pattern Anal. Mach. Intell. **45**(3), 3539–3553 (2022)
20. Wang, J., et al.: Objectformer for image manipulation detection and localization. In: Proceedings of the IEEE/CVF Conference on Computer Vision and Pattern Recognition, pp. 2364–2373 (2022)
21. Bayar, B., Stamm, M.C.: Constrained convolutional neural networks: a new approach towards general purpose image manipulation detection. IEEE Trans. Inf. Forensics Secur. **13**(11), 2691–2706 (2018)
22. Fridrich, J., Kodovsky, J.: Rich models for steganalysis of digital images. IEEE Trans. Inf. Forensics Secur. **7**(3), 868–882 (2012)
23. Huang, G., Liu, Z., Van Der Maaten, L., Weinberger, K.Q.: Densely connected convolutional networks. In: Proceedings of the IEEE Conference on Computer Vision and Pattern Recognition, pp. 4700–4708 (2017)
24. Nguyen, H.H., Fang, F., Yamagishi, J., Echizen, I.: Multi-task learning for detecting and segmenting manipulated facial images and videos. In: 2019 IEEE 10th International Conference on Biometrics Theory, Applications and Systems (BTAS), pp. 1–8. IEEE (2019)
25. Dang, H., Liu, F., Stehouwer, J., Liu, X., Jain, A.K.: On the detection of digital face manipulation. In: Proceedings of the IEEE/CVF Conference on Computer Vision and Pattern recognition, pp. 5781–5790 (2020)
26. Jia, G., et al.: Inconsistency-aware wavelet dual-branch network for face forgery detection. IEEE Trans. Biometrics, Behav. Identity Sci. **3**(3), 308–319 (2021)
27. Huang, Y., Juefei-Xu, F., Guo, Q., Liu, Y., Pu, G.: Fakelocator: robust localization of GAN-based face manipulations. IEEE Trans. Inf. Forensics Secur. **17**, 2657–2672 (2022)
28. Woo, S., Park, J., Lee, J.Y., Kweon, I.S.: Cbam: convolutional block attention module. In: Proceedings of the European conference on computer vision (ECCV), pp. 3–19 (2018)
29. He, K., Zhang, X., Ren, S., Sun, J.: Deep residual learning for image recognition. In: Proceedings of the IEEE Conference on Computer Vision and Pattern Recognition, pp. 770–778 (2016)
30. Zeng, Y., Fu, J., Chao, H., Guo, B.: Aggregated contextual transformations for high-resolution image inpainting. IEEE Trans. Vis. Comput. Graph. (2022)
31. Karras, T., Aila, T., Laine, S., Lehtinen, J.: Progressive growing of gans for improved quality, stability, and variation. In: 6th International Conference on Learning Representations, ICLR 2018, Vancouver, BC, Canada, April 30 - May 3, 2018, Conference Track Proceedings. OpenReview.net (2018). https://openreview. net/forum?id=Hk99zCeAb
32. Li, W., Lin, Z., Zhou, K., Qi, L., Wang, Y., Jia, J.: Mat: mask-aware transformer for large hole image inpainting. In: Proceedings of the IEEE/CVF Conference on Computer Vision and Pattern Recognition, pp. 10758–10768 (2022)
33. Zhou, B., Lapedriza, A., Khosla, A., Oliva, A., Torralba, A.: Places: a 10 million image database for scene recognition. IEEE Trans. Pattern Anal. Mach. Intell. **40**(6), 1452–1464 (2017)

Data Hiding

Neural Network Steganography Using Extractor Matching

Yunfei Xie⬤ and Zichi Wang(✉)⬤

School of Communication and Information Engineering, Shanghai University, Shanghai 200444,
China
{xieyunfei,wangzichi}@shu.edu.cn

Abstract. Neural networks have been applied in various fields, including
steganography (called neural network steganography). The network used for secret
data extraction is called the extractor. This paper proposes a neural network
steganography scheme using extractor matching. In our scheme, the extractor
is a publicly available normal network possessed by the receiver, which is used
for conventional intelligent tasks. Sender connects extractor to another neural
network (called cover network), and then trains the connected network to guarantee correctly data extraction without decreasing the performance of the original
task of cover network. During the process of training, the parameters of extractor remain unchanged. Specifically, these network parameters are obtained using
an extraction key. The receiver can correctly extract secret data with the help of
correct extraction key, while an incorrect key will fail to extract secret data. The
feasibility of our scheme is demonstrated in experiments.

Keywords: Steganography · Neural networks · Extractor matching

1 Introduction

As neural network continues to evolve, their capabilities have become more powerful and
diverse. Neural networks are now extensively utilized in various fields, demonstrating
their wide-ranging applications. Neural networks not only exhibit excellent performance
in many traditional tasks, e.g., target track [1], image recognition [2], natural language
processing [3], but also demonstrate strong performance in steganography.

Many researchers, like the authors in [4], use neural networks to achieve image
steganography. With the rapid advancement of neural network, it has become increasingly common to directly implement steganography within neural networks. Wu et al.
in [5] proposed a digital neural network watermarking framework, which can both perform the original task of the network and embed data within it. The authors in [6] used
a backdooring technique to train a model that deliberately outputs specific data with a
trigger. The specific data have no noticeable impact on the primary task for which the
model is designed.

In neural network steganography, the sender aims to embed secret data into neural
network model with tiny impact on the original model. The receiver can extract secret

data using an extraction network, which is designed for data extraction specially. The extraction network is typically transmitted to receiver with the stego network, as shown in Fig. 1. The problem with this framework is the risk of interception by a third party during the transmission of extraction network.

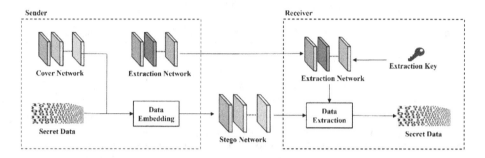

Fig. 1. Neural network steganography.

To mitigate this risk, this paper proposes a novel neural network steganography scheme using extractor matching. Extractor is a publicly available normal neural network possessed by receiver. The sender performs data embedding based on extractor in order to make sure the secret data must use extractor for extraction. Matching training can achieve it. Extraction key is agreed upon in advance by the sender and receiver. This means that extractor and extraction key using for data extraction are available to the receiver before the secret data is transmitted.

By doing so, the sender only needs to transmit the stego network because receiver already has extractor and extraction key. In this way, even if extraction key is accessed by a third-party during transmission, he cannot extract the secret data because secret data has not been transmitted at that time. If both extraction key and stego network are obtained by the third party, he cannot know which one is extractor because there are a huge number of publicly available normal networks on the Internet, and therefore he cannot get the secret data. Similarly, even if the third parties have extractor and stego network, they cannot extract valid secret data without extraction key agreed upon by the sender and receiver in advance. But receiver can correctly extract secret data by extractor with the help of correct extraction key. Wang et al. in [7] proposed a similar functionality, which involves extracting additional data without transmitting corresponding key. In this way, the security of steganography can be enhanced. Neural network steganography using extractor matching is designed precisely to avoid the issue of data leakage caused by extraction network.

For certain scenarios, e.g., a general steganographic framework for neural networks to achieve covert communication proposed in [8], a multi-source data hiding scheme which multiple senders can simultaneously transmit different secret data to a receiver in [9]. Utilizing steganography using extractor matching can omit the step of transmitting extracting network to receiver. The non-specific transmission of extraction network greatly reduces the risk of data leakage, as it is very difficult for third parties to find extractor in the massive public normal network.

In this paper, ResNet [10] and AlexNet [11] are used as an example. Neither network is necessary. Different networks need to design different embedding and extraction schemes. However, the use of a normal neural network is necessary because a publicly available neural network with normal functionality is less detectable by third parties than a special extractor or cover network, which is used for communication. In this paper, AlexNet serves as the extractor owned by receiver and ResNet acts as the cover of secret data, as shown in Fig. 2.

Fig. 2. Steganography scheme using extractor matching

During the data embedding process, the method described in [12] was employed, which involves embedding during the training process. Because AlexNet is a public network, the sender can get it directly without the receiver specifically transmitting it. Adding the weights of AlexNet to the loss function calculation of ResNet links the two networks. The loss function of ResNet is modified to two parts: extraction loss and original network loss. AlexNet is a pre-trained network. The weights of AlexNet are used in training to calculate the extraction loss with the weights of ResNet, and its weights do not change throughout the training process, which ensures that the receiver can use the same network to correctly extract secret data. Extraction key is used to obtain the parameters of AlexNet. The weights of ResNet are constantly changed during training to make the loss small enough. This approach helps to maintain the neural network's original detection accuracy without significant degradation. Receiver can extract the secret data from the stego network by AlexNet. Extraction key is agreed upon in advance by the sender and receiver, and it is only used to help extractor for extraction. The innovation and contributions of this paper can be summarized as follows:

1. This paper proposes a novel neural network steganography method using extractor matching. It involves incorporating extractor into cover network to participate in data embedding and model training, thereby achieving personalized neural network steganography tailored to the receiver.
2. This paper addresses the transmission security issue of transmission. By considering a public normal network possessed by the receiver as extractor, a customized neural network steganography scheme is proposed. This scheme achieves the correctly data extraction without specially transmitting extractor by the sender.
3. The proposed method in this paper does not significantly affect the detection accuracy of cover network by embedding secret data during the training process. Satisfactory results are shown in the experiments section.

2 Method

ResNet and AlexNet are both classic models of Convolutional Neural Networks (CNNs). In this paper, these two networks are used as examples to verify the feasibility of our scheme. The reason for using ResNet is that, in steganography, it is required that the cover network be common and that there be a large number of such networks. Such a network is secure in transmission, and ResNet is one such commonly available network. AlexNet is used because of its simple structure, which allows the receiver to perform as few operations as possible during the data extraction process. The requirement for the extractor is to make it easier for the receiver to use it to extract secret data.

In this paper, the proposed steganographic scheme is implemented on the MNIST dataset. Initially, the training of AlexNet is performed, simulating a scenario where the receiver possesses a normal network which is capable of recognizing handwritten digit images.

2.1 Data Embedding

This paper aims to embed data into the weights of ResNet during training process. The embedding function is

$$V_M = f(W, K), \tag{1}$$

where W is the weights of cover network (ResNet in this paper) using for data embedding, and K is the weights of extractor (AlexNet in this paper). Since the sizes of two networks' weights are different, it needs to be standardized during the embedding process so that their sizes are the same with secret data. The secret data is a string of binary sequences that can be thought of as a two-dimensional tensor. In this paper, tensor is standardized by increasing the dimension, retaining the axis while translating the input tensor, and modifying the number of neurons. The output $V_M = [V_m(1), V_m(2), \ldots, V_m(t)]^T \in [0, 1]^t$, "$t$" represents the length of secret data. It is also the number of neurons in the last dimension after standardization.

Let secret data $M = [m(1), m(2), \ldots, m(t)]^T \in (0, 1)^t$. The value of V_M should close to M as much as possible. This makes the embedded data easy to extract in subsequent operations. To achieve the function, the total loss of ResNet in (2) needs to be divided into two parts. One is the extraction loss of secret data, and the other part is the original loss of ResNet.

$$Loss_R = \alpha \cdot Loss_M + Loss_r \tag{2}$$

$Loss_r$ is the original loss of ResNet using for ensuring satisfactory detection accuracy. $Loss_M$ is the mean square error of the value of V_M and the real embedded data, which is defined as

$$Loss_M = \frac{1}{n} \sum_{i=1}^{n} (M_i - V_{M_i})^2 \tag{3}$$

n is the number of samples in the dataset. The extraction error and extraction loss are not calculated in the same way. However, only if the extraction loss is small enough during training can sender guarantee that receiver will extract secret data correctly. Extraction error is the basis for judging whether secret data is extracted correctly. Extraction loss needs to use the weights of AlexNet to participate in the calculation. Although the weights of AlexNet do not change during training, it does affect the training results of ResNet. Modifications under the guidance of loss function connect the two networks.

In (2), α is a very significant parameter which balances extraction error and detection accuracy by controlling extraction loss and original loss of ResNet during training. As the value of α increases, the extraction error diminishes while the detection precision decreases. The discussion about the value of α will be addressed in Sect. 3.1.

2.2 Decoding Network

In the above subsection, it was stated that the secret data is embedded into the weights of ResNet. The data is embedded in the second fully connected layer of ResNet. This result was obtained through multiple experimental attempts, and the use of this layer has less impact on the original task. Based on the network structure of AlexNet in Fig. 3, this paper selects the second and third Conv (convolutional) layers, as well as the last FC (fully connected) layer for joint training. Again, this selection is the result of experimentation. The experimental result is mentioned in Sect. 3.2.

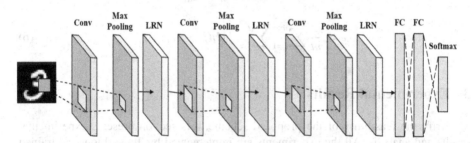

Fig. 3. Architecture of AlexNet for MNIST.

For the receiver, it only needs to use the pre-agreed extraction key to get the weights of extractor, and the perform some simple calculations on the weights of cover network and extractor. Mathematically, receiver only needs to get the value of V_m and round it to get the secret data. The value of V_M can be obtained by computing the Hadamard product between W_{dr} and K_{dr}, as shown in Fig. 4. W_{dr} and K_{dr} are the parameters W and K after standardization referred to in Sect. 2.1 and normalization referred to below. W_{dr} and K_{dr} come in different sizes with W and K.

In Fig. 4, the value of W_{dr} and K_{dr} is any real number, but the value of V_M is a real number between 0 and 1. To guarantee the value of $V_M \epsilon (0, 1)^t$, it is also necessary to normalize the Hadama product of W_{dr} and K_{dr} using the sigmoid function. Cause the output range of sigmoid is 0 to 1, after processing the Hadama product with sigmoid, the value of V_M is maintained between 0 and 1. Hadamard product is defined as

$$\mathbf{A} \odot \mathbf{B} = a_{ij} \times b_{ij} \tag{4}$$

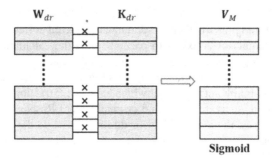

Fig. 4. Decoding network.

A and **B** are both m × n-order matrices. $A = [a_{ij}]_{m \times n}$ and $B = [b_{ij}]_{m \times n}$. Since the value of V_M is any real number between 0 and 1, and the value of M is 0 or 1, and the value of V_M is designed as close to M as possible by continuously adjusting the weights of ResNet during training process, the extraction data can be recovered by rounding the values of V_M in (5). Thus, the extraction error can be calculated using (6). In (6), $e = 0$ means that the secret data are extracted correctly.

$$M_r = round(W_{dr} \odot K_{dr}) \tag{5}$$

$$e = \frac{1}{nt} \sum_{i=1}^{n} \sum_{j=1}^{t} |M_i(j) - M_{ri}(j)| \tag{6}$$

3 Experimental Results

To verify the feasibility of the proposed scheme, this section presents experimental results and analysis. All the experiments are implemented by TensorFlow and trained under the environment of Python 3.6 on a Windows 10 system with an NVIDIA GeForce GTX 1660 SUPER GPU with 14 GB of memory. The Adam optimizer [13] is used for optimization.

3.1 Parameter Determination

In Sect. 2.1, it was mentioned that α can balance the relationship between extraction error and detection accuracy by controlling extraction loss and original loss of ResNet, ensuring both are satisfactory. This section will verify this point through experiments. In the experiment, a batch size of 20 and a capacity of 2000 bits were set. The second and third Conv layers, as well as the last FC layer were used for joint training. The detection accuracy of pre-trained AlexNet is hovered around 0.92. The extraction error and detection accuracy with different α are shown in Fig. 5.

In Fig. 5(a), when $\alpha = 0$, the extraction error $e = 0.4803$, which is significantly higher compared to $\alpha \neq 0$. This shows that it is indeed possible to control the magnitude of to alter the network's emphasis of training.

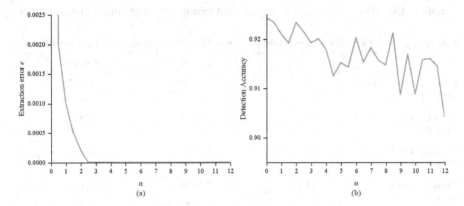

Fig. 5. Extraction error and detection accuracy with different α, (a) extraction error; (b) detection accuracy

The experimental results indicate that increasing reduces the extraction error while simultaneously decreasing the detection accuracy, consistent with the description in Sect. 2.1 and confirming the rationality of the design.

According to the Fig. 5, when $\alpha = 2$, the detection accuracy starts to noticeably decline, whereas at $\alpha = 2$, the extraction error is relatively small. This suggests that the value of α should be taken as small or equal to 2 as possible. Considering these factors, it can be concluded that $\alpha = 2$ satisfies the requirements for both detection accuracy and extraction error.

3.2 Determination of Extractor Layers

Layers of AlexNet using for joint training affects the efficiency of embedding and extracting. To obtain a good solution, the following experiments were conducted to decide which layer to use. The experiments were conducted with a batch size of 20, a capacity of 100 bits, and the parameter α in (2) being 2. The extraction error, detection accuracy and training time with different layers or keys are shown in Table 1.

In Table 1, "None" represents the original ResNet without being connected AlexNet for data embedding. "Random Key" refers to using a randomly generated extraction key to get the weights of AlexNet. This is to verify whether an erroneous key can assist the extractor in data extraction.

Compared with the original ResNet, it can be observed that embedding secret data using AlexNet has an impact on the training time of ResNet. The magnitude of this impact depends on which and how many layers of AlexNet are used. The more layers are used, the longer training time will be. The training time with the FC layer evidently increasing when compared to the results of the experiments without FC layer. This is because when FC layers are connected to ResNet, they establish more neural connections than Conv layers, resulting in greater computational requirements. The experimental results also showed that if only Conv layers are used, the network would sacrifice detection accuracy to ensure the correct extraction of secret data. This elucidates the rationale

Table 1. Extraction error, detection accuracy and training time with different layers or keys

Layers or keys of Extractor	Extraction Error e	Detection Accuracy	Training Time(s)
None	/	0.9371	7.898
Conv1, Conv2, Conv3 FC1, FC2, FC3	0	0.9395	86.807
Conv1, Conv2, Conv3 FC2, FC3	0	0.9259	26.516
Conv2, Conv3 FC2, FC3	0	0.9217	26.572
Conv2, Conv3 FC3	0	0.9349	11.770
Conv2, Conv3	0	0.8940	11.696
Random Key I	0.171	0.9242	8.020
Random Key II	0.183	0.9234	8.130

behind selecting the second and third Conv layers, along with the final FC layer in Sect. 2.2. By doing so, the training time is effectively reduced without compromising the extraction error of secret data and preserving the network's detection accuracy. The last two sets of experiments also demonstrate that only with the use of correct extraction key can the secret data be successfully recovered, thereby confirming the security.

3.3 Embedding Capacity and Security

Embedding capacity is an important evaluation metric in steganography. This subsection will experimentally explore the embedding capacity of the proposed steganography scheme.

The length of secret data is set as {500, 1000, ..., 6000}. The maximum capacity of the network is determined by extraction error and detection accuracy. When both values of extraction error and detection accuracy fall within the acceptable range, the maximum length of secret data is maximum embedding capacity. When the extraction error $e = 0$, it indicates that the secret data can be embedded and extracted correctly. The batch size is set to 20, and the parameter α in (2) being 2. The second and third Conv layers, as well as the last FC layer were used for training. The extraction error and detection accuracy with different capacity are shown in Fig. 6.

In Fig. 6(a), when capacity $t > 4000$, the extraction error $e \neq 0$. This can be considered as the embedding capacity of proposed scheme being 4000 bits, which is satisfactory for steganography.

In terms of the security of the proposed method, this paper proves that it is secure enough by comparing the histograms of the parameter distribution of ResNet before and after data embedding, as shown in Fig. 7. The histogram shows that the similarity of the parameter distribution before and after data embedding is very high, and it is difficult to distinguish between stego and normal networks, so the security is also guaranteed.

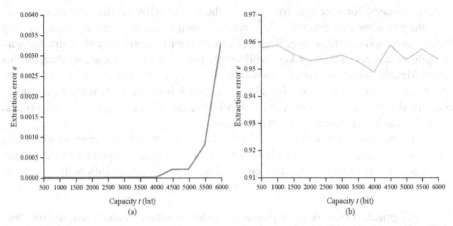

Fig. 6. Extraction error and detection accuracy with different capacity, (a) extraction error; (b) detection accuracy

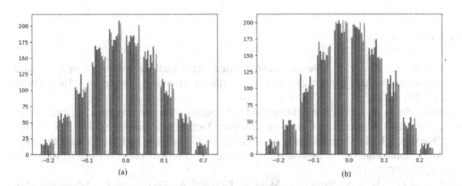

Fig. 7. Histogram of parameter distribution, (a) before data embedding, (b) after data embedding

The experiments in this section demonstrated the feasibility of the scheme and its successful embedding performance. However, further exploration is needed in the future to achieve better embedding performance. For example, how to achieve a larger embedding capacity with no impact on the detection accuracy of cover network.

4 Conclusion

This paper proposes a neural network steganography scheme using extractor matching. The sender and receiver agree in advance on an extraction key, and then sender connects a public normal network called extractor to the other network called cover network for data embedding. Extraction key is used to obtain the parameters of extractor. The proposed scheme does not modify the parameters of extractor during data embedding. Extractor is a publicly available normal network possessed by the receiver. Therefore, there is no need for specially transmitting the extractor by sender, and receiver can have the extractor and extraction key before transmission of stego network, making the steganography more secure. Receiver can obtain the secret data via extractor with the help of correct extraction key, while other keys fail to extract secret data. In this paper, AlexNet serves as extractor owned by receiver and ResNet acts as cover network. The detection accuracy of the original cover network is not affected while ensuring the correctly data extraction. This is achieved by embedding data during the training process instead of modifying parameters after training.

For further study, it is hoped to develop more neural network for steganography using extractor matching. Expand the choices for data embedding beyond just the weight of cover network, making the data embedding more covert. Additionally, a general framework of steganography using extractor matching can be considered.

Acknowledgment. This work was supported in part by the Natural Science Foundation of China under Grants 62376148 and 62002214, and supported in part by the Chenguang Program of Shanghai Education Development Foundation and Shanghai Municipal Education Commission under Grant 22CGA46.

References

1. Elhaki, O., Shojaei, K.: Neural network-based target tracking control of underactuated autonomous underwater vehicles with a prescribed performance. Ocean Eng. **167**(NOV.1), 239–256 (2018)
2. He K., Zhang X., Ren S., Sun J.: Deep residual learning for image recognition. In: Proceedings of the IEEE Conference on Computer Vision and Pattern Recognition, pp. 770–778 (2016)
3. Chowdhary, K.R.: Natural language processing. In: Chowdhary, K.R. (ed.) Fundamentals of Artificial Intelligence, pp. 603–649. Springer, New York (2020). https://doi.org/10.1007/978-81-322-3972-7_19
4. Devi A.G., Thota A., Nithya G., Majji S., Gopatoti A., Dhavamani L.: Advancement of digital image steganography using deep convolutional neural networks. In: 2022 International Interdisciplinary Humanitarian Conference for Sustainability (IIHC), Bengaluru, India, pp. 250–254 (2022)
5. Wu, H., Liu, G., Yao, Y., Zhang, X.: Watermarking neural networks with watermarked images. IEEE Trans. **31**(7), 2591–2601 (2021)
6. Adi Y., Baum C., Cisse M., Pinkas B., Keshet J.: Turning your weakness into a strength: watermarking deep neural networks by backdooring. In: 27th USENIX Security Symposium. pp. 1615–1631. {USENIX} Association, Baltimore (2018)
7. Wang, Z., Feng, G., Wu, H., Zhang, X.: Data hiding in neural networks for multiple receivers. IEEE Comput. Intell. Mag. **16**(4), 70–84 (2021)
8. Yang, Z., Wang, Z., Zhang, X.: A general steganographic framework for neural network models. Inf. Sci. **643**, 119250 (2023)
9. Yang Z., Wang Z., Zhang X., Tang Z.: Multi-source data hiding in neural networks. In: 2022 IEEE 24th International Workshop on Multimedia Signal Processing (MMSP), Shanghai, China, pp. 1–6 (2022)
10. He K., Zhang X., Ren S., Sun J.: Deep residual learning for image recognition. In: 2016 IEEE Conference on Computer Vision and Pattern Recognition (CVPR), Las Vegas, NV, USA, pp. 770–778 (2016)

11. Krizhevsky A., Sutskever I., Hinton G.: ImageNet classification with deep convolutional neural networks. Neural Information Processing Systems (NeurIPS), vol. 25, no. 2, pp. 84–90 (2012)

12. Uchida Y., Nagai Y., Sakazawa S., Satoh S.: Embedding watermarks into deep neural networks. In: Proceedings of the 2017 ACM International Conference on Multimedia Retrieval, pp. 269–277 (2017)

13. Kingma D.P., Ba, L.J.: Adam: a method for stochastic optimization. In: International Conference on Learning Representations (ICLR), Ithaca, NY. ArXiv, San Diego (2015)

Inversion Image Pairs for Anti-forensics in the Frequency Domain

Houchen Pu[1,2], Xiaowei Yi[1,2](✉), Bowen Yang[1,2], Xianfeng Zhao[1,2],
and Changjun Liu[1,2]

[1] Institute of Information Engineering, Chinese Academy of Sciences,
Beijing 100193, China
{puhouchen,yixiaowei,yangbowen,zhaoxianfeng,liuchangjun}@iie.ac.cn
[2] School of Cyber Security, University of Chinese Academy of Sciences,
Beijing 101408, China

Abstract. Recent studies have demonstrated that generative models, such as Generative Adversarial Networks (GANs), leave discernible traces in their results. Based on these traces, several forensic methods have achieved remarkable detection accuracy and strong generalization across different generative models. To counter forensic methods and identify potential vulnerabilities in detectors, existing anti-forensics methods primarily focus on embedding adversarial noises into spacial images. In addition, most methods design distinct noise patterns to each image, making it challenging to generate many adversarial samples within a short time. To address these limitations, this paper proposes a novel anti-forensics method in the frequency domain via using image pairs generated with GAN inversion technology. The objective is to design a universally effective approach that avoids introducing noticeable spatial traces. The proposed method introduces a fresh perspective by applying GAN inversion technology to the field of frequency-domain anti-forensics and only requires 100 images, which is effective to handle all the outputs of the target generator and to generate numerous adversarial samples in turn to help enhance the performance of the detector. Our experiment results show a significant reduction of the detection performance. Specially, when two target models detect both generated and edited images based on the StyleGAN, the area under the receiver-operating curve (AUC) decreases by 9.0%.

Keywords: Frequency-domain anti-forensics · GAN inversion · Spectral amplitude modification

1 Introduction

With the rapid advancement of generative models [12,20] such as Generative Adversarial Networks (GANs) and diffusion models, even non-expert users have the ability to produce a large number of highly realistic synthetic images and manipulated images. However, these fake images may trigger severe social problems or political threats over the world because of its high-quality and strong misleading content.

In order to reduce the risk brought by synthetic images, researchers have carried out numerous studies on the task of detecting images generated by GANs, which shows that GANs leave artifacts in both spacial and frequency domains. Considering these astonishing performance of detection methods, even for images from the different type of GANs and operations, such as resize and JPEG compression, we may raise a question whether GAN generated images can be detected easily?

In this paper, we focus on the detection of GAN-based entire image synthesis and propose an attack method in the frequency domain. To the best of our knowledge, we are the first to use GAN inversion technology to the task of GAN generated image anti-forensics in the frequency domain. We get image pairs from GAN inversion networks and analyse frequency-domain differences, which we add to the images generated by the same generator. We find that post-processed images become not only more real and detailed, but also less likely to be detected. Inspired by the phenomenon, we further try to design an approach to attacking the state-of-the-art detector by increasing the spectral amplitude of the target image, while keeping the spacial domain from additional visual fingerprints. Overall, we make the following contributions:

- Introduce inversion image pairs into the field of anti-forensics and promote the image quality as well by compensating with the average difference.
- Analyze the impact of spectral amplitude modification on the state-of-the-art detector and propose an anti-forensics method in the frequency domain.
- Only require a small number of images to achieve the anti-forensics objective, which is effective to all the images based on the target generator.

The remainder of the paper is organized as follows: In Sect. 2 we briefly review the related works about forensic and anti-forensic methods based on GAN generated images. Then in Sect. 3, we propose our anti-forensic method in detail and clarify the idea of our design. The experimental results are reported and discussed in Sect. 4. Lastly, Sect. 5 leads a conclusion of our work.

2   Related Work

In this section, we briefly review several forensics and anti-forensics techniques for detecting images generated by GAN. On one side, anti-forensics techniques can hide or eliminate the traces in the generated images. On the other side, it can find the vulnerability of existing detectors [25] and promote the development of forensic techniques in the process of confrontation. Besides, we will also briefly introduce GAN inversion, which serves as the starting point of this paper.

2.1 Forensics Techniques

GAN generated images have artifacts that differ from real images in both spatial and frequency domains. It is pointed in [16] that GANs leave artifacts in generated images similar to camera fingerprints, which can be used for source identification. Co-occurrence matrices are computed in [17] on the RGB channels of an image and inputs them to the convolutional neural network. Checkerboard artifacts have been observed in the Fourier spectrum of GAN generated images, which is explained as zero-padding around coefficient values during the up-sampling process in [29]. Therefore, the amplitude of the spectrum is used as a feature to train the classifier. Similar artifacts are found in the mean DCT spectrum of the generative dataset in [7] and different up-sampling operations in the GAN architecture lead to different checkerboard artifacts.

Due to the diversity of GAN architectures, even for models with the same architecture, their training data and objectives may differ. Therefore, the detectors' generalization ability needs to be considered. Besides, considering the varying sizes of images propagated on the Internet, the detector also needs to maintain good performance for manipulated images which are routinely compressed and resized. Data augmentation techniques, such as JPEG compression and image blurring, influence detection performance and the detector is able to generalize well towards unseen CNN-generated images [24]. A method training on image patches is proposed in [3], which has better detection performance for fake facial regions compared to the entire image. With critical analysis on the papers above in [8], down-sampling operations are avoided in the first block of the network and achieves the state-of-the-art performance.

2.2 Anti-forensics Techniques

Anti-forensics methods for GAN detection can be sorted into two categories based on their process: one is to eliminate the generative artifacts by improving the GAN architecture, which usually occurs during the image generation process; the other is to add perturbations to the generated images to degrade the detector's performance after the generation process. Inspired by the work on convolutional networks' shift-invariance in [28], the convolution kernels are optimized in [7] in the downsampling step of StyleGAN [12], significantly reducing the checkerboard artifacts in the DCT spectrum. A novel spectral regularization term is added to the generator loss in [6] to compensate spectral distortions.

Regarding the second category of anti-forensic methods, our research focuses on black-box attacks [15], which means the architecture and internal parameters of the detector can not be accessed. Interactions with the detector are limited to the input and output interfaces. This scenario aligns more closely with real-world scenarios. A local model is trained in [2] to substitute for the target detector with a large amount of data and conducts white-box attacks. Specific noise is added to the GAN generated images in [10] to disrupt the original periodic noise and to dodge detection via implicit spatial-domain notch filtering. An autoencoder is trained in [18] to eliminate the artifacts introduced by GAN in the generated images, rendering the detectors based on these artifacts ineffective. Camera fingerprints are incorporated into generated images in [5], aiming to deceive the detector. It is found in [27] that common adversarial examples, which is added noise in the spacial domain, exhibit highly identifiable artifacts in the frequency domain. Therefore, an improved algorithm is proposed that removes certain high-frequency components to reform the inherent drawbacks.

2.3 GAN Inversion

GAN can accept latent code inputs to generate images which fit a specific distribution. Correspondingly, GAN inversion aims to reverse a given image back into the latent space of a pretrained GAN model, allowing the image to be accurately reconstructed from the inverted latent code by the generator [26]. It should be noted that the GAN inversion technique has the same structure as the autoencoder, both of which have an "encoder-decoder" structure. However, in the former, the decoder is the fixed generator of GAN, and the generative fingerprints in the reconstructed image are consistent with those in the GAN-generated results. In contrast, both the encoder and decoder of the latter need to be trained, and reconstructing the image will introduce fingerprints from the new decoder, not the target generator. Due to the good disentanglement property of certain latent space [12], which means modifying the latent code can intuitively reflect the specific change of the generated image, inversion techniques are widely used in image editing [21], image restoration [1], and multimodal learning [14].

3 Methodology

In this section, we propose our anti-forensic method in detail and clarify the idea of our design. We firstly prepare GAN inversion image pairs to execute the first stage of the attack and then conduct frequency analysis for the second stage of the attack. A complete algorithm is described in steps. Figure 1 graphically shows the overall pipeline of our method.

3.1 GAN Inversion Image Pairs

Reasons for Using Inversion. When analyzing the spectral features of GAN-generated images, it is common practice to take the logarithmic difference of the Fourier spectrum amplitude between generated and real images, and then to calculate the average. There is no inherent correlation between the content of the generated images and the real images. Although the averaging process largely eliminates the influence of different contents on the results and highlights systemic differences, the problem lies in the lack of a "correct answer" for modification of each image. The way to assess the effectiveness of modification is through the detector's indirect reflection in prior work.

To tackle the challenges mentioned above, we adopt the GAN inversion technique proposed in [23]. Firstly, the real image is mapped back to a generated

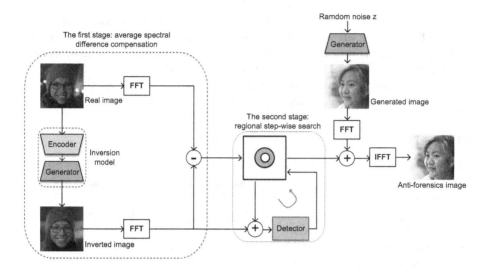

Fig. 1. The framework of the frequency-domain anti-forensics method. Inverted images are generated from real images through the inversion model and the average spectral difference is calculated in the first stage. Regional step-wise search based on the average spectral difference is repeated to reach the optimal solution. Then, according to the optimal solution, other generated images are processed to get anti-forensics results.

image through the inversion network, creating a pair of images with consistency in content. Secondly, after comparing the spectral differences between the two images, we conduct an analysis on how to modify the generated image in the frequency domain to minimize the spectral differences and to degrade the target detector's performance, without introducing visible changes in the spatial domain. Lastly, by combining the exploration process above, a frequency-domain anti-forensics method is designed.

Usage for Image Pairs. An intuitive idea is to compensate generated images with the difference. Compared to adding perturbations in the spatial domain, the challenges in the frequency domain not only lie in the fact that the Fourier spectrum is a complex matrix, which includes both amplitude and phase factors, but also in the difficulty of intuitively reflecting the modification's impact on the spatial domain. The phase of the spectrum reflects the image's structural information, while the amplitude reflects the grayscale value information. The amplitude exhibits visually interpretable features in prior work, so we choose to add perturbations in the spectrum amplitude of the image.

Assuming that $\boldsymbol{F}_i^{\mathrm{r}}$ and $\boldsymbol{F}_i^{\mathrm{f}}$ are the Fourier spectra of the i-th pair of the real image $\boldsymbol{I}_i^{\mathrm{r}}$ and the inverted image $\boldsymbol{I}_i^{\mathrm{f}}$ and that $\boldsymbol{A}_i^{\mathrm{r}}$ and $\boldsymbol{A}_i^{\mathrm{f}}$ are natural logarithms of corresponding spectrum amplitudes, we compute the average logarithmic difference $\overline{\Delta \boldsymbol{A}} = \frac{\sum_{i=1}^{N}(\boldsymbol{A}_i^{\mathrm{r}} - \boldsymbol{A}_i^{\mathrm{f}})}{N}$. Then, we prepare extra K generated images and

Table 1. Evaluating the spectral difference compensation from two aspects: the difficulty in identifying authenticity and the clarity of images. $\varnothing$ means not attacking. $+A$ means average spectral difference compensation.

Method	ΔLogits $\downarrow$	Var $\uparrow$
$\varnothing$	11.28	52.94
$+A$	7.90	121.69

compensate $\overline{\Delta A}$ to their Fourier spectrum amplitudes. Equation (1) explains the process of compensation. Noted that amp(F) and angle(F) are respectively the amplitude and phase of F, we can deduce Eq. (3) based on Eq. (1) and Eq. (2). These modified spectrum amplitudes are then transformed back to the spatial domain to obtain the modified images. In the training process, F stands for the inverted image which is generated by the target generator, and F^{adv} stands for the real image which can be treated as the correct answer of anti-forensics for F. In the testing process, F can be any image generated by the target generator, and F^{adv} is the anti-forensics result. In this paper, $\log(F)$ and e^A respectively represent the natural logarithm and exponent of each element in matrix F and A. The dot multiplication sign indicates the multiplication of the elements at the corresponding positions of two matrices of the same size.

$$\mathrm{amp}(F^{\mathrm{adv}}) = \mathrm{e}^{A+\Delta A} = \mathrm{e}^A \cdot \mathrm{e}^{\Delta A} = \mathrm{amp}(F) \cdot \mathrm{e}^{\Delta A} \tag{1}$$

$$\mathrm{angle}(F^{\mathrm{adv}}) = \mathrm{angle}(F) \tag{2}$$

$$F^{\mathrm{adv}} = F \cdot \mathrm{e}^{\Delta A} \tag{3}$$

In this section, we use logits, the unnormalized probability of the detector's output, as a reference. Keeping the real image unchanged, if the logits' difference between real and fake images become larger, it indicates that the fake images are more likely to be detected by the detector. We use the Laplace operator to evaluate image quality. The Laplace operator is used to calculate the second derivative of the image, reflecting the edge information. For images of the same content, the higher the clarity is, the greater the variance is after the Laplace operator filtering. Our experimental setups are listed in Sect. 4.1 and 20 real images are used to evaluate the spectral difference compensation. The visualization results of $\varnothing$ and $+A$ in Table 1 correspond to the second and third columns in Fig. 2. From the results in Fig. 2 and in Table 1, we find an interesting point that both anti-forensics ability and image quality are improved by compensating generated images with the average spectral amplitude difference. This point is different from other anti-forensics methods which add noise in the spacial domain and damage the original image's quality.

Further Thinking. The experiment above shows that compensating with the average difference is a forward optimization for both anti-forensics and image quality. We wonder if the spectrum amplitude of the generated image is directly replaced with that of the corresponding real image, will it perform better? We

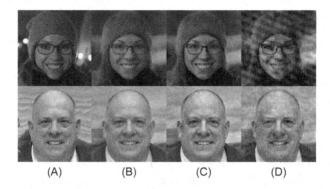

Fig. 2. From left to right: (A) real images; (B) inverted images; (C) images with the average spectral difference compensation; (D) images with spectrum amplitude totally replaced.

conduct such experiment, but obvious raster patterns appear in all modified images and the logits' difference become even larger, as shown in Fig. 2(D). We think one of the reasons for this phenomenon is that during the training process of the inversion encoder, the focus is placed on modeling and feature extraction of the target object (e.g., a person's face), while the reconstruction on the background is poor. The operation of directly replacing the amplitude values introduces additional background information into the generated image, which manifests as raster patterns in the spatial domain. This result indicates that the idea trying to train a network from the spectrum amplitude of the inverted images to that of the corresponding real images is quite wrong. The average difference compensation can only serve as one part for anti-forensics.

3.2 Frequency Analysis

Sensitivity to Different Frequencies. Inspired by [11], we first analyze the sensitivity of the target detector to different frequency bands of images. In the rest of the paper, we shift the spectrum so that the low frequency components are at the center of the spectrum. By gradually decreasing the distance from the points on the Fourier spectrum to the center and filtering higher frequency components out of this distance, the result in Fig. 3 shows that for filtered images, the detector's performance only slightly improves. However, as the filtering range expands, especially when the distance is less than 192, the probability of detector increases significantly. This indicates that although the target detector has strong ability to resist image post-processing, it is more affected by the low-frequency part of the image. Considering frequency amplitude variations below 32 have a significant impact on the image itself, we conduct perturbations within the frequency range of 32 to 192.

Limitations of Modification. From the analysis above, we choose to adjust the frequency amplitude in the distance of 32 to 192. Due to the high complexity

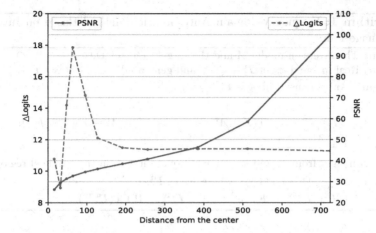

Fig. 3. Changes of logits and PSNR with low pass filtering. In the range of 32 to 192, the detector is more sensitive to frequency changes, and the image quality remains good (PSNR > 30 dB).

of optimizing all points in the region, which still can not guarantee effectiveness for all images, we adopt a regional step-wise amplitude increasing approach to find a universally applicable anti-forensic method. Considering the peculiarity of frequency-domain anti-forensics methods, we can not introduce obvious visible traces in the spatial domain, so we set two limitations. Firstly, we set an upper limit on the amplitude increase, with lower frequency regions having lower limits. This is because both detectors and people's eyes are more sensitive to low-frequency changes. Secondly, we impose limitations on the amplitude variations in adjacent regions. Without such limitation, distinct traces appear near regions with drastic brightness changes in the images. Although these images can cheat the detector, it is easy for people to detect that the image has been manipulated, which obviously does not meet the purpose of anti-forensics.

3.3   Algorithm Design

Our method consists of two stages: the stage of average spectral difference compensation and the stage of regional step-wise search. The experiment sets the modification region in the range of 32 to 192, the distance away from the center point, with a maximum upper limit of amplitude variation being e^x times the original value, where x is a linear function ranging from 0.25 to 1.5. Regions are divided by distance intervals of 8, with x changing at interval of 0.25. The optimal solution will retain and update from each search. To prevent visible artifacts in the spatial domain caused by excessively large changes, the x variation between adjacent intervals can not exceed 0.25.

Algorithm 1: Frequency-domain Anti-forensics Method Based on Inversion Image Pairs

---

Input: The inversion model M and the target detector D.
Input: Real image dataset $\boldsymbol{I}^{\mathrm{r}}_{i=1,2,\ldots,N}$ and generated image dataset $\boldsymbol{I}^{\mathrm{g}}_{j=1,2,\ldots,K}$.
Output: Anti-forensics images $\boldsymbol{I}^{\mathrm{adv}}_{j=1,2,\ldots,K}$.

1   $\boldsymbol{I}^{\mathrm{f}}_i \leftarrow \mathrm{M}(\boldsymbol{I}^{\mathrm{r}}_i)$;

2   $\boldsymbol{F}^{\mathrm{r}}_i \leftarrow \mathrm{FFT}(\boldsymbol{I}^{\mathrm{r}}_i)$; $\boldsymbol{F}^{\mathrm{f}}_i \leftarrow \mathrm{FFT}(\boldsymbol{I}^{\mathrm{f}}_i)$; $\boldsymbol{A}^{\mathrm{r}}_i \leftarrow \log(\mathrm{abs}(\boldsymbol{F}^{\mathrm{r}}_i))$; $\boldsymbol{A}^{\mathrm{f}}_i \leftarrow \log(\mathrm{abs}(\boldsymbol{F}^{\mathrm{f}}_i))$;

3   $\overline{\Delta A} \leftarrow \dfrac{\sum_{i=1}^{N}(\boldsymbol{A}^{\mathrm{r}}_i - \boldsymbol{A}^{\mathrm{f}}_i)}{N}$;

4   select S images from $\boldsymbol{I}^{\mathrm{f}}_i$ and search $\boldsymbol{x}$ at equal interval in the specified region to minimize $\sum_{i=1}^{S} \mathrm{D}(\mathrm{IFFT}(\boldsymbol{F}^{\mathrm{t}}_i))$, where $\boldsymbol{F}^{\mathrm{t}}_i \leftarrow \boldsymbol{F}^{\mathrm{f}}_i \cdot e^{\overline{\Delta A} + \boldsymbol{x}}$;

5   $\boldsymbol{F}^{\mathrm{g}}_j \leftarrow \mathrm{FFT}(\boldsymbol{I}^{\mathrm{g}}_j)$; $\boldsymbol{F}^{\mathrm{adv}}_j \leftarrow \boldsymbol{F}^{\mathrm{g}}_j \cdot e^{\overline{\Delta A} + \boldsymbol{x}\min}$; $\boldsymbol{I}^{\mathrm{adv}}_j \leftarrow \mathrm{IFFT}(\boldsymbol{F}^{\mathrm{adv}}_j)$.

---

4   Experiments

4.1   Experimental Setups

Datasets. In the experiments, we use 1000 real images in the FFHQ [12] dataset to make inversion image pairs base on the inversion network proposed in [23]. We prepare 1000 generated images based on StyleGAN2 [13] and 1000 edited images using techniques proposed in GANSpace [9] and SeFa [22] to evaluate our anti-forensics method.

Target Detectors. To verify the performance of our method, we choose two detectors Grag2021 [8] and Wang2020 [24]. We treat both of the two detectors as a black box and only use the output of the models. We follow the procedure used in [4]. For each image to be detected, a crop with random size and position is chosen, resized to 200×200 pixels and compressed in a random JPEG quality factor from 65 to 100.

Parameters and Evaluation Metrics. The parameters in our algorithm are as follows: $N = 100, K = 1000, S = 20$. We use AUC to evaluate the detector's performance instead of accuracy, because the threshold does not work well on images of different origin. Accuracy can be improved using Platt scaling method [19] if it is necessary.

4.2   Anti-forensics Results

We first compare the performance of the state-of-the-art detector Grag2021 before and after our attack. From the results in Table 2, AUC of Grag2021 has decreased from 90.33% to 81.66% when detecting generated images. Considering our training objective is generated images' anti-forensics, we test our method on edited images. The decrease in AUC approximately keep the same as that of

Fig. 4. Visual display of experimental results. First two rows: Generated images and their anti-forensics results. Last two rows: Edited images and their anti-forensics results.

the generated images. Additionally, we choose another detector Wang2020 with good detection performance. Our method still successfully attacks the detector without additional training and achieves 9.45% average decrease of AUC, which means the method has strong generalization ability. Some examples are displayed in Fig. 4. These results prove the effectiveness of our anti-forensics method. The method is universally effective to most of the generator's results. We just need 100 inversion image pairs to achieve it.

Table 2. Evaluating the performance of our anti-forensics method. Our method is trained only on generated images for attacking the detector Grag2021. Our method is tested on both generated images (different from training data) and edited images, evaluated by the detection performance of Grag2021 and Wang2020.

AUC% ↓ before/after attack	Grag2021 [8]	Wang2020 [24]
Generated images	90.33/81.66	80.43/68.99
Edited images	91.66/83.17	80.85/73.40
Average decrease	8.58	9.45

Table 3. Evaluating the effectiveness of different steps in the method. ∅ means not attacking. +A means average spectral difference compensation. +S means regional step-wise search.

Method	AUC% ↓
∅	90.33
+A	84.80
+S	85.37
+A+S	81.66

4.3   Ablation Study

We conduct ablation study to prove the effectiveness of two stages in our method. Two experiments are conducted in comparison with the detection performance before and after the complete attack. One experiment is only compensating generated images with average spectral difference, and the other is only searching in the certain region with the step-wise strategy. Results in Table 3 show that any stage can decrease the detection performance, but the best anti-forensics ability is achieved when two stages work together.

4.4   Spacial-Domain Influence

The difficulty of directly perturbing the frequency domain lies in evaluating the impact of perturbations on the spatial domain. We increase the spectral amplitude of frequencies ranging from 32 to 192 away from the center by $e^{0.5}$ times. The detector's result shows a decrease from 11.28 to 9.41, and slight watermark-like traces are observed in some images. The impact of perturbations on the images is evaluated using the PSNR (Peak Signal-to-Noise Ratio) metric. The metric 34.24dB indicates that the distortion is in an acceptable range.

We further explore the impact of watermark-like traces' transparency and positional distribution on the target detector. Different values of transparency are tested to evaluate the anti-forensics performance. Regarding the positional distribution, we select half of the watermark area and examine its placement in the left, center, edges, and uniformly dispersed across the image. As shown in Table 4, even when a somewhat transparent watermark-like trace or a relatively intact portion of the trace is selected, there is still an improvement in the images' anti-forensics ability. However, distributing watermark-like traces along the edges leads to a decrease in the anti-forensics ability. We are inspired to choose some regions clustered together to add watermark-like traces or make the traces more transparent in the future work so that traces can hardly be found by people's eyes.

Table 4. Evaluating the impact of traces' transparency and positional distribution.

Transparency	ΔLogits $\downarrow$	Positional distribution	ΔLogits $\downarrow$
0	9.41	Whole	9.41
0.2	9.99	Left	10.44
0.4	10.54	Center	10.48
0.6	10.88	Edges	11.44
0.8	11.31	Uniform	10.47
1.0	11.28	Nowhere	11.28

5   Conclusion

In this paper, we propose a novel anti-forensics method in the frequency domain based on inversion image pairs. First, we explain the reason and the usage for image pairs which come from GAN inversion techniques. We find that compensating with average spectral difference can promote both anti-forensics ability and image quality at the same time. Then, we analyze the target detector's sensitivity to changes in different frequencies and design a two-stage method of perturbing images' spectral amplitude. Experimental results show that our method is universally effective to both generated and edited images from the same generator at a low computational cost. The anti-forensics images not only leave few visible traces in the spacial domain but also remain effectiveness to the unseen detector.

Acknowledgments. This work was supported by National Key Technology Research and Development Program under 2020AAA0140000.

References

1. Abdal, R., Qin, Y., Wonka, P.: Image2stylegan++: how to edit the embedded images? In: Proceedings of the IEEE/CVF Conference on Computer Vision and Pattern Recognition (CVPR), pp. 8296–8305. IEEE (2020). https://doi.org/10.1109/CVPR42600.2020.00832
2. Carlini, N., Farid, H.: Evading deepfake-image detectors with white- and black-box attacks. In: Proceedings of the IEEE/CVF Conference on Computer Vision and Pattern Recognition Workshops (CVPRW), pp. 658–659. IEEE (2020). https://doi.org/10.1109/CVPRW50498.2020.00337
3. Chai, L., Bau, D., Lim, S.-N., Isola, P.: What makes fake images detectable? Understanding properties that generalize. In: Vedaldi, A., Bischof, H., Brox, T., Frahm, J.-M. (eds.) ECCV 2020. LNCS, vol. 12371, pp. 103–120. Springer, Cham (2020). https://doi.org/10.1007/978-3-030-58574-7_7
4. Corvi, R., Cozzolino, D., Zingarini, G., Poggi, G., Nagano, K., Verdoliva, L.: On the detection of synthetic images generated by diffusion models. In: Proceedings of the IEEE International Conference on Acoustics, Speech and Signal Processing (ICASSP), pp. 1–5. IEEE (2023). https://doi.org/10.1109/ICASSP49357.2023.10095167

5. Cozzolino, D., Thies, J., Rossler, A., Nießner, M., Verdoliva, L.: SpoC: spoofing camera fingerprints. In: Proceedings of the IEEE/CVF Conference on Computer Vision and Pattern Recognition Workshops (CVPRW), pp. 990–1000. IEEE (2021). https://doi.org/10.1109/CVPRW53098.2021.00110

6. Durall, R., Keuper, M., Keuper, J.: Watch your up-convolution: CNN based generative deep neural networks are failing to reproduce spectral distributions. In: Proceedings of the IEEE/CVF Conference on Computer Vision and Pattern Recognition (CVPR), pp. 7890–7899. IEEE (2020). https://doi.org/10.1109/CVPR42600.2020.00791

7. Frank, J., Eisenhofer, T., Schönherr, L., Fischer, A., Kolossa, D., Holz, T.: Leveraging frequency analysis for deep fake image recognition. In: Proceedings of the 37th International Conference on Machine Learning, pp. 3247–3258. ICML 2020, JMLR.org (2020)

8. Gragnaniello, D., Cozzolino, D., Marra, F., Poggi, G., Verdoliva, L.: Are GAN generated images easy to detect? a critical analysis of the state-of-the-art. In: Proceedings of the IEEE International Conference on Multimedia and Expo (ICME), pp. 1–6. IEEE (2021). https://doi.org/10.1109/ICME51207.2021.9428429

9. Härkönen, E., Hertzmann, A., Lehtinen, J., Paris, S.: Ganspace: Discovering interpretable GAN controls. In: Proceedings of the 34th International Conference on Neural Information Processing Systems. NIPS 2020, vol. 33, pp. 9841–9850 (2020)

10. Huang, Y., Juefei-Xu, F., Guo, Q., Liu, Y., Pu, G.: Dodging deepfake detection via implicit spatial-domain notch filtering. IEEE Trans. Circ. Syst. Video Technol. (2023). https://doi.org/10.1109/TCSVT.2023.3325427

11. Jiang, L., Dai, B., Wu, W., Loy, C.C.: Focal frequency loss for image reconstruction and synthesis. In: Proceedings of the IEEE/CVF International Conference on Computer Vision (ICCV), pp. 13919–13929. IEEE (2021). https://doi.org/10.1109/ICCV48922.2021.01366

12. Karras, T., Laine, S., Aila, T.: A style-based generator architecture for generative adversarial networks. In: Proceedings of the IEEE/CVF Conference on Computer Vision and Pattern Recognition (CVPR), pp. 4401–4410. IEEE (2019). https://doi.org/10.1109/CVPR.2019.00453

13. Karras, T., Laine, S., Aittala, M., Hellsten, J., Lehtinen, J., Aila, T.: Analyzing and improving the image quality of stylegan. In: Proceedings of the IEEE/CVF Conference on Computer Vision and Pattern Recognition, pp. 8110–8119. IEEE (2020). https://doi.org/10.1109/CVPR42600.2020.00813

14. Lee, S.H., et al.: Sound-guided semantic image manipulation. In: Proceedings of the IEEE/CVF Conference on Computer Vision and Pattern Recognition (CVPR), pp. 3377–3386. IEEE (2022). https://doi.org/10.1109/CVPR52688.2022.00337

15. Mahmood, K., Mahmood, R., Rathbun, E., van Dijk, M.: Back in black: a comparative evaluation of recent state-of-the-art black-box attacks. IEEE Access 10, 998–1019 (2022). https://doi.org/10.1109/ACCESS.2021.3138338

16. Marra, F., Gragnaniello, D., Verdoliva, L., Poggi, G.: Do GANs leave artificial fingerprints? In: Proceedings of the IEEE Conference on Multimedia Information Processing and Retrieval (MIPR), pp. 506–511. IEEE, IEEE (2019). https://doi.org/10.1109/MIPR.2019.00103

17. Nataraj, L., et al.: Detecting GAN generated fake images using co-occurrence matrices. Electronic Imaging 31(5), 532–1–532-7 (2019). https://doi.org/10.2352/ISSN.2470-1173.2019.5.MWSF-532

18. Neves, J.C., Tolosana, R., Vera-Rodriguez, R., Lopes, V., Proença, H., Fierrez, J.: GANprintr: improved fakes and evaluation of the state of the art in face manipu-

lation detection. IEEE J. Sel. Top. Sig. Process. **14**(5), 1038–1048 (2020). https://doi.org/10.1109/JSTSP.2020.3007250

19. Platt, J., et al.: Probabilistic outputs for support vector machines and comparisons to regularized likelihood methods. Adv. Large Margin Classifiers **10**(3), 61–74 (1999)

20. Rombach, R., Blattmann, A., Lorenz, D., Esser, P., Ommer, B.: High-resolution image synthesis with latent diffusion models. In: Proceedings of the IEEE/CVF Conference on Computer Vision and Pattern Recognition (CVPR), pp. 10674–10685. IEEE (2022). https://doi.org/10.1109/CVPR52688.2022.01042

21. Shen, Y., Gu, J., Tang, X., Zhou, B.: Interpreting the latent space of GANs for semantic face editing. In: Proceedings of the IEEE/CVF Conference on Computer Vision and Pattern Recognition (CVPR), pp. 9243–9252. IEEE (2020). https://doi.org/10.1109/CVPR42600.2020.00926

22. Shen, Y., Zhou, B.: Closed-form factorization of latent semantics in GANs. In: Proceedings of the IEEE/CVF Conference on Computer Vision and Pattern Recognition (CVPR), pp. 1532–1540. IEEE (2021). https://doi.org/10.1109/CVPR46437.2021.00158

23. Tov, O., Alaluf, Y., Nitzan, Y., Patashnik, O., Cohen-Or, D.: Designing an encoder for styleGAN image manipulation. ACM Trans. Graph. **40**(4), 1–14 (2021). https://doi.org/10.1145/3450626.3459838

24. Wang, S.Y., Wang, O., Zhang, R., Owens, A., Efros, A.A.: CNN-generated images are surprisingly easy to spot... for now. In: Proceedings of the IEEE/CVF Conference on Computer Vision and Pattern Recognition (CVPR), pp. 8695–8704. IEEE (2020). https://doi.org/10.1109/CVPR42600.2020.00872

25. Wei, W., Feng, Z., Min, T., Junjie, C., Hongjun, L.: Survey on anti-forensics techniques of digital image. J. Image Graph. **21**(12), 11 (2016). https://doi.org/10.11834/JIG.20161201

26. Xia, W., Zhang, Y., Yang, Y., Xue, J.H., Zhou, B., Yang, M.H.: Gan inversion: a survey. IEEE Trans. Pattern Anal. Mach. Intell. **45**(3), 3121–3138 (2023). https://doi.org/10.1109/TPAMI.2022.3181070

27. Ye, D., Jie, W., Qi, W., Qing, L.: Analysis of adversarial examples from frequency domain. Inf. Technol. Netw. Secur./Xinxi Jishu yu Wangluo Anquan **41**(5) (2022). https://doi.org/10.19358/J.ISSN.2096-5133.2022.05.009

28. Zhang, R.: Making convolutional networks shift-invariant again. In: Proceedings of the 36th International Conference on Machine Learning. ICML 2019, vol. 97, pp. 7324–7334. PMLR (2019)

29. Zhang, X., Karaman, S., Chang, S.F.: Detecting and simulating artifacts in GAN fake images. In: Proceedings of the IEEE International Workshop on Information Forensics and Security (WIFS), pp. 1–6. IEEE (2019). https://doi.org/10.1109/WIFS47025.2019.9035107

Cross-channel Image Steganography Based on Generative Adversarial Network

Bin Ma[1,2], Haocheng Wang[1,2], Yongjin Xian[1,2(✉)],
Chunpeng Wang[1,2], and Guanxu Zhao[3]

[1] Key Laboratory of Computing Power Network and Information Security, Ministry
of Education, Shandong Computer Science Center (National Supercomputer Center
in Jinan), Qilu University of Technology (Shandong Academy of Sciences), Jinan,
China
matxyj@163.com
[2] Shandong Provincial Key Laboratory of Computer Networks, Shandong
Fundamental Research Center for Computer Science, Jinan, China
[3] Jinan Art School, Jinan 250002, China

Abstract. Traditional steganographic algorithms often suffer from
issues such as low visual quality and limited resilience against steganalysis at high-capacity data embedding. To address these limitations, this
paper proposes a cross-channel image steganography algorithm based
on generative adversarial networks. In contrast to conventional image
steganography techniques that directly embed secret data into original carrier images, the proposed algorithm embeds the secret data into
the difference-plane of the two similar color channels. The proposed
data embedding scheme involves a U-Net structure based generator for
steganography, an adversarial network for steganalysis, and an optimization network for enhancing anti-steganalysis capabilities. In addition, a
newly introduced Lion optimizer is introduced to effectively optimize the
convergence speed of the proposed networks by adaptively setting learning rates and weight decay values. At the same time, the mean square
error loss, structural similarity loss, and adversarial loss are employed to
progressively enhance the visual quality of generated stego images. Consequently, a color image can be seamlessly embedded into the same-sized

This work was supported by National Natural Science Foundation of China
(62272255, 62302248, 62302249); National key research and development program of China (2021YFC3340600, 2021YFC3340602); Taishan Scholar Program
of Shandong (tsqn202306251); Shandong Provincial Natural Science Foundation
(ZR2020MF054, ZR2023QF018, ZR2023QF032, ZR2022LZH011), Ability Improvement Project of Science and Technology SMES in Shandong Province (2022TSGC2485,
2023TSGC0217); Jinan "20 Universities"-Project of Jinan Research Leader Studio
(2020GXRC056); Jinan "New 20 Universities"-Project of Introducing Innovation Team
(202228016); Youth Innovation Team of Colleges and Universities in Shandong Province
(2022KJ124); The "Chunhui Plan" Cooperative Scientific Research Project of Ministry
of Education (HZKY20220482); Achievement transformation of science, education and
production integration pilot project (2023CGZH-05), First Talent Research Project
under Grant (2023RCKY131, 2023RCKY143), Integration Pilot Project of Science
Education Industry under Grant (2023PX006, 2023PY060, 2023PX071).

B. Ma et al. (Eds.): IWDW 2023, LNCS 14511, pp. 194–206, 2024.
https://doi.org/10.1007/978-981-97-2585-4_14

color image, and achieving high perceptual quality. Experimental results demonstrate that the proposed algorithm achieves a peak PSNR of 41.6 dB for color stego images, significantly reducing the distortion caused by secret image embedding.

Keywords: Steganography · Cross-channel differences · Difference plane · Lion optimizer · Generative adversarial network

1   Introduction

Steganography is an important means in the field of information hiding, which makes people not perceive the existence of secret information by embedding it into the minute details of the carrier. With the continuous development of steganography algorithms and steganography models, from the earliest text steganography to the current multimedia steganography such as images, audio, video, and so on. In today's era of many digital media technologies, the mainstream steganographic media is still dominated by images. Spatial domain image steganography, as the most common steganography, mainly modifies the pixel values of the carrier image and embeds secret information, which can be categorized into traditional steganography and deep learning based steganography in its steganography method. The classical steganography algorithms in the spatial domain include the least significant bit (LSB) algorithm [1], the spatially generalized wavelet relative distortion algorithm (S-UNIWARD) [2], the wavelet weighting algorithm (WOW) [3], etc. Among them, the S-UNIWARD algorithm is the most effective algorithm for steganography. Among them, S-UNIWARD and WOW algorithms minimize the embedding distortion function by choosing the appropriate modification location to avoid embedding into texture-smooth regions, which reduces the detectability of the steganographic image.

The common steganography methods in deep learning based steganography are deep neural network based steganography, residual network based steganography, generative adversarial network based steganography, and style migration based steganography, etc. Baluja [4] et al. Mentioned in their paper that image preprocessing, image hiding, and image extraction tasks have been accomplished using multiple deep neural subnets, respectively, which has greatly utilized the advantages of deep neural networks in the image steganography field. steganography field. Thanks to the advantages of jump connection of residual network, Liu [5] and others combined the residual network with image steganography and constructed an encoder and decoder based on the residual network, which is more suitable for steganography of RGB three-channel images. With the development of deep learning technology, more and more fields are combined with deep learning technology. The proposal of the generative adversarial network, which is a deep learning model that can be counteracted by its own game has also attracted more attention in the field of information hiding. Duan [6] implemented steganography of image hidden image using u-net as a generator. More and more steganographic models based on generative adversarial networks have been proposed [7–10], and the ASDL-GAN [11] model proposed by Tang et al. can adaptively compute the embedding distortion that thereby avoiding embedding into smooth regions of the image, and Xu-Net [12] was introduced as a steganography analyzer for adversarial training to further improve the quality of the generated steganographic images. Style migration can change the overall style of the image, so that the carrier image is converted to the picture style of the stylized image, in the steganography of style migration, image stylization can be a good way to mask the image distortion caused by the embedding of the secret image. Li [13] et al. combined style migration and quadratic moments with image steganography to improve the amount of image embedding at the same time, further solving the problem of distortion of the image due to embedding.

In contrast, steganalysis [14,15], as the antithesis of steganography, has a thatch and shield relationship with steganography. Steganalysis, as a means of detecting and discriminating steganography, analyzes the changes in the feature information of an image to determine whether it contains secret information. For example, the classical spatial rich model (SRM) [16] steganalysis algorithm detects secret information by analyzing the statistical properties of the image. With the development of deep learning technology, the rapid integration of the field of steganalysis with deep learning was born. Deep learning based steganalysis models learn the deep features of an image by designing multiple neural network structures to distinguish between carrier images and steganographic images. Common deep learning based steganalysis models are Xu'Net [13], Ye'Net [17], and SRNet [14].

Based on the above problems, this paper proposes a cross-channel image steganography algorithm based on GAN. The network model consists of four networks: steganography network, steganalysis network, extraction network, and steganalysis optimization network. Through inter-network confrontation and iterative optimization of the loss function in each network, the steganography network gradually improves the stego image generation capability and the steganalysis detection capability of stego images, and the two complement each other to promote the generation of stego images with high visual quality and strong resistance to steganography detection capability. At the same time, this algorithm adopts the U-Net [6,18] structure of the generative network, through the skip conncetion of the U-Net network will be fused with the features in the lower sampling layer and the upper sampling layer, which effectively combines the distribution of the shallow and deep content features of the image and reduces the loss of details, to generate high-quality stego images. In the training process of the generative model, multiple loss functions are used to iteratively optimize the generator. Based on the fact that steganography is done on images in the spatial domain, this algorithm weights the discriminative loss of the joint steganalysis network with two pixel-level loss functions, namely, the mean square error (MSE) and the structural similarity (SSIM), to form the total loss of the generator network for iterative optimization. High-quality images are generated while ensuring that the network can converge quickly and stably. The main contributions of this paper are as follows:

(1) A cross-channel image steganography algorithm based on the generative adversarial network is first proposed in this paper. The secret image is embedded into the difference-plane of two similar color channels, thereby the relationship between two similar color channels is comprehensively used and the image distortion caused by large-capacity data embedding is minimized.

(2) The Lion optimizer is introduced to improve the performance of the generative adversarial networks, thereby, the convergence speed and the data embedding capability are both enhanced compared with the networks that use Adam optimizer.

(3) The mean square error (MSE), structural similarity (SSIM), and discriminative loss are involved with adaptive weighting parameters, solving the problem of gradient vanishing during the network training process. Consequently, the data embedding performance of the proposed scheme is greatly improved.

The rest of the paper is composed as follows. In Sect. 2, we introduce our proposed steganography algorithm and steganography model based on generative adversarial network and expand in detail on the other sub-networks and loss functions included in the model. In Sect. 3, we give the experimental setup, experimental results, and performance evaluation of the proposed method with other graph steganography methods in the same experimental setting. The last section summarizes the conclusions and the outlook for future research.

2   Proposed Method

2.1   Cross-channel Image Steganography Model Based on Generative Adversarial Network

This algorithm is based on the basic model of the generative adversarial network, which is improved in four aspects: embedding algorithm, network structure, model optimizer, and loss function. A cross-channel image steganography model based on generative adversarial network is proposed. The algorithm model as a whole is shown in Fig. 1. Visually the R and G channels are the most similar. For the single-channel grayscale image, visually similar means that the luminance average is more similar, and the difference plane made by using R and G is embedded and then summed with the original carrier image, the pixel overflow range is smaller, so the pixel loss is smaller in the pixel overflow processing after embedding of the secret image. Therefore, according to the principle of minimizing the loss in pixel overflow processing, the R and G channels are differenced to obtain a more suitable difference plane for embedding secret information and achieve better information hiding capability. This model consists of three sub-networks: steganography network (SN), steganalysis adversarial network (SAN), and steganalysis optimization network (SON).

In this study, a generator based on the U-Net structure is designed in the Steganography Network (SN) of Fig. 1. The encoding and decoding network of the whole generator is divided into down-sampling layer and up-sampling layer.

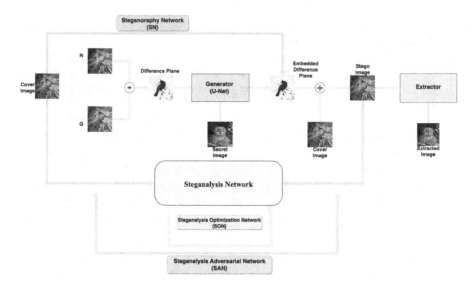

Fig. 1. Cross-channel image steganography Model Structure based on Generative Adversarial Network

The features of the down-sampling layer and the up-sampling layer are spliced by skip connection to enhance the accuracy of the reconstructed image.

In the Steganalysis Adversarial Network (SAN) shown in Fig. 1, the stego image generated by the generator in the steganographic network SN and the original carrier image Cover are jointly input into the optimized steganalysis analyzer, and the adversarial training is carried out synchronously to improve the s analysis ability of the generated stego image. In the training process, to improve the steganalysis resistance of the stego image generated after embedding the secret image, this algorithm designs the steganalysis confrontation loss SD_loss based on the steganalysis analyzer, and takes it as a part of the loss function of the generator in the steganographic network SN, and jointly weights it with the loss functions such as MSE_loss, SSIM_loss, and so on, to carry out the training. Thus, the stego image generated after the steganalysis adversarial network SAN adversarial training is guaranteed to have not only high visual quality but also high resistance to steganalysis, to achieve the purpose of improving security.

Finally, as shown in Fig. 1, the Steganalysis Optimization Network (SON), when the Steganalysis Network discriminates the steganographic images, its discriminative loss will be used as its loss to optimize itself after each iteration and update the network parameters to synchronously improve its own discriminative ability.

2.2   Steganography Network

Generator Based on U-Net Structure. U-Net is a convolutional neural network model based on the encoder-decoder structure, which adopts the Ushaped structure of the down-sampling layer and up-sampling layer to realize the reconstruction of the original image. On the left side of the constracting path of the U-network, the sensory field is continuously enhanced by stacking and convolution, and the number of output feature maps is gradually increased, to extract the underlying features of the original image; while on the right side of the expansive path, the convolution operation after upsampling gradually approximates the original input image. At the same time, jump connections are used to fuse shallow and deep features between feature maps of the same size, so that the whole network can fully utilize the features to reconstruct the image. In this scheme, the U-Net structure is used to transfer the detailed information of different sensory fields in the encoding stage to the decoding stage, and the secret image is stacked and fused with the carrier image to finally generate a high-quality stego image, and the structure of the generator is shown in Fig. 2. The U-Net structure makes full use of the local and overall feature information of the input original carrier image and the secret image by splicing the shallow information with the deep information. The U-Net structure gradually constructs high-precision stego images and guarantees that the generated stego images have excellent visual performance.

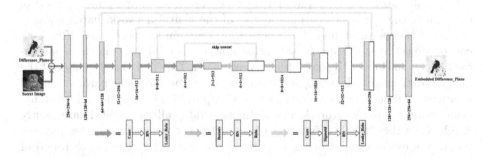

Fig. 2. Generator Model Based on U-Net

2.3 Steganalysis Adversarial Networks and Steganalysis Optimization Networks

Steganalysis adversarial network is used to discriminate whether the input image hides secret information or not. With the development of deep learning, the combination of steganalysis and deep learning poses a great threat to steganography algorithms. In this study, XU-Net and SR-Net, which are commonly used in the field of steganalysis of images in the airspace, are used to jointly discriminate stego images and output the corresponding discriminative loss, so as to improve the steganalysis resistance of stego images. As shown in the steganalysis optimization network SAN in Fig. 1, the original carrier image Cover and its corresponding stego image are firstly used as inputs to the steganalyzer network, and the corresponding discriminative loss SD_loss is generated, and the joint mean square error loss and the structural similarity loss are used as the generator's total loss to carry out the iterative optimization, so as to ensure that the resulting stego image has a high visual quality again and also has a high resistance to steganography. Also has a high resistance to steganalysis.

The steganalysis optimization network SON then uses the output SD_loss of the SAN network for the iterative optimization of the steganalyzer network itself, as shown in Fig. 1. The SD_loss output from the SAN network is used to synchronously update the network parameters of XU-Net and SR-Net, which improves the steganalyzer's detection capability and further improves the stego image's steganographic detection resistance.

2.4 Extractor Structure

The extractor network is used to extract the embedded extracted image from the stego image. We have designed the extractor network for extracting the secret image by using a Convolutional Neural Network (CNN) with the structure of the encoding and decoding network for the extraction of the secret image, which guarantees accurate and efficient recovery of the secret image from the stego image while controlling a smaller number of network parameters. This decoder network all uses 3×3 convolutional layers with step size 1 and padding 1. Also, to enhance the nonlinear learning ability of the neural network, batch normalization (BN) operation and ReLU activation function are performed after each convolutional layer. A sigmoid activation function is used after the last convolutional layer to extract the embedded secret image. The high visual quality of the image is extracted by learning nonlinear features to better fit the parameters and to achieve the mapping between input and output. The two pixel-level loss functions of mean square error (MSE) and structural similarity (SSIM) are jointly weighted as the total loss function of the extractor network, and the parameters of the extractor network are progressively updated through each round of iterative optimization to generate extracted images with higher visual quality.

2.5 Loss Function

Generative adversarial networks are powerful but sometimes suffer from slow convergence and unstable training. The gradient of the generator in the generative adversarial network is generated by the discriminator through the loss function, which calculates the degree of consistency between the generated image and the original image, and provides the gradient for iterative optimization of the generator. The convergence speed and stability of the generative adversarial network, as well as the quality of the generated image, are largely determined by the loss function. Designing a loss function with good performance is a prerequisite to ensure the stable convergence of the generative adversarial network and the generation of high-quality images. Therefore, in this study, we dynamically superimpose the mean square error loss MSE_loss, the structural similarity loss SSIM_loss, and the steganalysis loss SD_loss as the total loss of the steganographic network generator, to ensure that the generative adversarial network converges quickly and stably and generates stego images with higher visual quality and higher steganalysis resistance. The total loss function of the generator in the training phase is:

$$loss = MSE\_loss + SSIM\_loss + \alpha L\_D \tag{1}$$

where MSE_loss is the sum of MSE_loss of stego image and cover image, extract image and secret image. SSIM_loss is the sum of SSIM_loss of stego image and cover image, extract image and secret image. L_D is the adversarial loss generated by adversarial networks. The value of α will be discussed later.

Generator confronts the steganalyzer SD and the steganalysis adversarial network SAN. The generator tries to confuse the discriminator so that it outputs wrong judgments. In order to reduce the distance between the carrier image and the stego image, a pixel-by-pixel computation with pixel-level mean square error loss (MSE_loss) is used. The mean square error of the stego image and the original carrier image Cover is calculated, and the MSE_Sloss between the carrier image and the stego image is obtained by calculating the mean square error loss. The formula used to calculate the mean squared error loss is given by:

$$MSE\_loss = \sum_{i=1}^{n} \frac{\left(x_i - x_i'\right)^2}{n} \tag{2}$$

Where x_i is the value of the pixel point of the original image, x_i' is the value of the pixel point of the target image, and n is the number of image pixel points. Meanwhile, in order to further improve the structural similarity of the generated images, we designed the structural similarity loss (SSIM_loss) in a targeted way while using MSE and used it on both the carrier image and stego image, secret image, and extracted image. The structural similarity loss is calculated as:

$$SSIM\_loss = \frac{(2\mu_x\mu_{x'} + C_1)(2\sigma_{xx'} + C_2)}{(\mu_x{}^2 + \mu_{x'}{}^2 + C_1)(\sigma_x{}^2 + \sigma_{x'}{}^2 + C_2)} \tag{3}$$

$\mu_x, \mu_{x'}$ and $\sigma_x, \sigma_{x'}$ denote the mean and variance of the original image and the target image, respectively. The value of C_i (i = 1, 2, 3, 4) is the square of the product of the coefficients much less than 1 and the dynamic range of the image grayscale.

It is also necessary to discriminate the original carrier image from the generated stego image by the steganalyzer. When the steganalyzer replaces the function of the discriminator, its training goal is: when the input is a stego image, the expected output probability is close to 0; when the input is the original carrier image, the expected output probability is close to 1. In this experiment, we combine the classical null-domain steganalyser, XU-Net, with the SR-Net in a joint-weighted adversarial, so as to further improve the anti-steganography analysis ability of the generated stego image. Its loss function can be expressed as:

$$L_{SD} = -\sum_{i=1}^{2} x_i' \log(x_i) \tag{4}$$

where x_1, x_2 is the output of the steganalyzer finally passing through the softmax layer, which is the probability of the original image and the stego image, respectively, and $x_{1'}$, $x_{2'}$ are the labels corresponding to the input original carrier image and the generated stego image, respectively.

3   Experimental Component

3.1   Experimental Setup

The dataset selected for this experiment is the ImageNet dataset, which is divided into a training set and a test set. The training set is divided into two parts, trainA and trainB, each containing 35000 RGB color natural images of 256 × 256 size, and the test set test contains 10000 RGB color natural images of 256 × 256 size. This experiment uses the Lion optimizer newly proposed by Google-Brain to optimize the training of the model during the model training process. The Lion optimizer can take into account the effectiveness and efficiency of the training, thanks to the fact that the Lion optimizer only tracks the momentum and therefore saves more memory than Adam. In the training phase, we set the inputs of the model as 16 carrier images and 16 secret images and generate the corresponding stego images through iterative optimization. After completing the training phase, the trained model is built on a test set of 10,000 natural images in ImageNet for model performance evaluation.

For this experiment, Dell T630, 128 GB RAM, RTX3090 GPU with 24 GB display memory, python 3.7, and Tensorflow 2.4.0 were used for hardware.

3.2   Discussion of Experimental Results

In the experiment, as the number of model iterations increases, the stego images generated by the steganographic network through the generator are more and more visually similar to the original carrier images. Figure 3 shows the frequency histogram comparison of three randomly selected original images and their corresponding stego images. From the comparison, it can be seen that the histogram distribution of the original carrier image and the stego image after embedding the secret image are basically the same.

Figure 4 shows the distribution of PSNR and SSIM for generating stego images. Where the horizontal coordinate is the index of the number of images, and the vertical coordinate is the calculated PSNR or SSIM value, a total of 200 images were selected for testing in the test experiment design. From the test results, the average PSNR of 200 stego images is 38.78 dB; of which 25.2% of the images have PSNR values greater than 40 dB, 71.3% of the images have PSNR greater than 38 dB, and 97.1% of the images have PSNR greater than 35 dB; on the other hand, the average SSIM value of the stego images is greater than 0.96, of which 23.1% of the images have SSIM value greater than 0.98, 61.2% of the images had SSIM values greater than 0.96, and 96.2% of the images had SSIM values greater than 0.9

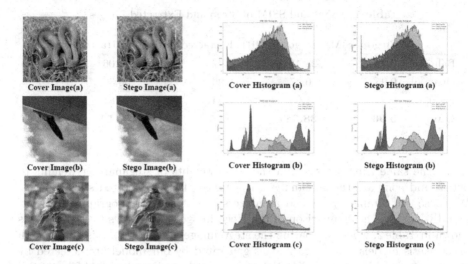

Fig. 3. Frequency Histograms of the Original Cover Image and Its Corresponding Stego Image

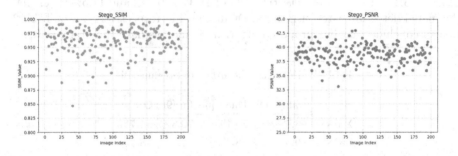

Fig. 4. SSIM and PSNR Scatterplot Distribution of Stego Images

3.3   Comparison of Experimental Results

In this section, we compare the experiments of Balujia, Duan, and Fu. As shown in Table 1, We mainly compare the visual metrics (PSNR, SSIM) and the resistance to steganalysis of the stego images with the extracted images, and in making the comparison of stego images, the PSNR of the stego images generated by our model is 10.37 dB, 2.11 dB, and 7.73 dB higher than that of the stego images generated by the models of Balujia, Duan, and Fu, respectively. The SSIM of the stego image is 0.04, 0.01, and 0.02 higher than that of the Balujia, Duan, and Fu, respectively. From the data presented in Table 1, we can see that the stego image generated by our proposed algorithm has higher values in PSNR and SSIM metrics than the model we compared.

Table 1. PSNR and SSIM of Stego and Extracted Images

	Stego (SSIM)	Stego (PSNR)	Extracted (SSIM)	Extracted (PSNR)
Baluja [4]	0.92	28.41	0.92	28.06
Duan [6]	0.95	36.67	0.96	36.32
Fu [9]	0.94	31.05	0.94	29.83
Ours	**0.96**	**38.78**	**0.95**	**34.51**

At the same time, we selected the same weights of steganalyzer weights to detect and compare the steganalysis resistance of the generated stego images. We chose two commonly used steganalyzers in airspace steganography, XU-Net, and SR-Net, for the comparison of the stego image's steganalysis resistance. As shown in Table 2, when comparing the stego image anti-steganalysis ability with other thesis models, the stego images generated by our model are detected by the steganographic analyzer XU-Net at a lower detection rate of 44.2%, 30.4%, and 16.1% lower than those of the models of Balujia, Duan, and Fu, respectively. The stego images generated by our model are detected by the steganographic analyzer SR-Net at a lower detection rate of 41.4%, 30.3%, and 17.5% lower than the model detection rates of Balujia, Duan, and Fu, respectively. From Table 2, we can conclude that our proposed algorithmic models have the best results.

Table 2. Detection Accuracy of Steganalysis

	Baluja [4]	Duan [6]	Fu [9]	**Ours**
Xu'Net	98.9%	85.1%	70.8%	**54.7%**
SR'Net	99.5%	88.4%	75.6%	**58.1%**

4   Conclusion

In this paper, we proposed a Cross-channel Image steganography Model Based on Generative Adversarial Network. In the generator, the shallow features of the image are effectively fused with the deep features through the skip connection layer in the U-Net network structure, to reconstruct and generate high-quality images. The classical spatial steganalysis tools XU-Net and SR-Net are used as the discriminator to form a steganalysis adversarial network, which is used to carry out steganalysis on the stego images generated by the steganalysis network, and the anti-steganalysis ability of the generated stego image is improved through adversarial training. In the meantime, the mean square error loss (MSE_loss), structural similarity loss (SSIM_loss), and steganalysis adversarial loss generated by the steganalysis adversarial network (SD_loss) were used for weighted joint iteration as the total loss of the steganographic network generator, and the dynamic allocation of weight coefficients was carried out in the model training process. At the same time, the newly proposed Lion optimizer is used to optimize the appropriate learning rate and weight decay coefficient, to ensure that the network can converge quickly and smoothly. Experimental results show that the stego images generated by the proposed algorithm have better visual quality and higher anti-steganalysis ability than those generated by other mainstream image steganography methods.

References

1. Mielikainen, J.: LSB matching revisited. IEEE Signal Process. Lett. **13**(5), 285–287 (2006). https://doi.org/10.1109/LSP.2006.870357
2. Holub, V., Fridrich, J., Denemark, T.: Universal distortion function for steganography in an arbitrary domain. EURASIP J. Inf. Secur. **2014**(1), 1 (2014). https://doi.org/10.1186/1687-417X-2014-1
3. Holub, V., Fridrich, J.: Designing steganographic distortion using directional filters. In: 2012 IEEE International Workshop on Information Forensics and Security (WIFS), Costa Adeje - Tenerife, Spain: IEEE, Dec. 2012, pp. 234-239 (2022). https://doi.org/10.1109/WIFS.2012.6412655.
4. Baluja, S.: Hiding Images in Plain Sight: Deep Steganography
5. Liu, L., Meng, L., Wang, X., Peng, Y.: An image steganography scheme based on ResNet. Multimed. Tools Appl. **81**(27), 39803–39820 (2022). https://doi.org/10.1007/s11042-022-13206-2
6. Duan, X., Jia, K., Li, B., Guo, D., Zhang, E., Qin, C.: Reversible image steganography scheme based on a U-Net structure. IEEE Access **7**, 9314–9323 (2019). https://doi.org/10.1109/ACCESS.2019.2891247
7. Yang, J., Ruan, D., Huang, J., Kang, X., Shi, Y.-Q.: An embedding cost learning framework using GAN. IEEE Trans. Inf. Forensics Secur. **15**, 839–851 (2020). https://doi.org/10.1109/TIFS.2019.2922229
8. Duan, X., Gou, M., Liu, N., Wang, W., Qin, C.: High-capacity image steganography based on improved Xception. Sensors **20**(24), 7253 (2020). https://doi.org/10.3390/s20247253
9. Duan, X., Wang, W., Liu, N., Yue, D., Xie, Z., Qin, C.: StegoPNet: image steganography with generalization ability based on pyramid pooling module. IEEE Access **8**, 195253–195262 (2020). https://doi.org/10.1109/ACCESS.2020.3033895
10. Fu, Z., Wang, F., Cheng, X.: The secure steganography for hiding images via GAN. EURASIP J. Image Video Process. **2020**(1), 46 (2020). https://doi.org/10.1186/s13640-020-00534-2
11. Tang, W., Tan, S., Li, B., Huang, J.: Automatic steganographic distortion learning using a generative adversarial network. IEEE Signal Process. Lett. **24**(10), 1547–1551 (2017). https://doi.org/10.1109/LSP.2017.2745572
12. Xu, G., Wu, H.-Z., Shi, Y.-Q.: Structural design of convolutional neural networks for steganalysis. IEEE Signal Process. Lett. **23**(5), 708–712 (2016). https://doi.org/10.1109/LSP.2016.2548421
13. Li, Q., et al.: Image steganography based on style transfer and quaternion exponent moments. Appl. Soft Comput. **110**, 107618 (2021). https://doi.org/10.1016/j.asoc.2021.107618
14. Boroumand, M., Chen, M., Fridrich, J.: Deep residual network for steganalysis of digital images. IEEE Trans. Inf. Forensics Secur. **14**(5), 1181–1193 (2019). https://doi.org/10.1109/TIFS.2018.2871749

15. Zhou, L., Feng, G., Shen, L., Zhang, X.: On security enhancement of steganography via generative adversarial image. IEEE Signal Process. Lett. **27**, 166–170 (2020). https://doi.org/10.1109/LSP.2019.2963180

16. Fridrich, J., Kodovsky, J.: Rich models for steganalysis of digital images. IEEE Trans. Inf. Forensics Secur. **7**(3), 868–882 (2012). https://doi.org/10.1109/TIFS.2012.2190402

17. Ye, J., Ni, J., Yi, Y.: Deep learning hierarchical representations for image steganalysis. IEEE Trans. Inf. Forensics Secur. **12**(11), 2545–2557 (2017). https://doi.org/10.1109/TIFS.2017.2710946

18. Ma, B., Han, Z., Li, J., Wang, C., Wang, Y., Cui, X.: A high-capacity and high-security generative cover steganography algorithm. In: Sun, X., Zhang, X., Xia, Z., Bertino, E. (eds.) ICAIS 2022. CCIS, vol. 1588, pp. 411–424. Springer, Cham (2022). https://doi.org/10.1007/978-3-031-06764-8_32

A Reversible Data Hiding Algorithm for JPEG Image Based on Paillier Homomorphic Encryption

Chunxin Zhao[1,2] , Ruihe Ma[3] , and Yongjin Xian[1,2(✉)]

[1] Key Laboratory of Computing Power Network and Information Security, Ministry of Education, Shandong Computer Science Center (National Supercomputer Center in Jinan), Qilu University of Technology (Shandong Academy of Sciences), Jinan, China
matxyj@163.com
[2] Shandong Provincial Key Laboratory of Computer Networks, Shandong Fundamental Research Center for Computer Science, Jinan, China
[3] School of Economics, Jilin University, Changchun 130015, China

Abstract. With the rapid advancement of cloud computing, great security concern regarding data storage has been brougth forth. This paper proposes a reversible data hiding algorithm for JPEG images based on Paillier homomorphic encryption. The secret data is embedded by modifying quantized DCT coefficients. The AC coefficients are modified through homomorphic encryption during the embedding process, while the DC coefficients, which significantly impact image quality, are subjected only to Arnold transform. Additionally, an algorithm that can control the distribution range of ciphertext of Paillier homomorphic encryption is frist presented, thereby, minimizing the ciphertext expansion. Extensive experimental results demonstrate that the proposed algorithm achieves larger data embedding capacity, higher security of ciphertext images, and more stable image quality in comparison with other advanced algorithms.

Supported by National Natural Science Foundation of China (62272255, 62302248, 62302249); National key research and development program of China (2021YFC3340600, 2021YFC3340602); Taishan Scholar Program of Shandong (tsqn202306251); Shandong Provincial Natural Science Foundation (ZR2020MF054, ZR2023QF018, ZR2023QF032, ZR2022LZH011), Ability Improvement Project of Science and Technology SMES in Shandong Province (2022TSGC2485, 2023TSGC0217); Jinan "20 Universities"-Project of Jinan Research Leader Studio (2020GXRC056); Jinan "New 20 Universities"-Project of Introducing Innovation Team (202228016); Youth Innovation Team of Colleges and Universities in Shandong Province (2022KJ124);The "Chunhui Plan" Cooperative Scientific Research Project of Ministry of Education (HZKY20220482); Achievement transformation of science, education and production integration pilot project (2023CGZH-05), First Talent Research Project under Grant (2023RCKY131, 2023RCKY143), Integration Pilot Project of Science Education Industry under Grant (2023PX006, 2023PY060, 2023PX071).

B. Ma et al. (Eds.): IWDW 2023, LNCS 14511, pp. 207–222, 2024.
https://doi.org/10.1007/978-981-97-2585-4_15

Keywords: Reversible data hiding · Homomorphic encryption · JPEG image · Ciphertext range

1 Introduction

With the development of cloud platforms, cloud storage has become the main method of data storage. The public does not need to purchase special storage devices to store a large amount of personal data [1]. Instead, they choose cloud storage to store a large amount of personal data, which greatly saves the cost of storing data. However, at the same time, data uploaded to cloud platforms also carries incalculable risks, such as the leakage, tampering, and dissemination of personal privacy data [2]. In order to solve such problems, reversible data hiding in the ciphertext domain has emerged [3]. Cryptographic reversible data hiding refers to the process in which content owners encrypt the file data before uploading it to make it invisible for safer storage of data files [4]. Cloud platform managers, in order to facilitate the management and storage of encrypted multimedia data, embed additional data in the encrypted multimedia data, making it more convenient to manage cloud platform storage resources [5].

Most reversible data hiding methods in the ciphertext domain are applied to uncompressed images [6–8], which have more embedding space and larger embedding capacity, and have good performance in embedding additional data. Nowadays, the more popular one is the Joint Photographic Experts Group (JPEG) image, which compresses the image in a certain way to reduce the additional data embedded in the JPEG image [9]. In order to embed more additional data in the JPEG image, Huang et al. [10] proposed embedding data in the DCT coefficients of the JPEG image, using a histogram shift method to embed data with communication coefficient values of 1 and −1, And propose a block selection strategy based on zero coefficients to achieve high embedding capacity and visual quality on JPEG graphics. Li et al. [11] constructed a two-dimensional histogram of non zero AC coefficient pairs in JPEG images, forming a reversible two-dimensional map and modifying the AC coefficient pairs to embed data. Qian et al. [12] encrypted the JPEG bitstream, reducing the size of the JPEG image to meet the requirements of the JPEG encoder, and embedding more bit data using Huffman encoding mapping and histogram shifting. Reference [13] proposes an asymmetric and exchangeable elliptic curve separable RDHEI method, which can resist plaintext attacks (CPA) and ensure the security of encrypted images.

When content owners encrypt uploaded multimedia files, most of them use symmetric encryption or stream encryption. The encryption method is too simple, which makes it easy for attackers to crack the encrypted content and obtain private data uploaded by users [14]. At the same time, these traditional encryption methods cannot modify the encrypted data and embed additional data after encryption [15]. Therefore, experts have started to adopt asymmetric encryption, For example, homomorphic encryption algorithms encrypt data [16]. The advantage of homomorphic encryption algorithms is that they are not easily cracked and can operate on the encrypted ciphertext without affecting the decryption result [17]. Reference [18] adopts homomorphic encryption technology, which embeds additional data into encrypted images by moving the histogram of absolute difference, achieving high-capacity data hiding. Reference [19] applied homomorphic encryption to three types of RDH: difference extension, histogram shift, and pixel value sorting, which improved the security of encryption. However, its embedding data method is relatively simple, and the embedding performance needs to be improved. Ke et al. [20] used a fully homomorphic encryption encapsulation differential expansion method to hide data in the least significant bit of key conversion, achieving fully homomorphic encryption and extracting data in the encryption domain without the need for a key. Reference [21] proposes the method of absolute moment block truncation coding (AMBTC) combined with homomorphic encryption, which can embed variable-size additional data and avoid overflow issues. Zhou et al. [22] implemented a separable RDH method based on the combination of NTRU and homomorphic encryption, but the security and embedding performance still need to be improved.

Therefore, this article proposes a scheme for reversible data hiding in image ciphertext based on the Paillier homomorphic encryption algorithm.

(1) The proposal suggests utilizing the Paillier homomorphic encryption algorithm to encrypt JPEG images and embed secret data during image encryption process. Specifically, it only applies homomorphic encryption to the suitable quantized AC coefficients of JPEG images for secret data embedding.

(2) A method that can control the range of generated ciphertext values from Paillier homomorphic encryption is frist proposed in this paper, thereby preventing excessively large ciphertext values that would occupy excessive storage space. This approach ensures that the ciphertext values of homomorphic encryption fall within a certain range and avoids an overly wide numerical range of ciphertext values. As a result, it enables a uniform distribution of ciphertexts after homomorphic encryption and enhances data security.

(3) During the process of encrypting JPEG images, separate data encryption algorithms are applied to DCT coefficient. Only Arnold transform is employed for DC coefficients while AC coefficients undergo homomorphic encryption for secret data embedding. Consequently, The proposed scheme not only enhances the security of encryption but also enables more sexret data embedding and increases data embedding capacity.

2   Related Work

2.1   Paillier Homomorphic Encryption

The paillier homomorphic encryption algorithm involves the inverse of modular multiplication and the extended euclidean algorithm, which plays a crucial role in encryption, decryption, and computation.

Module multiplication inverse element: two coprime integers x and n, with one integer present φ meet:

$$\varphi x \equiv 1 \ (\bmod\ n) \tag{1}$$

From the formula, it can be seen that φx divided by n is 1, and the necessary and sufficient conditions for the existence of a multiplication inverse are φ interprime with n. Namely, the maximum common divisor $\gcd(\varphi, n) = 1$, where x is called φ the inverse of multiplication can be written as φ^{-1}. And φ^{-1} can be obtained using the extended euclidean algorithm.

$$ax + by = m \tag{2}$$

Provide the values of a, b, and m, and solve for the values of x and y. The linear equation only has a solution when $m \bmod (\gcd(a, b)) == 0$, that is, when and only when m is a multiple of the greatest common divisor of both a and b. When m is 1, only a and b are mutually prime to have a solution.

Key Generation. Randomly generate two large prime numbers p and q, where needs to be satisfied is that the maximum common divisor of $p \times q$ and $(p-1) \times (q-1)$ is 1. Namely:

$$\gcd\big(p \cdot q, (p-1) \cdot (q-1)\big) = 1 \tag{3}$$

Generate public keys n and g, where n is the product of p and q, randomly select the integer g, and g needs to satisfy $g \in Z_{n^2}^*$, where the value of g is between $[0, n^2)$ and the maximum common divisor $\gcd(g, n^2) = 1$, g is coprime with n^2. To improve the computational efficiency of homomorphic encryption, $g = n + 1$, $g = n + 1$ satisfies the above conditions and is coprime with n^2. The public key is (n, g).

To generate the private keys λ and μ, calculate λ first:

$$\lambda = lcm((p-1), (q-1)) \tag{4}$$

lcm is the calculated minimum common multiple.

Construct the L function as follows:

$$L(x) = \frac{(x-1)}{n} \tag{5}$$

Calculation coefficient μ as follows:

$$\mu = L(g^{\lambda(n)} \bmod n^2)^{-1} \bmod n \tag{6}$$

mod is a modular operation and the remainder is obtained. The private key is (λ, μ).

Data Encryption. Assuming that the plaintext that the data encryption party needs to encrypt is m and receives the public key (n, g) sent by the user in the cloud. The data encryption party randomly generates a random number r, where r satisfies $0 \leq r < n$ and $r \in Z_n^*$, that is, r and n are mutually prime.

Calculate ciphertext:

$$c = g^m r^n mod\ n^2 \tag{7}$$

Data Decryption. Based on the ciphertext c obtained from the cloud and the private key (λ, μ), perform ciphertext decryption. According to the following formula, clear text m can be obtained.

$$m = L(c^\lambda\ mod\ n^2)\ \mu\ mod\ n \tag{8}$$

Additive Homomorphism. Firstly, encrypt and generate ciphertext for two data types, c_1 and c_2. Among $c_1 = g^{m_1} r^n\ mod\ n^2$, $c_2 = g^{m_2} r^n\ mod\ n^2$.

According to the formula:

$$c = c_1\ c_2\ mod\ n^2 \tag{9}$$

The property of paillier addition homomorphic encryption is that for two ciphertexts c_1 and c_2, multiplication in the ciphertext domain is equal to addition in the plaintext domain. Namely:

$$\begin{aligned} c &= (g^{m_1} r^n\ mod\ n^2)(g^{m_2} r^n\ mod\ n^2) mod\ n^2 \\ &= (g^{m_1} g^{m_2} r^n r^n) mod\ n^2 \\ &= g^{m_1+m_2} r^{2n} mod\ n^2 \\ &= g^{m_1+m_2} (r^2)^n mod\ n^2 \end{aligned} \tag{10}$$

Among them, $m_1 + m_2$ is equivalent to m in Eq. 7, that is, the plaintext result obtained by decrypting c is $m_1 + m_2$.

3   A Reversible Data Hiding Algorithm Based on Paillier Homomorphic Encryption in Image Ciphertext Domain

To ensure the security of the media data of the supercomputing cloud platform and ensure the safe and reliable storage and forwarding of data, before uploading the image data, the user will encrypt the image first to prevent others from seeing the uploaded data. To facilitate the management of the data on the cloud platform, the administrator of the supercomputing cloud platform will embed additional data in the encrypted image data of the user. When the user needs to download the original image, the system extracts the embedded data and recovers the original image data, to realize the safe and reliable storage and forwarding of data. Therefore, a JPEG image reversible data hiding algorithm in encrypted domain based on quantized DCT coefficient operation was proposed.

3.1   Algorithm Theoretical Framework

In this paper, paillier homomorphic encryption technology is used to realize reversible data hiding in JPEG images. Using the property of paillier homomorphic encryption: ciphertext multiplication is equal to plaintext addition, in JPEG images, using the RDH method of quantizing DCT coefficients, the secret data is encrypted while embedding. At the same time, when the AC component of the DCT coefficient is encrypted, the distribution range of the generated ciphertext value is controlled and the size of the ciphertext value is controlled. At the same time, the ciphertext image is formed and stored in the cloud platform.

As shown in Fig. 1, in the homomorphic encryption embedding process, the original JPEG image is first processed, the JPEG image is partitioned and quantized, the quantized DCT coefficient blocks are processed, and the DCT coefficients are encrypted respectively. Arnold transform is used for DC coefficient, and the AC coefficient is encrypted by homomorphic encryption. While the AC coefficient is paillier homomorphic encrypted, the ciphertext λ_0 and λ_1 of the binary data 0 or 1 are embedded to form the ciphertext DCT coefficient. In the process of homomorphic encryption, a method is used to control the range of the ciphertext value of the AC coefficient. The range of the ciphertext value of the AC coefficient containing the embedded data is controlled to be range (x_1, x_2), and the range of the ciphertext value of the other AC coefficient is controlled to be range (x_3, x_4), forming the ciphertext image.

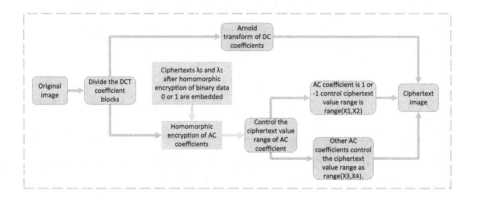

Fig. 1. Homomorphic encryption and embedding graph.

As shown in Fig. 2, for image restoration and extraction of embedded secret data, it is a separable operation. The first method can directly decrypt the ciphertext image first, and then extract the embedded secret data after decryption. The second method is to extract the secret data from the ciphertext image first, and then decrypt the ciphertext image to restore the original image. For both methods, the embedded secret data can be extracted and the original ciphertext image can be restored. the first method is to obtain the key of homomorphic encryption to decrypt the ciphertext image and form plaintext. By using the property of homomorphic encryption: ciphertext multiplication is equal to plaintext addition, the plaintext data are analyzed to extract the embedded secret data and restore the original image. The second method is to find the ciphertext value range of the embedded secret data according to the distribution range of the ciphertext value controlled by the previous homomorphic encryption, and calculate the inverse element of the modular multiplication according to the embedded key λ_0 and λ_1, and multiply it with the ciphertext to analyze and extract the embedded secret data. The ciphertext image is decrypted according to the encryption key, and then the original image is restored.

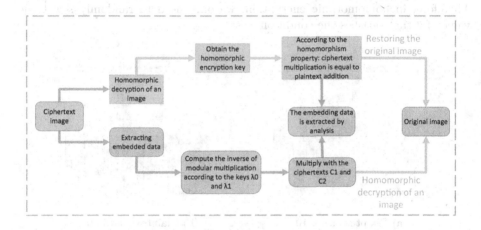

Fig. 2. Homomorphic decryption and extraction graph.

3.2   Control the Distribution Range of Ciphertext

According to the above paillier homomorphic encryption algorithm, it can be known that the parameters related to ciphertext generation are p, q, g, r, n, λ. According to the formula of homomorphic encryption, it is known that the size of ciphertext data generation is controlled by the parameters p, q, r, and other parameters are related to it. Where r is randomly generated data, r satisfies $0 \leq r < n$, and $r \in Z_n^*$, that is, r and n are coprime. The large primes p and q are the most important parameters that control the size range of ciphertext values. Figure 3 show that after fixing the values of p and q, no matter how the r value is randomly generated, the size range of the ciphertext value is determined after encrypting the plaintext. For example, in the figure, $p = 11$, $q = 19$, and the range of ciphertext values does not exceed 4.5×10^4. It can filter the qualified ciphertext values, control the range of ciphertext values, and ensure the uniform distribution of the histogram of ciphertext values after encrypting plaintext data, which ensures the security of data and is not easy to be found by monitors.

Influence of Parameter r on Ciphertext Value Range in Homomorphic Encryption. In the process of Paillier homomorphic encryption, according to the formula Eq. 7 When the plaintext is encrypted, the parameter r will affect the range of the ciphertext value generation. The influence of parameter r on the ciphertext value is observed by controlling the influence of other parameters on the ciphertext value when the plaintext is encrypted. Where $n = p \times q$, and $g = n + 1$, the parameters associated with the two are p and q. By controlling p and q as fixed values, the influence of the parameter r on the ciphertext value is observed. As can be seen in Fig. 3, fixing p as 11, q as 19, encrypting different plaintext data, and randomly generating the value of parameter r from 10000, it can be seen that no matter how r changes, the ciphertext value range is up to 4.5×10^4, and the choice of r value will not affect the ciphertext value range. Therefore, in homomorphic encryption, we only need to randomly select the value of r that satisfies the condition.

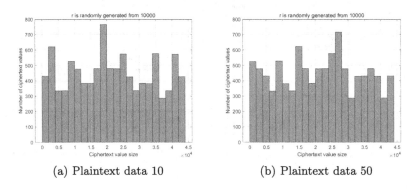

(a) Plaintext data 10   (b) Plaintext data 50

Fig. 3. Encrypt different plaintext data (p=11, q=19)

Influence of Parameters p and q on the Range of Ciphertext Values in Homomorphic Encryption. According to the formula of paillier homomorphic encryption, the determination of parameters in the encryption process is related to large prime numbers p and q. By keeping other parameters fixed, the influence of p and q on the ciphertext value range is studied. As can be seen in Fig. 4, when Paillier homomorphic encryption is carried out on plaintext data 1, r is randomly generated from 10000. When $p = 17$ and $q = 19$, the ciphertext value range is up to 1.05×10^5. When $p = 23$ and $q = 29$, the ciphertext value range is up to 4.5×10^5. At the same time, we find that when the values of parameters p and q are larger and larger, the range of ciphertext value is larger and larger. Therefore, p and q can be controlled within a certain data range, so that the range of ciphertext value is not too large, resulting in too much data and saving storage space.

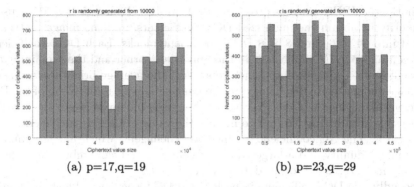

(a) p=17,q=19   (b) p=23,q=29

Fig. 4. Encrypt the plaintext with different parameters

In this paper, the process of controlling the size of the ciphertext range is selected as follows:

Firstly, the range of ciphertext value is controlled between 1.0×10^4–4.5×10^4 to ensure that the encrypted ciphertext value data are all 5-digit. For 1 and -1 in the AC coefficient, the ciphertext value data after encrypting 1 and -1 and embedding secret information (binary 0 or 1) is controlled in the range of 1.0×10^4–2.0×10^4. For the other data in the AC coefficient, the ciphertext data is controlled between 2.0×10^4–4.5×10^4. The embedded secret messages 0 and 1 are encrypted by paillier homomorphic encryption, and the encrypted ciphertext values are recorded as λ_0 and λ_1. If for 1 encrypted ciphertext data values are a_1 and a_2 respectively, where a_1 and a_2 meet the range of ciphertext value data, for 1 embedded binary data 0, according to paillier homomorphic encryption, then:

$$C_1 = a_1 \times \lambda_0 \, mod \, N^2 \tag{11}$$

For 1 embedded binary data 1, according to paillier homomorphic encryption, that is:

$$C_2 = a_2 \times \lambda_1 \, mod \, N^2 \tag{12}$$

3.3 Paillier Homomorphic Encryption and Embedding Process

As shown in Fig. 1, the image homomorphic encryption and embedding process. Firstly, the original image is divided into 64×64 and 8×8 DCT coefficient blocks, and an 8×8 DCT coefficient block is selected to encrypt the DC coefficients respectively. The DC coefficients are used to Arnold transform, and the AC coefficients are encrypted by Paillier homomorphic encryption. Moreover, binary data 0 or 1 is embedded at the encrypted AC coefficients 1 and -1, and the ciphertext range of AC coefficients is controlled to facilitate the subsequent accurate extraction of secret information and image recovery. The encrypted DCT coefficient block is recovered and the ciphertext image is formed, which is described in detail in the following five steps.

Firstly, the JPEG image is processed, and the JPEG image 512×512 is used to quantify it, obtain the quantized DCT coefficients, and the image matrix is divided into 64×64 and 8×8 DCT coefficient blocks. Each DCT coefficient block contains one DC coefficient in the upper left corner and the remaining 63 AC coefficients, DC coefficient which represents the average brightness value of the image and has a greater impact on the image quality. Modification of the DC coefficient may cause loss and irreversible change to the image, so in the process of reversible data hiding, try not to modify the value of the DC coefficient, only transform it. The AC coefficient is called AC coefficient, which has less impact on the image quality and a larger range of changes, and can embed more data, including most zeros and other data that tend to 0.

Secondly, the DC coefficients in each 8×8 DCT coefficient block are transform by Arnold to ensure data security. Then, 1 and -1 of the AC coefficients are selected for Paillier homomorphic encryption and the secret information is embedded at the same time, and the rest AC coefficients are only encrypted by Paillier homomorphic encryption.

In the encryption process of AC coefficient 1 and -1, the range of ciphertext value distribution is allocated according to the proportion of AC coefficient value 1 and -1 in each 8×8 DCT coefficient block with other non-zero values, so that the histogram distribution of encrypted data value is more uniform after image encryption. It ensures that the encrypted ciphertext image is not easy to be detected by attackers. Keep the information secure.

For the embedded secret data, only binary data 0 or 1 is selected for embedding, and paillier homomorphic encryption is performed on 0 and 1 respectively. The encrypted ciphertext of 0 and 1 is selected and recorded, denoted as λ_0 and λ_1 respectively, which is used as the key in the subsequent extraction of secret information.

3.4 Image Decryption and Extraction Process

When managing ciphertext image data, the administrator of the supercomputing cloud platform needs to view the flag information previously embedded in the ciphertext image for classification management of ciphertext image data. The administrator of the supercomputing cloud platform can directly extract the secret information from the ciphertext image without decrypting the ciphertext image. After obtaining the ciphertext image uploaded by the user from the supercomputing cloud platform, the user can decrypt the original image and parse out the secret data.

As shown in the process of homomorphic decryption extraction in Fig. 2, the separable reversible data hiding algorithm is proposed in this paper, that is, extracting secret data and decrypting the original image can be carried out separately. The original image can be completely recovered and the true homomorphic invertibility can be achieved.

Without knowing the decryption key, cloud platform managers can use the embedded keys λ_0 and λ_1 to extract the embedded binary data, calculate the inverse modular multiplication of the embedded key and multiply it with the ciphertext value to calculate and analyze the embedded binary data. And restore the original image. The user can directly use the decryption key to homomorphic decrypt the ciphertext image into the original image containing the embedded data. According to the special properties of homomorphic encryption, the secret data in the embedded image can be extracted and the original image can be recovered.

4 Experimental Results and Analysis

4.1 Paillier Homomorphic Encryption Parameter Analysis

For the method of controlling the distribution range of ciphertext proposed in Sect. 3, we focus on the influence of the change of parameters in Paillier homomorphic encryption on the range of ciphertext value, among which, there are three parameters related to the size of ciphertext value data, namely p, q and r parameters. In order not to make the parameters affect each other and cause the change of ciphertext value size, the changes of the three parameters on the size of the generated ciphertext value will be discussed separately.

For r parameter, in the ciphertext generation formula Eq. 7, r is generated randomly, only $0 \leq r < n$, and $r \in Z_n^*$, that is, r and n are prime. For the parameters p and q, are randomly generated prime numbers satisfying the formula Eq. 3, that is, the product of p and q and the product of $p - 1$ and $q - 1$ have the greatest common divisor of 1.

As can be seen from Table 1, this experiment screens the ciphertext value of AC coefficient encrypted by Paillier homomorphic encryption. When the values of three groups of p and q are given, r is randomly generated from 1000 to 2000. It can be seen that no matter how r changes, the minimum and maximum values of AC coefficient encryption do not change. And it can be found from

Table 1. Parameter r affects the change of ciphertext value size.

Parameter p and q values (p and q are randomly generated prime numbers)	The value of the parameter r	The minimum ciphertext value after the AC coefficient is encrypted	The maximum ciphertext value after the AC coefficient is encrypted
p = 11 q = 19	r = 1000	10132	43577
	r = 1500	10132	43577
	r = 2000	10132	43577
p = 13 q = 23	r = 1000	10094	89101
	r = 1500	10094	89101
	r = 2000	10094	89101
p = 23 q = 29	r = 1000	11106	444221
	r = 1500	11106	444221
	r = 2000	11106	444221

the values of three groups of p and q that with the increase of p and q values, the value range of the maximum ciphertext value of AC coefficient encryption greatly increases.

4.2   Algorithm Performance Indicators

Embedding Capacity. In the image ciphertext domain reversible data hiding, most use uncompressed image, the image encryption embedded secret data, because the image has not been compressed, so the image embedding capacity is larger, the embedded secret data is more. And in order to improve the embedding capacity in JPEG image, and embed the secret data, it is necessary to improve the encryption algorithm, this paper proposed a method to control the distribution of ciphertext, greatly reduces the embedding secret data caused by the large value of the ciphertext, improve the embedding capacity.

Table 2. Embedding capacity of four images under different QF.

Image	Method	QF = 50	QF = 60	QF = 70	QF = 80	QF = 90
Baboon	Ref. [23]	3775	4472	5311	6344	6520
	Ref. [24]+ Ref. [25] (HuaEnc)	4387	5239	6148	7153	8589
	Ref. [25]	6581	7433	8888	11315	13477
	Proposed	32860	36210	40650	48020	60780
Couple	Ref. [23]	1184	1601	2347	3107	4772
	Ref. [24]+ Ref. [25] (HuaEnc)	1653	2032	2819	4237	6568
	Ref. [25]	1931	2740	3950	5737	9327
	Proposed	21080	23570	27150	32420	43740
Boat	Ref. [23]	1728	2173	2726	4332	6913
	Ref. [24]+ Ref. [25] (HuaEnc)	2540	3043	3832	5368	7660
	Ref. [25]	3032	3736	4963	7710	11782
	Proposed	19678	21930	25340	31500	46080
Lena	Ref. [23]	889	1171	1843	2762	5660
	Ref. [24]+ Ref. [25] (HuaEnc)	1577	1975	2679	4134	7082
	Ref. [25]	1973	2588	3966	5993	9997
	Proposed	14318	16616	20170	25980	38590

Firstly, the embedding capacity of the following four images under different QF(Quality Factor) between QF = 50 and QF = 90 is measured, as shown in the table. The algorithm proposed in this paper is compared with literature [23–25] (HuaEnc) and [25]. The largest number of AC coefficients 1 and −1 are encrypted by paillier homomorphic encryption and embedded data. Table 2 shows that the image embedding capacity of the proposed algorithm is the maximum value from QF = 50 to QF = 90, and the embedding capacity value is more than 10000. With the increase of the QF, the embedding capacity increases. This indicates that the proposed algorithm has good embedding performance.

PSNR. Peak signal noise ratio (PSNR) is one of the indicators to measure the quality of the image. The larger the PSNR value is, the more similar the restored image is to the original image, and the closer it is to the original image. For the encrypted image using Paillier homomorphic encryption and embedded data, the PSNR is used to measure the difference between the encrypted image and the original image. The smaller the PSNR value is, the better it is. This shows that the encrypted image has a smaller similarity with the original image, and the security of the encrypted image is better guaranteed.

As shown in the following four line charts, the PSNR of the four images under different QF and payloads. It can be found that with the increase of the quality factor QF, the PSNR value is also increasing, and the encrypted image is more similar to the original image. With the increase of the embedded payload, the PSNR value of the image is decreasing, and the encrypted image is more similar to the original image. The PSNR value of the encrypted image is below 17.5, and the similarity between the encrypted image and the original image is small, and it cannot be distinguished. With the increase of embedded data, the PSNR changes little, which also shows the stability of the quality of the ciphertext image, which is not easy to be analyzed by the monitor, and ensures the security of the ciphertext image (Fig. 5).

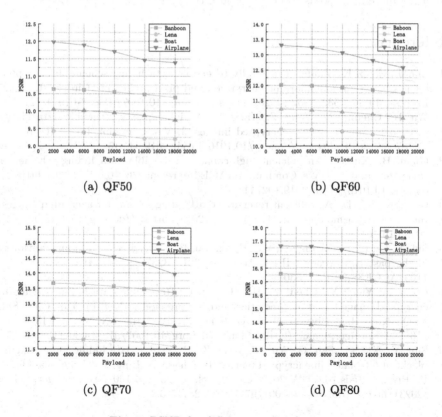

(a) QF50 (b) QF60

(c) QF70 (d) QF80

Fig. 5. PSNR for different quality factors

5   Conclusion

This paper proposes a reversible data hiding method for JPEG images based on paillier homomorphic encryption. After experimental analysis and verification, the algorithm has certain advantages compared with other reversible data hiding methods for JPEG images in encryption domain. The main contributions of this paper are as follows:

The use of Paillier homomorphic encryption algorithm in JPEG images is proposed to realize reversible data hiding. It can not only encrypt the image by using the Paillier homomorphic encryption algorithm in asymmetric encryption, fully guarantee the security of the image, but also maintain a high embedding capacity in JPEG compressed images. A method to control the ciphertext range of paillier homomorphic encryption algorithm is proposed, which can reduce the length of the ciphertext, and by encrypting the ciphertext range embedded with secret data, the ciphertext image of the encrypted embedded data can be restored losslessly, and the additional data embedded can be extracted, which can realize true homomorphic reversibility.

In the future, the ciphertext generation range method of paillier homomorphic encryption can be optimized to achieve higher embedding capacity and lower file increment, so as to better ensure the security of digital images.

References

1. Liu, J., Zhao, K., Zhang, R.: A fully reversible data hiding scheme in encrypted images based on homomorphic encryption and pixel prediction. Circ. Syst. Signal Process. **39**(7), 3532–3552 (2020). https://doi.org/10.1007/s00034-019-01321-9
2. Wu, H., Cheung, Y., Yang, Z., Tang, S.: A high-capacity reversible data hiding method for homomorphic encrypted images. J. Vis. Commun. Image Represent. **62**, 87–96 (2019). https://doi.org/10.1016/j.jvcir.2019.04.015
3. Guan, B., Xu, D.: An efficient high-capacity reversible data hiding scheme for encrypted images. J. Vis. Commun. IMAGE Represent. **66**, 102744 (2020). https://doi.org/10.1016/j.jvcir.2019.102744
4. Chen, K., Xu, D.: An efficient reversible data hiding scheme for encrypted images. Int. J. Digit. Crime Forensics **10**(2), 1–22 (2018). https://doi.org/10.4018/IJDCF.2018040101
5. Shi, Y.-Q., Li, X., Zhang, X., Wu, H.-T., Ma, B.: Reversible data hiding: advances in the past two decades. IEEE Access **4**, 3210–3237 (2016). https://doi.org/10.1109/ACCESS.2016.2573308
6. Chen, F., Yuan, Y., Chen, Y., He, H., Qu, L.: Reversible data hiding scheme in encrypted-image based on prediction and compression coding. In: Yoo, C.D., Shi, Y.-Q., Kim, H.J., Piva, A., Kim, G. (eds.) IWDW 2018. LNCS, vol. 11378, pp. 216–229. Springer, Cham (2019). https://doi.org/10.1007/978-3-030-11389-6_17
7. Sun, Y., Xie, G., Fan, G., Yu, C., Tang, Z.: Reversible data hiding via arranging blocks of bit-planes in encrypted images. In: Zhao, X., Tang, Z., Comesaña-Alfaro, P., Piva, A. (eds.) IWDW 2022. LNCS, vol. 13825, pp. 190–204. Springer, Cham (2023). https://doi.org/10.1007/978-3-031-25115-3_13

8. Song, C., Zhang, Y., Lu, G.: Reversible data hiding in encrypted images based on image partition and spatial correlation. In: Yoo, C.D., Shi, Y.-Q., Kim, H.J., Piva, A., Kim, G. (eds.) IWDW 2018. LNCS, vol. 11378, pp. 180–194. Springer, Cham (2019). https://doi.org/10.1007/978-3-030-11389-6_14

9. Xiong, L., Xu, Z., Shi, Y.: An integer wavelet transform based scheme for reversible data hiding in encrypted images. Multidimens. Syst. Signal Process. 29(3), 1191–1202 (2018). https://doi.org/10.1007/s11045-017-0497-5

10. Huang, F., Qu, X., Kim, H.J., Huang, J.: Reversible data hiding in JPEG images. IEEE Trans. Circuits Syst. Video Technol. 26(9), 1610–1621 (2016). https://doi.org/10.1109/TCSVT.2015.2473235

11. Li, N., Huang, F.: Reversible data hiding for JPEG images based on pairwise nonzero AC coefficient expansion. Signal Process. 171, 107476 (2020). https://doi.org/10.1016/j.sigpro.2020.107476

12. Qian, Z., Xu, H., Luo, X., Zhang, X.: New framework of reversible data hiding in encrypted JPEG bitstreams. IEEE Trans. Circuits Syst. Video Technol. 29(2), 351–362 (2019). https://doi.org/10.1109/TCSVT.2018.2797897

13. Sheidani, S., Mahmoudi-Aznaveh, A., Eslami, Z.: CPA-secure privacy-preserving reversible data hiding for JPEG images. IEEE Trans. Inf. Forensics Secur. 16, 3647–3661 (2021). https://doi.org/10.1109/TIFS.2021.3080497

14. Anushiadevi, R., Amirtharajan, R.: Design and development of reversible data hiding- homomorphic encryption rhombus pattern prediction approach. Multimed. Tools Appl. 82, 46269–46292 (2023). https://doi.org/10.1007/s11042-023-15455-1

15. Xiang, S., Luo, X.: Efficient reversible data hiding in encrypted image with public key cryptosystem. EURASIP J. Adv. Signal Process. 2017, 1–13 (2017). https://doi.org/10.1186/s13634-017-0496-6

16. Jiang, C., Pang, Y.: Encrypted images-based reversible data hiding in Paillier cryptosystem. Multimed. Tools Appl. 79(1), 693–711 (2020). https://doi.org/10.1007/s11042-019-07874-w

17. Zhang, X., Long, J., Wang, Z., Cheng, H.: Lossless and reversible data hiding in encrypted images with public-key cryptography. IEEE Trans. Circuits Syst. Video Technol. 26(9), 1622–1631 (2016). https://doi.org/10.1109/TCSVT.2015.2433194

18. Xiong, L., Dong, D., Xia, Z., Chen, X.: High-capacity reversible data hiding for encrypted multimedia data with somewhat homomorphic encryption. IEEE Access 6, 60635–60644 (2018). https://doi.org/10.1109/ACCESS.2018.2876036

19. Anushiadevi, R., Praveenkumar, P., Rayappan, J.B.B., Amirtharajan, R.: Reversible data hiding method based on pixel expansion and homomorphic encryption. J. Intell. Fuzzy Syst. 39(3), 2977–2990 (2020). https://doi.org/10.3233/JIFS-191478

20. Ke, Y., Zhang, M.-Q., Liu, J., Su, T.-T., Yang, X.-Y.: Fully homomorphic encryption encapsulated difference expansion for reversible data hiding in encrypted domain. IEEE Trans. Circuits Syst. Video Technol. 30(8), 2353–2365 (2020). https://doi.org/10.1109/TCSVT.2019.2963393

21. Bhardwaj, R.: A high payload reversible data hiding algorithm for homomorphic encrypted absolute moment block truncation coding compressed images. Multimed. Tools Appl. 80(17), 26161–26179 (2021). https://doi.org/10.1007/s11042-021-10722-5

22. Zhou, N., Zhang, M., Wang, H., Ke, Y., Di, F.: Separable reversible data hiding scheme in homomorphic encrypted domain based on NTRU. IEEE Access 8, 81412–81424 (2020). https://doi.org/10.1109/ACCESS.2020.2990903

23. He, J., Chen, J., Luo, W., Tang, S., Huang, J.: A novel high-capacity reversible data hiding scheme for encrypted JPEG bitstreams. IEEE Trans. Circ. Syst. Video Technol. **29**(12), 3501–3515 (2019). https://doi.org/10.1109/TCSVT.2018.2882850
24. Ong, S., Wong, K., Tanaka, K.: Scrambling-embedding for JPEG compressed image. Signal Process. **109**, 38–53 (2015)
25. Hua, Z., Wang, Z., Zheng, Y., Chen, Y., Li, Y.: Enabling large-capacity reversible data hiding over encrypted JPEG bitstreams. IEEE Trans. Circuits Syst. Video Technol. **33**(3), 1003–1018 (2023). https://doi.org/10.1109/TCSVT.2022.3208030

Convolutional Neural Network Prediction Error Algorithm Based on Block Classification Enhanced

Hongtao Duan[1,2], Ruihe Ma[4], Songkun Wang[1,2], Yongjin Xian[1,2(✉)],
Chunpeng Wang[1,2], and Guanxu Zhao[3]

[1] Key Laboratory of Computing Power Network and Information Security, Ministry
of Education, Shandong Computer Science Center (National Supercomputer Center
in Jinan), Qilu University of Technology (Shandong Academy of Sciences), Jinan,
China
matxyj@163.com
[2] Shandong Provincial Key Laboratory of Computer Networks, Shandong
Fundamental Research Center for Computer Science, Jinan, China
[3] Jinan Art School, Jinan 250002, China
[4] School of Economics, Jilin University, Changchun 130015, China

Abstract. Reversible data hiding techniques can effectively solve the
information security problem, and One crucial approach to enhance the
level of reversible data hiding is to predict images with higher accuracy, thereby effectively increasing the embedding capacity. However,
conventional prediction methods encounter limitations in fully exploiting global pixel correlation. In this study, we propose a novel framework that combines image splitting and convolutional neural network
(CNN) techniques. Specifically, grayscale images are divided into non-
overlapping blocks and categorized into texture and smoothing groups
based on mean square error calculations for each block. This strategy not
only improves operational efficiency but also enhances prediction accuracy. Additionally, we leverage the multi-sensory field and global optimization capability of CNNs for image prediction. By assigning image
blocks to specific predictors targeting either texture or smoothing groups

This work was supported by National Natural Science Foundation of China
(62272255, 62302248, 62302249); National key research and development program of China (2021YFC3340600, 2021YFC3340602); Taishan Scholar Program
of Shandong (tsqn202306251); Shandong Provincial Natural Science Foundation
(ZR2020MF054, ZR2023QF018, ZR2023QF032, ZR2022LZH011), Ability Improvement Project of Science and Technology SMES in Shandong Province (2022TSGC2485,
2023TSGC0217); Jinan "20 Universities"-Project of Jinan Research Leader Studio
(2020GXRC056); Jinan "New 20 Universities"-Project of Introducing Innovation Team
(202228016); Youth Innovation Team of Colleges and Universities in Shandong Province
(2022KJ124); The "Chunhui Plan" Cooperative Scientific Research Project of Ministry
of Education (HZKY20220482); Achievement transformation of science, education and
production integration pilot project (2023CGZH-05), First Talent Research Project
under Grant (2023RCKY131, 2023RCKY143), Integration Pilot Project of Science
Education Industry under Grant (2023PX006, 2023PY060, 2023PX071).

B. Ma et al. (Eds.): IWDW 2023, LNCS 14511, pp. 223–232, 2024.
https://doi.org/10.1007/978-981-97-2585-4_16

according to their respective categories, more precise predicted images can be obtained. The image predictor is trained using 3000 randomly selected images from ImageNet. The experimental results show that the proposed method can predict images accurately and improve the prediction performance more effectively. Compared with other methods, the overall performance of the method is higher.

Keywords: convolutional neural network · Image dividing · reversible data hiding

1 Introduction

Reversible data hiding in image encryption [1] can correctly extract secret information in hidden carriers and restore the original carrier without any distortion [2], Can effectively protect information security [3], and this powerful feature makes it widely used in a series of sensitive fields such as medical field and military field and has achieved good results.

Since Barton [4] proposed the reversible data hiding algorithm, there have been many methods and approaches to optimize the reversible data hiding algorithm so far, in general, there are two main ways to optimize the reversible data hiding algorithm, one is to make optimization in embedding to reduce the image distortion after data embedding, such as the RDH method based on differential expansion proposed by Tian [5], the histogram proposed by Ni et al. in [6], the RDH method based on histogram translation, the prediction error expansion proposed by Thodi et al. [7] and the reversible data hiding algorithm based on code division multiplexing proposed by Ma [8]; the other one is to make improvements on predicting the original image and propose a higher accuracy prediction algorithm or predictor to improve the prediction accuracy of the image. The difference prediction algorithm (DP) proposed by Tian [5]. He calculated the adjacent pixel difference to extend the embedded secret data. Weinberger et al. [9] proposed the median edge prediction. This method uses the comparison between the target pixel and its surrounding four pixels and then to predict the target pixel. The lossless data hiding algorithm proposed by Fallahpour [10] is based on gradient adaptive prediction (GAP), he estimates the target pixel value by using the gradient data between adjacent pixels. This method operates on seven pixels adjacent to the target pixel and predicts the target pixel by the gradient change between adjacent pixels. Sachnev [11] proposed a reversible data hiding method based on rhombic prediction. Ma et al. proposed an adaptive error prediction algorithm based on multiple linear regression [12]. With the consistency of the pixel distribution in the local area of the natural image, the inherent connection between the pixels around the target pixel is learned adaptively, and the prediction of the target pixel is achieved by building a multiple linear regression function matrix, which greatly improves the prediction accuracy. Some methods utilize multi-predictors [14] and adaptive strategies [15–17].

With the rise of deep learning, convolutional neural networks are useful in the fields of image segmentation [13], natural language processing [18], and information hiding [19], and in the field of information hiding, Luo [20] et al. proposed a reversible data hiding method for stereo images based on convolutional neural networks, which predicts images by using the connection between the left and right views of stereo images. Hu et al. predicted images by CNN in [19]. The feature extraction and image prediction by CNN has greatly improved the prediction accuracy compared with the previous methods, but due to the first application of CNN to predict grayscale images, the structure of CNN predictor is too simple and the prediction accuracy needs to be improved. Hu et al. in [21] firstly the image is divided into four parts and then combined with CNN predictor for prediction, by using the other three parts as the context for Each part is predicted, and the prediction accuracy is improved.

In general, due to the fluctuation of texture complexity of the whole image, the pixel value change will affect the overall effect of the neural network training, resulting in the reduction of the accuracy of the trained predictor; reducing the learning efficiency of the affected neural network. Osamah et al. [22] proposed to divide the image into smooth and non-smooth regions to improve the performance of the reversible data hiding algorithm. Therefore, in this paper, we make improvements in image processing by dividing the image into m n * n size image blocks, and dividing the image blocks into texture blocks and smooth blocks according to the texture complexity of the image blocks according to the set threshold for training; then we design a convolutional neural network predictor with dilated convolution, which can capture multi-scale contextual information. Image dividing makes the predictor's classification ability and data extraction ability enhanced to some extent. We randomly selected 3000 images in Imagenet for training, and the experimental results show that the method in this paper can have better prediction performance than the previously proposed methods.

The rest of this paper is organized as follows. Section 2 describes the proposed RDH scheme in detail, and Sect. 3 reports the comparison experiments with other predictors. We conclude our work in Sect. 4.

2 Convolutional Neural Network Error Prediction Algorithm

2.1 Image Preprocessing

First, we divide the original image into a cross-set image and a dot-set image. For the cross-set image, the pixel value at the dot-set location is set to 0 and vice versa, in the way shown in Fig. 1. This ensures the reversibility of the image. Then the two subset images are divided, which can increase the prediction accuracy, reduce the prediction error, and decrease the distortion degree of the embedded image. Assuming that the original image is a grayscale image of size N * M, the size of each sub-block is n * m, and the original image is divided into K sub-blocks. Subsequently, a threshold T is set concerning the image texture complexity, and the method defines the mean square error (MSE) as a criterion to evaluate the roughness of the image, and uses the MSE to classify the smooth and textured blocks of the original image, and defines the image block as a textured block when the MSE of an image block ¿ T, and defines the image block as a smooth block when the MSE of an image block ¡= T. The MSE is calculated as follows:

$$\text{MSE} = \frac{1}{mn} \sum_{i=0}^{m-1} \sum_{j=0}^{n-1} [I(i,j) - I_{ave}(i,j)]^2 \tag{1}$$

In the equation MSE denotes the mean square error between I and I_{ave}, the former being the original image and the latter being the mean of the image.

After dividing the images into texture blocks and smooth sliders, according to their groups, they can be put into the predictor for texture blocks and the predictor for smooth blocks, respectively. In this way, more accurate predicted images can be generated.

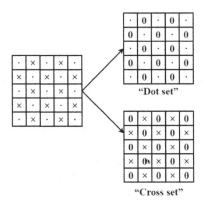

Fig. 1. The original image is divided into "cross set" and "dot set" images.

2.2   Convolutional Neural Network Model

As shown in Fig. 2, the whole network is divided into two main parts, where the number of channels is set to 32 and k is the convolutional kernel size. The first part has four parallel convolutional blocks for feature extraction, which contain a dilated convolution module. The size of convolution kernel k is 3 * 3, 5 * 5, 7 * 7, and 3 * 3 (dilation is set to 2), and each convolution block includes three operations of image convolution, activation layer, and image convolution, the first three convolution modules can comprehensively extract the feature information at different levels of the image; based on this, the dilated convolution module is added, which can expand the sensory field of the extracted features to refine the more subtle features of the image and four parallel The four parallel convolution blocks produce four feature matrices, which are superimposed by channels to generate a 128-channel full feature matrix for the next image prediction. The second part has two convolutional blocks that are connected in series to globally optimize the feature matrix and generate the final predicted image.

According to the proposed image division method, the dot set image block and the cross set image block can be predicted from each other. It is assumed that the cross image block is used to predict the dot image block as shown in Fig. 2. First, the image block is feature extracted, the cross image block is fed into four parallel convolution blocks, and the output of each convolution block is summed to generate the full feature matrix, and then the full feature matrix is input to the image prediction module, which passes the refinement of the feature image through jump connection and a serial connection to generate the final predicted dot set image.

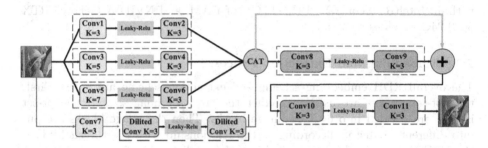

Fig. 2. Convolutional neural network framework.

2.3   Training

We randomly select 3000 images from ImageNet to train the predictor for convergence in this paper. In the training process, we convert the images into grayscale images of 512 * 512 size, set the threshold T to 3000, divide them into texture group and smooth group according to the chunking requirement, and then train the convolutional neural network separately using the two groups of images, the neural network for the texture area using texture group and the neural network for the smooth area using smooth group. Figure 3 shows an example of an image chunked and trained separately, where the image is classified according to the classification method in Sect. 2.1 so that it enters the CNN for training separately, which allows it to obtain predictors for textured and smoothed regions.

We use a back-propagation based ADAM optimizer with a batch size of 4 to optimize the following loss function:

$$loss = \frac{1}{N} \sum_{i=1}^{N} (\tilde{I}_D - I_D)^2 + \varphi \parallel \omega \parallel_2^2 \tag{2}$$

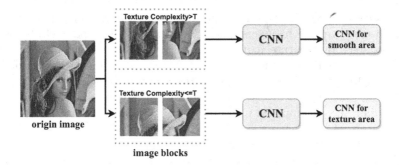

Fig. 3. Neural network training flow chart.

where N is the number of training data, φ is the bias, and ω is the weight in the network, and we set φ to 2.5 * 1e$-$5

All experiments were performed on a DeLL graphics workstation configured with an Intel(R) Xeon(R) CPU, 112 GB of RAM, and NVIDIA GeForce RTX 3090 24G graphics memory.

2.4   Embedding and Extraction Process

The overall RDH embedding scheme and extraction scheme are given. First, the dot-set images are used to predict the cross-set images, and the dot-set images are divided into 64 * 64 image blocks. Then the image blocks are put into different predictors according to the relationship with the threshold T. If the MSE is less than T, they are fed into the predictor for the smooth set; and vice versa, they are fed into the predictor for the texture set. The predicted cross set image block is obtained, and then the predicted cross set image blocks are merged according to the original position to obtain the predicted cross set image, and then the predicted cross set image and the original cross set image are reversible to hide one half of the data and produce the cross set image after embedding the information, and similarly the cross set image after embedding the information is divided according to the same steps and put into different predictors according to the relationship with the threshold value The predicted dot set image block is obtained, then the predicted dot set image block is merged into the predicted dot set image, the predicted dot set image and the original dot set image are reversible to hide the other half of the information and produce the dot set image after embedding the information, after embedding the complete information, the cross set image after embedding the information and the dot set image after embedding the information are merged into the data hiding image.

In the extraction process and image reduction process, the image after embedding information is first divided into subset images (cross-set and dot-set) of the same pattern. Then the data-hidden cross-set image is divided into 64 * 64 image blocks, which are put into different predictors according to the relationship with the threshold value to generate the predicted dot-set image blocks, and subsequently, the predicted dot-set image blocks are merged into the predicted dot-set image. And the second part of the information is extracted and the original dot set image is recovered in cooperation with the dot set image after embedding the information. Then the recovered dot set image is divided as above and the image blocks are put into different predictors according to the relationship with the threshold value to obtain the predicted cross-set image blocks, and subsequently, all the predicted cross-set image blocks are merged into the predicted cross set image for recovering the original cross set image and extracting the first part of the information. The recovered cross-set image and dot-set image are subsequently added together in the null field to recover the original image, and the two parts of information are also combined to recover the original information bits.

3 Experimental Settings

3.1 Comparison of Prediction Performance Between Different Methods

We use the mean square error, mean, and variance as criteria to analyze the prediction performance of this paper's algorithm in comparison with other algorithms.

Table 1. Image cross-correlation coefficient of the same source

Predictor/evaluation standard	MSE	Mean	Variance
DP [5]	280.5058	8.9459	173.9047
MEDP [9]	185.7578	6.9980	118.1088
CNNP [19]	52.0809	3.4063	34.2653
Proposed	49.1615	3.2114	33.4406

Table 1 compares the prediction error MSE, absolute mean, and variance of the proposed method with the other three prediction algorithms. The MSE of the proposed method is 49.1615, which is much lower than that of CNNP (52.0809), MEDP (185.7578), and DP (280.5058). It can be seen that the prediction accuracy of the method in this paper is higher than that of other methods.

Table 2 presents a comparison of the number of prediction errors "0" among various methods. Generally, a higher count of prediction errors "0" indicates a smaller difference between the predicted and original images, thereby reflecting greater accuracy. This index serves as an indicator to assess the level of prediction accuracy. In this study, we focused on analyzing the MISC standard image library comprising Lena, Barbara, Baboon, and Boat with dimensions of 512 × 512 pixels. Notably, our method exhibits the highest number of prediction errors "0". This can be attributed to our utilization of block classification-based prediction algorithms in conjunction with convolutional neural networks (CNNs) , the characteristics of different blocks are fully utilized, and the powerful features of CNN are exploited.

Table 2. Comparison Of The Number Of Pixel Values "0" In The Prediction Errors Of Different Methods

Predictor	DP [5]	MEDP [9]	CNNP [19]	Proposed
Lena	11483	28961	31263	**31721**
Barbara	8409	22295	28056	**29980**
Baboon	4220	8781	11004	**11238**
Boat	7414	17045	19259	**20188**

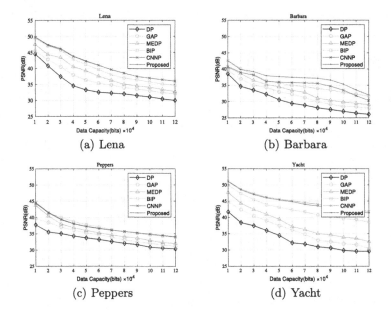

(a) Lena

(b) Barbara

(c) Peppers

(d) Yacht

Fig. 4. Comparison of five RDH methods by using four benchmark images.

3.2   Embedding Performance

To further validate the enhanced capability of the prediction algorithm proposed in this paper for image reversible data hiding, we utilize the prediction error planes generated by the prediction algorithm in this paper to embed the information by using the error extension embedding technique. Figures 4(a) to (d) show the PSNR values of the four standard graphs (Lena, Barbara, Yacht, and Pepper) when the embedding capacity is 10000 to 120000 bits. It can be seen in Fig. 4(a–d). The carrier-secret image after data embedding using the prediction error plane produced by image segmentation combined with the convolutional neural network error prediction algorithm produces a slightly higher PSNR than the other prediction methods. It turns out that the reason for the better performance is that image segmentation to train different weights of the network separately for different characteristics of the image is more targeted, coupled with the convolutional neural network has the characteristics of multiple sensory fields and global optimization ability, so that it can produce better results in image feature extraction and prediction. The experimental results in this section confirm that the proposed method outperforms the classical and other state of the art methods.

4   Conclusion

In this paper, a reversible data hiding error prediction algorithm based on the combination of block classification and convolutional neural network is proposed. Firstly, the image is divided into texture area and smooth area according to the threshold to train the network with different coefficients respectively, which makes full use of the correlation between adjacent pixels. Through the convolutional neural network, the image features can be extracted more accurately and the predicted image that is highly similar to the target image can be generated; meanwhile, the expansion convolution is added to enhance the expressive ability of the convolutional neural network and improve the convergence speed of the network. The experimental results show that the prediction accuracy of the method proposed in this paper is higher than several other classical prediction methods.

References

1. Tang, Z., Chen, L., Zhang, X., Zhang, S.: Robust image hashing with tensor decomposition. IEEE Trans. Knowl. Data Eng. **31**(3), 549–560 (2019). https://doi.org/10.1109/TKDE.2018.2837745
2. Hu, R., Xiang, S.: Lossless robust image watermarking by using polar harmonic transform. Sig. Process. **179**, 107833 (2021)
3. Shi, Y.-Q., et al.: Reversible data hiding: advances in the past two decades. IEEE Access **4**, 3210–3237 (2016)
4. Barton, J.M.: Method and apparatus for embedding authentication information within digital data. United States Patent, 5 646 997 (1997)
5. Tian, J.: Reversible data embedding using a difference expansion. IEEE Trans. Circ. Syst. Video Technol. **13**(8), 890–896 (2003)
6. Ni, Z., et al.: Reversible data hiding. IEEE Trans. Circ. Syst. Video Technol. **16**(3), 354–362 (2006)
7. Thodi, D.M., Rodríguez, J.J.: Expansion embedding techniques for reversible watermarking. IEEE Trans. Image Process. **16**(3), 721–730 (2007)
8. Ma, B., Shi, Y.Q.: A reversible data hiding scheme based on code division multiplexing. IEEE Trans. Inf. Forensics Secur. **11**(9), 1914–1927 (2016)
9. Weinberger, M.J., Seroussi, G., Sapiro, G.: The LOCO-I lossless image compression algorithm: principles and standardization into JPEG-LS. IEEE Trans. Image Process. **9**(8), 1309–1324 (2000)
10. Fallahpour, M.: Reversible image data hiding based on gradient adjusted prediction. IEICE Electron. Express **5**(20), 870–876 (2008)
11. Sachnev, V., et al.: Reversible watermarking algorithm using sorting and prediction. IEEE Trans. Circ. Syst. Video Technol. **19**(7), 989–999 (2009)

12. Ma, B., et al.: Adaptive error prediction method based on multiple linear regression for reversible data hiding. J. Real-Time Image Process. **16**, 821–834 (2019)

13. Lang, C., et al.: Learning what not to segment: a new perspective on few-shot segmentation. In: Proceedings of the IEEE/CVF Conference on Computer Vision and Pattern Recognition (2022)

14. Jafar, I.F., et al.: Efficient reversible data hiding using multiple predictors. Comput. J. **59**(3), 423–438 (2016)

15. Dragoi, I.-C., Coltuc, D.: Local-prediction-based difference expansion reversible watermarking. IEEE Trans. Image Process. **23**(4), 1779–1790 (2014)

16. Lee, B.Y., Hee, J.H., Hyoung, J.K.: Reversible data hiding using a piecewise autoregressive predictor based on two-stage embedding. J. Electr. Eng. Technol. **11**(4), 974–986 (2016)

17. Wang, X., et al.: High precision error prediction algorithm based on ridge regression predictor for reversible data hiding. IEEE Sig. Process. Lett. **28**, 1125–1129 (2021)

18. Dong, L., et al.: Question answering over freebase with multi-column convolutional neural networks. In: Proceedings of the 53rd Annual Meeting of the Association for Computational Linguistics and the 7th International Joint Conference on Natural Language Processing (Volume 1: Long Papers) (2015)

19. Hu, R., Xiang, S.: CNN prediction based reversible data hiding. IEEE Sig. Process. Lett. **28**, 464–468 (2021)

20. Luo, T., et al.: Convolutional neural networks-based stereo image reversible data hiding method. J. Vis. Commun. Image Representation **61**, 61–73 (2019)

21. Hu, R., Xiang, S.: Reversible data hiding by using CNN prediction and adaptive embedding. IEEE Trans. Pattern Anal. Mach. Intell. **44**(12), 10196–10208 (2021)

22. Al-Qershi, O.M., Khoo, B.E.: Authentication and data hiding using a hybrid ROI-based watermarking scheme for DICOM images. J. Digit. Imaging **24**, 114–125 (2011)

Dual-Domain Learning Network for Polyp Segmentation

Yan Li[1,2], Zhuoran Zheng[3], Wenqi Ren[4(✉)], Yunfeng Nie[5],
Jingang Zhang[6], and Xiuyi Jia[3]

[1] Institute of Information Engineering, Chinese Academy of Sciences,
Beijing 100195, China
liyan1999@iie.ac.cn
[2] School of Cyber Security, University of Chinese Academy of Sciences,
Beijing 100195, China
[3] School of Computer Science and Engineering, Nanjing University of Science
and Technology, Nanjing 210094, China
[4] School of Cyber Science and Technology, Sun Yat-sen University,
Shenzhen 510006, China
renwq3@mail.sysu.edu.cn
[5] Brussel Photonics, Department of Applied Physics and Photonics,
Vrije Universiteit Brussel and Flanders Make, 1050 Brussels, Belgium
[6] School of Future Technology, University of Chinese Academy of Sciences,
Beijing 100039, China

Abstract. Automatic polyp segmentation is a crucial application of
artificial intelligence in the medical field. However, this task is challenging due to uneven brightness, variable colors, and blurry boundaries.
Most current polyp segmentation methods focus on features extracted
from the spatial domain, ignoring the valuable information contained in
the frequency domain. In this paper, we propose a Dual-Domain Learning Network (D^2LNet) for polyp segmentation. Specifically, we propose
a Phase-Amplitude Attention Module, which enhances the details in the
phase spectrum, while reducing interference from brightness and color
in the amplitude spectrum. Moreover, we introduce a Spatial-Frequency
Fusion Module that utilizes parameterized frequency-domain features to
adjust the style of spatial-domain features and improve polyp visibility.
Extensive experiments demonstrate that our method outperforms the
state-of-the-art approaches both visually and quantitatively.

Keywords: Polyp segmentation · Dual-domain learning · Artificial
intelligence and applications

1 Introduction

Colorectal cancer (CRC) has the third-highest incidence rate and the second-
highest mortality rate worldwide. Most colorectal cancer cases are caused by the

B. Ma et al. (Eds.): IWDW 2023, LNCS 14511, pp. 233–247, 2024.
https://doi.org/10.1007/978-981-97-2585-4_17

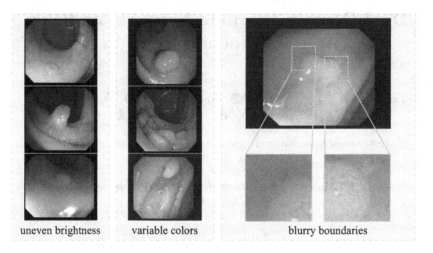

uneven brightness   variable colors   blurry boundaries

Fig. 1. This figure demonstrates the three types of image degradation that can occur in colonoscopy. These three types of degradation information influence the performance of polyp segmentation algorithms.

malignant transformation of polyps [17]. Therefore, early diagnosis and treatment of polyps are of great significance. Colonoscopy is the gold standard procedure for CRC screening. However, manual screening by doctors relies on their experience, often leading to missed or misdiagnosed polyps. Nowadays, many fields are driven by artificial intelligence methods, and automatic polyp segmentation as an important application of intelligent medicine has also been developed by leaps and bounds.

Unfortunately, images captured during colonoscopy exhibit uneven brightness, varying colors and blurry boundaries, as illustrated in Fig. 1. Training neural network models with such images will implicitly bring brightness and color information to the models, which weakens the segmentation performance and affects the generalization ability. Additionally, the blurry boundaries between polyps and normal tissues make it challenging to accurately segment the shapes of polyps. Currently, several automated polyp segmentation methods based on deep learning have been developed and made significant progress [4,21,23,33,34]. However, these methods miss valuable information i.e. the frequency domain signal (the phase and amplitude spectrums). In this study, we reveal the image information represented by the phase and amplitude spectrums. By employing them, we can achieve superior segmentation results.

Here, we conduct two experiments (see Fig. 2). The first experiment is shown in Fig. 2(a), and we convert spatial domain images into the amplitude and phase spectrums of the frequency domain using the Fast Fourier Transform (FFT) and then reconstruct the images using only the amplitude or phase spectrum. The second experiment is shown in Fig. 2(b), and we exchange the amplitude spectrums of two polyp images, then obtain the processed image by performing the Inverse Fast Fourier Transform (IFFT). Through the results, we can see that: i)

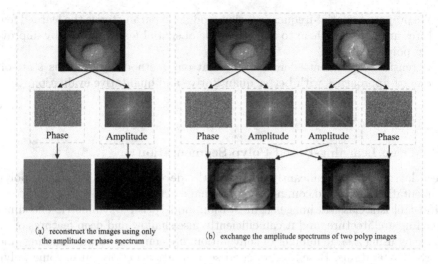

(a) reconstruct the images using only the amplitude or phase spectrum

(b) exchange the amplitude spectrums of two polyp images

Fig. 2. We show the role of the phase and amplitude spectrums of polyp images in the frequency domain.

When using only the phase spectrum for image reconstruction, the image retains details and structure but loses brightness information. ii) When reconstructing images using only the amplitude spectrum, the image appears completely black. This is because the phase spectrum contains spatial location information of different frequency components. Disregarding the phase leads to the reconstructed image being perceived as an unknown and irregular image. iii) After exchanging the amplitude spectrum, the brightness and color of the two images are swapped, while the image structure and boundaries remain unchanged. This indicates that the phase primarily represents the structure and details such as boundary, while the amplitude reflects the brightness and color. This conclusion is consistent with the observations from [12,19].

Based on the above observations and conclusions, we propose a new frequency and spatial learning network (dual-domain learning network) for polyp segmentation. It aims to address the problems of polyp segmentation by jointly exploring information in both frequency and spatial domains. We carefully developed two modules, the Phase-Amplitude Attention Module (PAAM) and the Spatial-Frequency Fusion Module (SFFM). Specifically, in the frequency domain branch, we cascade multiple PAAMs to capture multi-level amplitude and phase features. At the same time, we employ an attention mechanism to enhance the valuable details and suppress the brightness that cause interference. SFFM transforms the frequency domain global features into parameter feature maps and then integrates them with spatial features.

In conclusion, the main contributions of our work are as follows:

- We propose a phase-amplitude attention module to capture multi-level amplitude and phase features and utilize an attention mechanism to enhance global details while attenuating global illumination.

- We propose a spatial-frequency fusion module to parameterize the learned features and leverage them to alter the style of spatial features, thereby improving polyp visibility.
- Extensive experiments demonstrate that our method outperforms state-of-the-art approaches with better quantitative and qualitative evaluation.

2   Related Work

2.1   Deep Learning-Based Polyp Segmentation

Deep learning has significantly advanced the development of automatic polyp segmentation. It started coming from a U-shape network for polyp segmentation. UNet [16] is a classical image segmentation model with a symmetric encoding-decoding architecture, and it can efficiently fuse shallow and deep features of an image by using a large number of convolution, up-sampling, down-sampling, and shortcut operations. Based on its good segmentation performance, some polyp segmentation methods [13,14,29,30] adopt the U-shape network.

In addition, some methods use attention mechanisms and carefully designed strategies to improve segmentation accuracy. PraNet [7] uses a parallel partial decoder to roughly predict the polyp region and then introduces a reverse attention mechanism to refine the boundaries. SANet [27] designs a color-swapping data augmentation strategy to improve generalization ability and employs shallow attention to filter out noise. ACSNet [32] proposes a local context attention to focus on the hard region and utilizes channel-wise attention to select and aggregate features. CFA-Net [34] improves the segmentation performance by extracting boundary features and fusing cross-level features. Since the success of the Transformer in multiple fields, some approaches [3,4,23] have replaced the backbone of networks from CNN with Transformer and obtained good results.

Although the above models have made significant progress, these methods only focus on the spatial domain and overlook the frequency domain information. We have discovered that the amplitude and phase spectrums contain abundant information, which can effectively complement spatial features. As a result, we propose a network that explores features in both the spatial and frequency domains.

2.2   Learning in the Frequency Domain

In recent years, learning information from the Fourier domain of images has shown promising results in various tasks, which provides a new perspective for polyp segmentation. For instance, FFCNet [24] proposes complex convolutional networks for colon disease classification, which utilize brightness insensitivity and displacement invariance of the frequency domain. FSDGN [31] designs an amplitude-guided phase module and applies frequency domain features to guide local features for image dehazing tasks. FFC [2] proposes a new convolutional operator Fast Fourier Convolutions (FFC) that combines spatial and frequency domains for convolution, and LaMa [20] applies FFC to image restoration tasks.

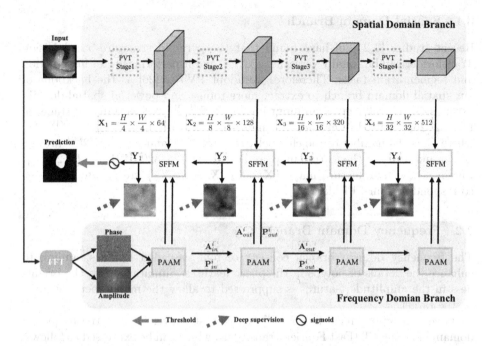

Fig. 3. The structure of our framework (D^2LNet). First, our method extracts spatial-domain feature maps and frequency-domain feature maps from the input image, respectively. Next, the extracted spatial-domain feature maps are enhanced to align with the extracted frequency-domain feature maps. Finally, these feature maps are normalized and go through residual concatenation to output a high-quality segmented image.

However, most methods only take into account the real and imaginary parts, neglecting further considerations of the amplitude and phase spectrum, or solely focusing on amplitude or phase, overlooking their combined effect. Besides, these methods are designed for other computer vision tasks. In contrast, our approach specifically addresses the difficulties in polyp segmentation. We propose a phase-amplitude attention module that applies attention to both amplitude and phase features, suppressing interference from brightness and colors in the amplitude spectrum, while enhancing the details of polyps in the phase spectrum. Moreover, we introduce a spatial-frequency fusion module that modifies spatial-domain features with parameterized frequency-domain features to improve polyp visibility, enabling effective polyp segmentation.

3 Method

D^2LNet (see Fig. 3) includes a spatial domain branch, a frequency domain branch, and a spatial-frequency fusion module. Four sets of feature maps are obtained in the frequency and spatial domains, respectively, and then fused by a spatial-frequency fusion module to obtain a high-quality segmented image.

3.1 Spatial Domain Branch

Recent studies [3,25,26] have found that the pyramid structure-based Vision Transformer can be equally versatile as CNN, and it performs better in detection and segmentation tasks. Therefore, we adopt PVTv2 [26] as the backbone of our spatial domain branch to extract more robust and powerful spatial-domain features. Specifically, given an input image $\mathbf{I} \in \mathbb{R}^{H \times W \times 3}$, four different stages of PVTv2 [26] are used to extract four multi-scale feature maps $\mathbf{X}_i \in \mathbb{R}^{H_i \times W_i \times C_i}$, where C_i is the number of the channels, $H_i = H/2^{i+1}$ and $W_i = W/2^{i+1}$ are the height and width of the feature map, and $C_i \in \{64, 128, 320, 512\}$, $i \in \{1, 2, 3, 4\}$. Finally, these four feature maps $\{\mathbf{X}_1, \mathbf{X}_2, \mathbf{X}_3$ and $\mathbf{X}_4\}$ are inputed into the corresponding four SFFMs.

3.2 Frequency Domain Branch

The frequency domain branch consists of four cascaded PAAMs, aiming to enhance the details of polyps in the phase features. Simultaneously, the brightness in the amplitude features is suppressed to allow the model focusing more on the target.

Given an input image $\mathbf{I} \in \mathbb{R}^{H \times W \times 3}$, we first convert it into the frequency domain via the FFT (Fast Fourier Transform), which can be expressed as follows:

$$\mathcal{F}(u, v) = \sum_{h=0}^{H-1} \sum_{w=0}^{W-1} \mathbf{I}(h, w) \cdot e^{-j2\pi(\frac{h}{H}u + \frac{w}{W}v)} \tag{1}$$

where $\mathbf{I}(h, w)$ and $\mathcal{F}(u, v)$ are the spatial coordinates of the pixel and the spectrum of the frequency domain, H and W indicate the height and width of the image, respectively.

The frequency-domain feature $\mathcal{F}(u, v)$ can further be denoted as $\mathcal{F}(u, v) = R(u, v) + jI(u, v)$, where $R(\cdot)$ and $I(\cdot)$ represent the real and imaginary part of $\mathcal{F}$. Then the two parts are converted to the amplitude and phase spectrums as follows:

$$A(u, v) = [R^2(u, v) + I^2(u, v)]^{1/2} \tag{2}$$

$$P(u, v) = \arctan[I(u, v)/R(u, v)] \tag{3}$$

where $A(u, v)$ is the amplitude spectrum, $P(u, v)$ is the phase spectrum. They serve as the inputs of our first PAAM. For simplicity, let A_{in}^i and P_{in}^i denote the inputs of the i-th PAAM, and A_{out}^i and P_{out}^i denote the outputs of the i-th PAAM, where $i \in \{1, 2, 3, 4\}$.

The structure of PAAM is shown in Fig. 4. Taking the i-th PAAM as an example. First, A_{in}^i and P_{in}^i are conducted the convolution operation with the convolution kernels of size 3×3, respectively, and the corresponding outputs are concatenated into U, denoted as:

$$U = \mathrm{Cat}((A_{in}^i \otimes k_3), (P_{in}^i \otimes k_3)) \tag{4}$$

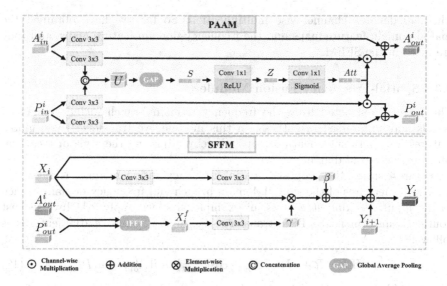

Fig. 4. This figure shows the details of the PAAM and SFFM pipelines. The role of PAAM is to selectively enhance and suppress the frequency-domain features of the image and the role of SFFM is to fuse the spatial-domain features with the frequency domain features.

where $\otimes$ denotes the convolution operator, k_n denotes the convolution filter with kernel size of $n \times n$, $\mathrm{Cat}(\cdot)$ denotes channel-wise concatenation.

Then, inspired by [8], we apply global average pooling (GAP) to U, and the feature vector $S \in \mathbb{R}^{1 \times 1 \times C}$ is obtained by calculating the average value of each channel. Next, we apply a channel-downscaling convolution layer with ReLU to generate a compact feature representation $Z \in \mathbb{R}^{1 \times 1 \times r}$, where $r = C/8$. Finally, we apply a channel-upscaling convolution layer with Sigmoid to Z, yielding attention activation Att, and this process can be expressed as:

$$S = \frac{1}{H \times W} \sum_i^H \sum_j^W U_C(i,j) \tag{5}$$

$$Z = \delta(S \otimes k_1) \tag{6}$$

$$Att = \sigma(Z \otimes k_1) \tag{7}$$

where U_C is the feature of the C-th channel of U, H and W are respectively the width and height of the feature, δ denotes ReLU, σ denotes Sigmoid.

We use Att to adaptively recalibrate amplitude feature A_{in}^i and phase feature P_{in}^i, the calibrated feature map A_{out}^i and P_{out}^i is calculated by:

$$A_{out}^i = (A_{in}^i \otimes k_3) \odot Att + A_{in}^i \otimes k_3 \tag{8}$$

$$P_{out}^i = (P_{in}^i \otimes k_3) \odot Att + P_{in}^i \otimes k_3 \tag{9}$$

where $\odot$ denotes channel-wise multiplication. So far, we have obtained the spatial-domain feature maps and the frequency-domain feature maps, and they are fused in the SFFM.

3.3   Spatial-Frequency Fusion Module

The features extracted from the frequency domain branch include enhanced details and suppressed brightness. In this module, we transform these global features into parameter feature maps, thereby adjusting the style of local features in the spatial domain.

Figure 4 shows the proposed spatial-frequency fusion module. It takes the same stage features of the spatial domain branch and frequency domain branch as the inputs. Taking stage i as an example, first we apply 2-D Inverse Fast Fourier Transformer (IFFT) to transform A_{out}^i and P_{out}^i into spatial features as follows:

$$X_i^f = \text{IFFT}(R + jI) = A_{out}^i \cdot \cos(P_{out}^i) + j(A_{out}^i \cdot \sin(P_{out}^i)) \qquad (10)$$

where $\cos(\cdot)$ denotes the cosine transform and $\sin(\cdot)$ denotes the sine transform.

Next, we adopt two sets of 3×3 convolution from the spatial feature X_i to learn parameter feature map β. A 3×3 convolution from the frequency feature X_i^f is adopted to learn parameter feature map γ, which contains illumination constraints and fine details learned in the frequency domain. By using them simultaneity, we can adjust the style of spatial features to meet the constraints learned in the frequency domain, thereby improving the visibility of polyps. Inspired by [15], the output feature is calculated by:

$$Y_i = \text{BN}((\gamma * \frac{X_i - \mu(X_i)}{\upsilon(X_i)} + \beta) \otimes k_3) + X_i + Y_{i+1} \qquad (11)$$

where BN is BatchNorm, $*$ denotes element-wise multiplication, $\mu(X_i)$ and $\upsilon(X_i)$ are the mean and variance of X_i, which are used to normalize the feature map. Y_{i+1} is the output of the SFFM corresponding to stage i + 1. It is worth noting that there is no Y_{i+1} in Eq. 11 in the last stage.

3.4   Loss Function

Our loss function can be formulated as:

$$\mathcal{L} = \mathcal{L}_{IoU}^w + \mathcal{L}_{BCE}^w \qquad (12)$$

where $\mathcal{L}_{IoU}^w$ and $\mathcal{L}_{BCE}^w$ represent the weighted intersection over union loss and weighted binary cross entropy loss [28], which are adopted for the global and local restriction. Deep supervision is performed on the four stage outputs (i.e., Y_1, Y_2, Y_3 and Y_4). Each output is up-sampled to the same size as the ground truth GT. The total loss can be formulated as:

$$\mathcal{L}_{total} = \sum_{i=1}^{i=4} \mathcal{L}(Y_i^{up}, GT) \qquad (13)$$

Table 1. Quantitative results on Kvasir-SEG [9] and CVC-ClinicDB [1] dataset. **Bold** indicates the best performance.

method	Year	Kvasir-SEG					
		mDice	mIoU	F_β^w	S_α	E_ξ^{max}	MAE
UNet [16]	2015	0.818	0.746	0.794	0.858	0.893	0.055
UNet++ [35]	2018	0.821	0.744	0.808	0.862	0.909	0.048
PraNet [7]	2020	0.898	0.841	0.885	0.915	0.948	0.030
ACSNet [32]	2020	0.898	0.838	0.882	0.920	0.952	0.032
SANet [27]	2021	0.904	0.847	0.892	0.915	0.953	0.028
MSNet [33]	2021	0.902	0.847	0.891	0.923	0.954	0.029
SSFormer [23]	2022	**0.925**	0.878	0.921	0.931	0.969	**0.017**
DCRNet [30]	2022	0.846	0.772	0.807	0.882	0.917	0.053
CFANet [34]	2023	0.915	0.862	0.903	0.924	0.962	0.023
D²LNet(Ours)	2023	**0.925**	**0.879**	**0.923**	**0.933**	**0.970**	**0.017**
method	Year	CVC-ClinicDB					
		mDice	mIoU	F_β^w	S_α	E_ξ^{max}	MAE
UNet [16]	2015	0.823	0.755	0.811	0.890	0.953	0.019
UNet++ [35]	2018	0.794	0.729	0.785	0.873	0.931	0.022
PraNet [7]	2020	0.899	0.849	0.896	0.937	0.979	0.009
ACSNet [32]	2020	0.882	0.826	0.873	0.928	0.959	0.011
SANet [27]	2021	0.916	0.859	0.909	0.940	0.976	0.012
MSNet [33]	2021	0.915	0.867	0.912	0.947	0.978	0.008
SSFormer [23]	2022	0.916	0.873	0.924	0.937	0.971	0.007
DCRNet [30]	2022	0.869	0.800	0.832	0.919	0.968	0.023
CFANet [34]	2023	0.933	0.883	0.924	0.951	**0.989**	0.007
D²LNet (Ours)	2023	**0.934**	**0.889**	**0.933**	**0.952**	0.982	**0.006**

In this paper, we use the output of the last stage as the segmented image generated by the model.

4   Experiment

4.1   Implementation Details

We implement our model in the PyTorch 1.8.1 framework. The graphics processing unit is NVIDIA TITAN RTX 3090. During training, all input images are uniformly resized to 352×352 and a multi-scale training strategy [7] is employed. The network is trained for 100 epochs with a batch size of 16. For optimizer, we use the AdamW [10] to update parameters. The learning rate is 1e−4 and the weight decay is 1e−4.

Table 2. Quantitative results on ColonDB [21], Endoscene [22] and ETIS [18]) dataset. **Bold** indicates the best performance.

method	Year	CVC-ColonDB					
		mDice	mIoU	F_β^w	S_α	E_ξ^{max}	MAE
UNet [16]	2015	0.504	0.436	0.491	0.710	0.781	0.059
UNet++ [35]	2018	0.481	0.408	0.467	0.693	0.763	0.061
PraNet [7]	2020	0.712	0.640	0.699	0.820	0.872	0.043
ACSNet [32]	2020	0.716	0.649	0.697	0.830	0.851	0.039
SANet [27]	2021	0.752	0.669	0.725	0.837	0.875	0.043
MSNet [33]	2021	0.747	0.668	0.733	0.837	0.883	0.042
SSFormer [23]	2022	0.772	0.697	0.766	0.844	0.883	0.036
DCRNet [30]	2022	0.661	0.576	0.613	0.767	0.828	0.110
CFANet [34]	2023	0.743	0.665	0.728	0.835	0.898	0.039
D²LNet (Ours)	2023	**0.820**	**0.736**	**0.798**	**0.870**	**0.932**	**0.028**

method	Year	Endoscene					
		mDice	mIoU	F_β^w	S_α	E_ξ^{max}	MAE
UNet [16]	2015	0.710	0.627	0.684	0.843	0.875	0.022
UNet++ [35]	2018	0.707	0.624	0.687	0.839	0.898	0.018
PraNet [7]	2020	0.871	0.797	0.843	0.925	0.972	0.010
ACSNet [32]	2020	0.863	0.787	0.825	0.923	0.968	0.013
SANet [27]	2021	0.888	0.815	0.859	0.928	0.972	0.008
MSNet [33]	2021	0.862	0.796	0.846	0.927	0.953	0.010
SSFormer [23]	2022	0.887	0.821	0.869	0.929	0.962	**0.007**
DCRNet [30]	2022	0.753	0.670	0.689	0.854	0.900	0.025
CFANet [34]	2023	0.893	**0.827**	0.875	0.938	0.978	0.008
D²LNet (Ours)	2023	**0.895**	**0.827**	**0.876**	**0.939**	**0.979**	**0.007**

method	Year	ETIS					
		mDice	mIoU	F_β^w	S_α	E_ξ^{max}	MAE
UNet [16]	2015	0.398	0.335	0.366	0.684	0.740	0.036
UNet++ [35]	2018	0.401	0.343	0.390	0.683	0.776	0.035
PraNet [7]	2020	0.628	0.567	0.600	0.794	0.841	0.031
ACSNet [32]	2020	0.578	0.509	0.530	0.754	0.764	0.059
SANet [27]	2021	0.750	0.654	0.685	0.849	0.897	0.015
MSNet [33]	2021	0.720	0.650	0.675	0.846	0.881	0.020
SSFormer [23]	2022	0.767	0.698	0.736	0.863	0.891	0.016
DCRNet [30]	2022	0.509	0.432	0.437	0.714	0.787	0.053
CFANet [34]	2023	0.733	0.655	0.693	0.846	0.892	**0.014**
D²LNet (Ours)	2023	**0.792**	**0.702**	**0.751**	**0.870**	**0.920**	**0.014**

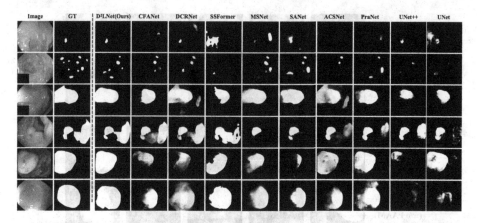

Fig. 5. Our method exhibits the best visual results, especially in conditions of highly uneven brightness and multiple polyps.

4.2   Datasets and Metrics

To evaluate the performance of the proposed D²LNet, we follow the popular experimental setups [7], which adopt five challenging public datasets: Kvasir-SEG [9], CVC-ClinicDB [1], ColonDB [21], Endoscene [22] and ETIS [18]. The training set comprises 900 images from Kvasir-SEG and 550 from CVC-ClinicDB. The remaining 100 images from Kvasir-SEG and 62 images from CVC-ClinicDB are used as test sets. Since they are already seen during training, we use them to assess the learning ability of our model. We perform testing using ColonDB, Endoscene, and ETIS, as these datasets are not used during the training stage, they are utilized to measure the generalization ability of our model.

Follow previous work [3, 7], We employ six widely-used metrics for quantitative evaluation, including Mean Dice Similarity Coefficient (mDice), mean Intersection over Union (mIoU), Mean Absolute Error (MAE), weighted F-measure (F_β^w) [11], S-measure (S_α) [5], and max E-measure (E_ξ^{max}) [6]. mDice and mIoU are similarity measures at the regional level, while MAE and F_β^w are utilized to evaluate the pixel-level accuracy. S_α and E_ξ^{max} are used to analyze global-level similarity. The lower value is better for the MAE and the higher is better for others.

4.3   Experiments with State-of-the-art Methods

We compare the proposed D²LNet with previous state-of-the-art methods, including CFANet [34], DCRNet [30], SSFormer [23], MSNet [33], SANet [27], ACSNet [32], PraNet [7], UNet++ [35], UNet [16]. For a fair comparison, we use their open-source codes to evaluate the same datasets or use the predictions provided by themselves.

Quantitative Comparison. As can be seen in Table 1 and Table 2, D²LNet achieves the best scores across five datasets on almost all metrics, which shows

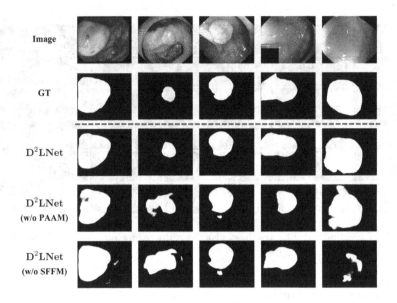

Fig. 6. This figure shows the role of the key modules in our approach.

the superior learning ability and generalization performance of our proposed model. On the CVC-ColonDB dataset, compared to the latest network, our model improves mDice by 1.1%, mIoU by 1.1%, and MAE is 0.011 lower. On the ETIS dataset, mDice is 1.08% higher, and mIou is 1.07% higher.

Visual Comparison. Figure 5 shows the visualization results of our model and the compared models. Our proposed model significantly outperforms other methods with better segmentation results, which are the closest to the ground truth. We discover that our model possesses the following advantages: 1) Our model can adapt to different lighting conditions, such as the overexposure observed in the third and sixth images in Fig. 5. Our method can effectively eliminate the interference caused by these lighting conditions and accurately locate the polyp regions. This is achieved by leveraging the amplitude features in our PAAM to diminish the effects of brightness and colors. 2) Our model has more accurate polyp boundaries, as shown in the fourth image of Fig. 5. It contains multiple polyps with complex boundaries. The boundaries segmented by our method are closer to the ground truth (GT) labels. This is due to our effective utilization of phase features to enhance details such as boundaries. 3) Our model is able to accurately locate the position of polyps, such as the second image in Fig. 5, which involves multiple small polyps and is highly challenging. With the help of our frequency domain branch, we can easily capture global features, thereby obtaining the regions where multiple polyps are located. 4) Our model can effectively distinguish between polyps and normal tissues. For example, in the first image of Fig. 5, polyps have strong camouflage abilities. Our SFFM is able to adjust the style of spatial features and improve polyp visibility.

Table 3. Ablation study for D^2LNet on the Kvasir-SEG [9] dataset.

	mDice	mIoU	F_β^w	S_α	E_ξ^{max}	MAE
D^2LNet (w/o PAAM)	0.905	0.846	0.894	0.913	0.955	0.028
D^2LNet(w/o SFFM)	0.912	0.857	0.901	0.917	0.960	0.026
D^2LNet	**0.925**	**0.879**	**0.923**	**0.933**	**0.970**	**0.017**

4.4 Ablation Study

We conduct an ablation study to validate the effectiveness of the modules designed in our D^2LNet. To analyze the effectiveness of the PAAM, we remove the frequency domain branch and train a version of "D^2LNet (w/o PAAM)". To analyze the effectiveness of the SFFM, we replace all SFFMs with a simple channel-wise concatenation operation to fuse features from spatial and frequency domains, and a version of "D^2LNet (w/o SFFM)" is trained. Table 3 demonstrates the model without the SFFMs and PAAMs drops sharply on Kvasir datasets. Figure 6 shows that the complete D^2LNet has the best predictions. To sum up, all the modules are necessary for the final predictions.

5 Conclusion

In this paper, we present a method that extracts information from both the spatial and frequency domains, achieving complementarity and communication between the two domains. Based on the analysis of the frequency domain, we propose a Phase-Amplitude Attention Module to focus on the details in the phase and effectively suppress the brightness and color in the amplitude. To integrate the spatial and frequency domain features, we propose a Spatial-Frequency Fusion Module that alter the style of spatial-domain features with parameterized frequency-domain features to enhance the visibility of polyps. Experimental results demonstrate the superiority of our method.

In future work, we consider adding an image enhancement module to overcome the signal interference problem associated with endoscopic videography.

Acknowledgements. This work was supported by the Shenzhen Science and Technology Program (Grant No. JCYJ20220530145209022), Chinese Academy of Sciences Cyber Security and Informatization Project (No. CAS-WX2022SF-0102), and Postgraduate Research & Practice Innovation Program of Jiangsu Province (KYCX22_0461).

References

1. Bernal, J., Sánchez, F.J., Fernández-Esparrach, G., et al.: WM-DOVA maps for accurate polyp highlighting in colonoscopy: validation vs. saliency maps from physicians. CMIG **43**, 99–111 (2015)

2. Chi, L., Jiang, B., Mu, Y.: Fast Fourier convolution. In: Advances in Neural Information Processing Systems, vol. 33, pp. 4479–4488 (2020)

3. Dong, B., Wang, W., Fan, D.P., Li, J., Fu, H., Shao, L.: Polyp-PVT: polyp segmentation with pyramid vision transformers. arXiv preprint arXiv:2108.06932 (2021)

4. Duc, N.T., Oanh, N.T., Thuy, N.T., Triet, T.M., Dinh, V.S.: ColonFormer: an efficient transformer based method for colon polyp segmentation. IEEE Access **10**, 80575–80586 (2022)

5. Fan, D.P., Cheng, M.M., Liu, Y., Li, T., Borji, A.: Structure-measure: a new way to evaluate foreground maps. In: Proceedings of the IEEE International Conference on Computer Vision, pp. 4548–4557. IEEE (2017)

6. Fan, D.P., Gong, C., Cao, Y., et al.: Enhanced-alignment measure for binary foreground map evaluation. In: Proceedings of the Twenty-Seventh International Joint Conference on Artificial Intelligence. International Joint Conferences on Artificial Intelligence Organization (2018)

7. Fan, D.-P., et al.: PraNet: parallel reverse attention network for polyp segmentation. In: Martel, A.L., et al. (eds.) MICCAI 2020. LNCS, vol. 12266, pp. 263–273. Springer, Cham (2020). https://doi.org/10.1007/978-3-030-59725-2_26

8. Hu, J., Shen, L., Sun, G.: Squeeze-and-excitation networks. In: Proceedings of the IEEE Conference on Computer Vision and Pattern Recognition, pp. 7132–7141. IEEE (2018)

9. Jha, D., et al.: Kvasir-SEG: a segmented polyp dataset. In: Ro, Y., et al. (eds.) MMM 2020. LNCS, vol. 11962, pp. 451–462. Springer, Cham (2020). https://doi.org/10.1007/978-3-030-37734-2_37

10. Loshchilov, I., Hutter, F.: Decoupled weight decay regularization. In: ICLR 2019. OpenReview.net (2018)

11. Margolin, R., Zelnik-Manor, L., Tal, A.: How to evaluate foreground maps? In: Proceedings of the IEEE Conference on Computer Vision and Pattern Recognition, pp. 248–255 (2014)

12. Oppenheim, A.V., Lim, J.S.: The importance of phase in signals. Proc. IEEE **69**(5), 529–541 (1981)

13. Patel, K., Bur, A.M., Wang, G.: Enhanced U-Net: a feature enhancement network for polyp segmentation. In: 2021 18th Conference on Robots and Vision (CRV), pp. 181–188. IEEE (2021)

14. Poudel, S., Lee, S.W.: Deep multi-scale attentional features for medical image segmentation. Appl. Soft Comput. **109**, 107445 (2021)

15. Ren, J., Hu, X., Zhu, L., et al.: Deep texture-aware features for camouflaged object detection. IEEE Trans. Circ. Syst. Video Technol. **33**(3), 1157–1167 (2023)

16. Ronneberger, O., Fischer, P., Brox, T.: U-Net: convolutional networks for biomedical image segmentation. In: Navab, N., Hornegger, J., Wells, W.M., Frangi, A.F. (eds.) MICCAI 2015. LNCS, vol. 9351, pp. 234–241. Springer, Cham (2015). https://doi.org/10.1007/978-3-319-24574-4_28

17. Shussman, N., Wexner, S.D.: Colorectal polyps and polyposis syndromes. Gastroenterology Rep. **2**(1), 1–15 (2014)

18. Silva, J., Histace, A., Romain, O., Dray, X., Granado, B.: Toward embedded detection of polyps in WCE images for early diagnosis of colorectal cancer. CARS **9**, 283–293 (2014)

19. Skarbnik, N., Zeevi, Y.Y., Sagiv, C.: The importance of phase in image processing. Technion-Israel Institute of Technology, Faculty of Electrical Engineering (2009)

20. Suvorov, R., Logacheva, E., Mashikhin, A., et al.: Resolution-robust large mask inpainting with Fourier convolutions. In: Proceedings of the IEEE/CVF Winter Conference on Applications of Computer Vision, pp. 2149–2159. IEEE (2022)

21. Tajbakhsh, N., Gurudu, S.R., Liang, J.: Automated polyp detection in colonoscopy videos using shape and context information. IEEE TMI **35**(2), 630–644 (2015)
22. Vázquez, D., Bernal, J., Sánchez, F.J., et al.: A benchmark for endoluminal scene segmentation of colonoscopy images. JHE **2017**, 9 (2017)
23. Wang, J., Huang, Q., Tang, F., Meng, J., Su, J., Song, S.: Stepwise feature fusion: local guides global. In: Wang, L., Dou, Q., Fletcher, P.T., Speidel, S., Li, S. (eds.) MICCAI 2022. LNCS, vol. 13433, pp. 110–120. Springer, Cham (2022). https://doi.org/10.1007/978-3-031-16437-8_11
24. Wang, K.N., et al.: FFCNet: Fourier transform-based frequency learning and complex convolutional network for colon disease classification. In: Wang, L., Dou, Q., Fletcher, P.T., Speidel, S., Li, S. (eds.) MICCAI 2022. LNCS, vol. 13433, pp. 78–87. Springer, Cham (2022). https://doi.org/10.1007/978-3-031-16437-8_8
25. Wang, W., Xie, E., Li, X., et al.: Pyramid vision transformer: a versatile backbone for dense prediction without convolutions. In: Proceedings of the International Conference on Computer Vision, pp. 568–578. IEEE (2021)
26. Wang, W., Xie, E., Li, X., et al.: PVT v2: improved baselines with pyramid vision transformer. Comput. Vis. Media **8**(3), 415–424 (2022)
27. Wei, J., Hu, Y., Zhang, R., Li, Z., Zhou, S.K., Cui, S.: Shallow attention network for polyp segmentation. In: de Bruijne, M., et al. (eds.) MICCAI 2021. LNCS, vol. 12901, pp. 699–708. Springer, Cham (2021). https://doi.org/10.1007/978-3-030-87193-2_66
28. Wei, J., Wang, S., Huang, Q.: F^3Net: fusion, feedback and focus for salient object detection. In: Proceedings of the AAAI Conference on Artificial Intelligence, vol. 34, pp. 12321–12328. AAAI Press (2020)
29. Wu, H., Zhong, J., Wang, W., Wen, Z., Qin, J.: Precise yet efficient semantic calibration and refinement in convnets for real-time polyp segmentation from colonoscopy videos. In: Proceedings of the AAAI Conference on Artificial Intelligence, vol. 35, pp. 2916–2924. AAAI Press (2021)
30. Yin, Z., Liang, K., Ma, Z., Guo, J.: Duplex contextual relation network for polyp segmentation. In: 2022 IEEE 19th International Symposium on Biomedical Imaging (ISBI), pp. 1–5. IEEE (2022)
31. Yu, H., Zheng, N., Zhou, M., Huang, J., Xiao, Z., Zhao, F.: Frequency and spatial dual guidance for image dehazing. In: Avidan, S., Brostow, G.J., Cissé, M., Farinella, G.M., Hassner, T. (eds.) ECCV 2022. LNCS, vol. 13679, pp. 181–198. Springer, Cham (2022). https://doi.org/10.1007/978-3-031-19800-7_11
32. Zhang, R., Li, G., Li, Z., Cui, S., Qian, D., Yu, Y.: Adaptive context selection for polyp segmentation. In: Martel, A.L., et al. (eds.) MICCAI 2020. LNCS, vol. 12266, pp. 253–262. Springer, Cham (2020). https://doi.org/10.1007/978-3-030-59725-2_25
33. Zhao, X., Zhang, L., Lu, H.: Automatic polyp segmentation via multi-scale subtraction network. In: de Bruijne, M., et al. (eds.) MICCAI 2021. LNCS, vol. 12901, pp. 120–130. Springer, Cham (2021). https://doi.org/10.1007/978-3-030-87193-2_12
34. Zhou, T., et al.: Cross-level feature aggregation network for polyp segmentation. Pattern Recogn. **140**, 109555 (2023)
35. Zhou, Z., Rahman Siddiquee, M.M., Tajbakhsh, N., Liang, J.: UNet++: a nested U-Net architecture for medical image segmentation. In: StoyanovD, D., et al. (eds.) DLMIA/ML-CDS -2018. LNCS, vol. 11045, pp. 3–11. Springer, Cham (2018). https://doi.org/10.1007/978-3-030-00889-5_1

Two-Round Private Set Intersection Mechanism and Algorithm Based on Blockchain

Yue Wang⬤, Zhanshan Wang⬤, Xiaofeng Ma⬤, and Jing He$^{(\boxtimes)}$ ⬤

Department of Control Science and Engineering, Tongji University, Shanghai 201804, China
2130719@tongji.edu.cn

Abstract. In the contemporary era of the digital economy, the significance of data as a fundamental production factor cannot be overstated, as it serves as a driving force behind the development of the digital economy. Despite the existence of numerous algorithms that have been proposed to achieve intersection of datasets while ensuring privacy, challenges related to computational efficiency and trust costs between institutions persist. This paper presents the BTPSI (Blockchain-based Two-round Private Set Intersection) algorithm, which aims to efficiently and reliably compute the intersection of datasets within the unbalanced private set intersection framework. The proposed BTPSI algorithm has been thoroughly tested and evaluated on various datasets of varying sizes.

Keywords: Blockchain · Private Set Intersection · Cuckoo Filter · Oblivious Transfer · Cuckoo Hash

1 Introduction

In recent years, there has been a significant increase in the development and utilization of advanced information technologies, such as big data and artificial intelligence. This exponential growth has led to a high demand for data, highlighting the crucial significance of this valuable resource [1]. Unrestricted access to data and its efficient utilization are widely acknowledged as essential catalysts for future innovation and growth. Nevertheless, the act of data sharing by institutions can lead to privacy breaches and the possibility of diminished interest, thereby raising concerns about the ambiguity of data ownership. Consequently, the urgent imperative of addressing the challenge of achieving data sharing while simultaneously safeguarding data privacy, clarifying data ownership, and fully unlocking the potential value of data has emerged in the era of the digital economy [2].

The decentralized nature of blockchain and its distributed data storage mechanism ensure the traceability and immutability of data stored on the chain [3]. By shifting reliance from third-party intermediaries to the technology itself, blockchain enables transparent transactions on the chain, creating a trusted ecosystem for inter-organizational data exchange. Nevertheless, the adoption of blockchain technology also

B. Ma et al. (Eds.): IWDW 2023, LNCS 14511, pp. 248–261, 2024.
https://doi.org/10.1007/978-981-97-2585-4_18

presents challenges, including privacy breaches and ambiguity surrounding data owner-ship. In response to these concerns, various techniques for privacy-preserving computation have been developed. These techniques facilitate the circulation and sharing of data while simultaneously safeguarding the confidentiality of privacy-related information.

Blockchain-based data-sharing schemes have been extensively researched [4–7]. Sultana et al. proposed a novel data-sharing framework that utilizes blockchain technology and incorporates smart contracts to regulate access to the shared data [8]. Rouhani et al. employed access control methodologies, such as attribute-based encryption, to effectively regulate data access permissions on the blockchain, thereby guaranteeing secure and privacy-preserving data sharing [9, 10]. Wang et al. introduced a dual-chain structure as a means to securely store both original data and transaction data, thereby facilitating secure data sharing and traceability [11]. Zheng et al. employed the Paillier cryptosystem as a means to ensure the confidentiality of shared data, thereby facilitating data trading [12]. Feng et al. introduced a novel data privacy access control scheme for blockchain, which is built upon searchable attribute-based encryption [13]. Li et al. proposed a privacy protection model that utilizes a dual-chain structure incorporating CP-ABE (Ciphertext-Policy Attribute-Based Encryption) and zk-SNARK (Zero-Knowledge Succinct Non-Interactive Argument of Knowledge). This model allows for fine-grained access control, sensitive access policies, and attribute privacy protection [14].

In the financial industry, characterized by the abundance of data and the need for privacy protection, there is a significant requirement for data sharing. As a result, several financial institutions, including banks, securities firms, and funds, have made significant advancements in implementing privacy-preserving computation [15]. Consequently, this paper presents a case study on cross-financial institutions and suggests a privacy data-sharing solution that combines blockchain technology with PSI technology. Specifically, a two-round private set intersection algorithm has been developed to address the issue of non-balanced data sets. The proposed solution seeks to aid financial institutions in addressing the issue of data silos, promoting improved data collaboration, and safeguarding privacy. The primary contributions of this paper are outlined as follows:

- We propose a two-round private set intersection algorithm, which includes a pre-intersection step utilizing the cuckoo filter prior to the OT-based private set intersection to improve efficiency of unbalanced private set intersection.
- Blockchain-based Two-round Private Set Intersection algorithm, namely BTPSI, is proposed by integrating blockchain and two-round private set intersection algorithm above. This integration aims to ensure the traceability of records and streamline collaboration among all parties involved.
- We implement a special list query scheme based on BTPSI. Extensive experiments and evaluations have been conducted to demonstrate the practicality of BTPSI across diverse datasets of different sizes.

The subsequent sections of this paper are structured as follows. Section 2 provides a brief overview of the related work. Section 3 outlines the architecture of the data-sharing system based on blockchain technology, and introduces the Blockchain-based Two-round Private Set Intersection algorithm (BTPSI). Experimental results and analysis are presented in Sect. 4, and ultimately, Sect. 5 concludes this work.

2   Related Work

Research on private set intersection (PSI) was initially initiated by Freedman et al. [16]. Over the years, substantial advancements have been achieved in the development of PSI protocols. In this section, we discuss and analyze four common PSI Schemes: PSI Based on Naive Hashing, PSI Based on Public-key Encryption, PSI Based on Oblivious Transfer, and Outsourced PSI.

2.1   PSI Based on Naive Hashing

The protocol for PSI based on naive hashing involves two parties, namely S and R. At the beginning, each party independently applies a pre-agreed hash function to compute hash values on their respective sets. Next, the computed hash values are transmitted from R to S. Subsequently, S performs a comparison between the received hash values from R and its own set of hash values in order to determine the intersection. Although this protocol demonstrates a high level of efficiency, it is important to note that a vulnerability arises when the input domain is limited in size. In instances of this nature, there exists the possibility of a brute-force attack, wherein one party can compute the hash values for every conceivable element within the domain and subsequently execute the intersection operation to acquire all the inputs of the opposing party.

2.2   PSI Based on Public-Key Encryption

Freeman [16] proposed a PSI protocol that utilizes polynomial interpolation and homomorphic encryption. During the implementation of this scheme, interpolation computations are conducted to generate a polynomial of high degree, where the degree of the polynomial corresponds to the cardinality of the client's set. When the set of elements in the client's dataset is extensive, the polynomial degree increases, leading to substantial computational burden in generating the high-degree polynomial and conducting computations on it. In 2010, Cristofaro and Tsudik [17] presented a PSI protocol based on blind signatures. However, the computational efficiency of this protocol is diminished as a result of the extensive use of exponentiation operations.

2.3   PSI Based on Oblivious Transfer

Oblivious Transfer (OT) is a cryptographic primitive that allows two parties to exchange information in a way that the sender remains oblivious to the information chosen by the receiver, while the receiver remains unaware of any information beyond what he has received. As an extension protocol of Oblivious Transfer, Oblivious Pseudo-Random Function (OPRF) was introduced by Freedman and Ishai et al. [18]. In the context of OPRF, the sender, denoted as Sender A, selects a random seed q, while the receiver, referred to as Receiver B, provides input data y. Receiver B then obtains the output of a pseudorandom function $F(q, y)$. At the same time, the sender is unable to access the input data, while the receiver remains unaware of the seed q.

OPRF can be effectively utilized in the construction of a PSI protocol. Specifically, receiver B takes as input data y and performs an OPRF $F(q, y) = Q_y$. Sender A provides input data x and executes a regular pseudorandom function $F(q, x) = Q_x$, subsequently transmitting Q_x to receiver B. By comparing Q_x and Q_y for equality, it is possible to establish a private equality test for inputs x and y, which can be further extended to PSI for datasets X and Y.

2.4 Outsourced PSI

With the emergence of cloud computing, a substantial quantity of storage resources can be allocated to external cloud servers for computational purposes. Consequently, the utilization of cloud-assisted PSI has garnered significant attention among researchers in recent times. In these protocols, the primary computational tasks are performed by the cloud server, without accessing any plaintext information. Compared to traditional PSI protocols, these protocols typically require the property of collusion resistance. This is because if the server colludes with a participant, the protocol would deteriorate into a secure multiparty computation protocol with varying computational capabilities among the participants [19].

Kerschbaum [20] introduced two outsourced PSI protocols that rely on the utilization of one-way hash functions and RSA public-key encryption. However, the first protocol, akin to naive hashing, is still susceptible to brute-force attacks, while the second protocol demonstrates reduced efficiency. Liu et al. [21] proposed a relatively simple PSI protocol by combining symmetric and asymmetric encryption techniques. The aforementioned approach, however, inadvertently reveals the cardinality of the intersection and poses a potential vulnerability to brute-force attacks. Kamara et al. [22] presented a PSI protocol that utilizes the Advanced Encryption Standard (AES) and pseudorandom functions. The algorithm demonstrates notable computational efficiency within the semi-honest model; however, it is susceptible to collusion attacks, thereby lacking collusion resistance.

3 Blockchain-Based Two-Round Private Set Intersection

3.1 System Architecture

The financial industry exhibits a significant degree of digitization and is currently facing a pressing demand for cross-institutional data collaboration. However, the prevalence of "data islands" remains relatively high due to ongoing concerns regarding data privacy. The paper presents a data-sharing system based on blockchain technology.

This system utilizes a two-round private set intersection algorithm to facilitate secure and compliant data sharing among financial institutions. Additionally, the system incorporates smart contracts to streamline the cooperation process and minimize management costs.

In order to alleviate the storage and computational load on the blockchain, each institution performs the data storage and primary calculation processes locally. Only select interactive data points in the calculation process are stored on the blockchain. When a dispute arises, parties involved have the option to submit an application to the arbitral agency. This agency will then render a verdict on the dispute by considering the evidence stored on the blockchain. The organizational architecture of the data-sharing system based on blockchain technology is illustrated in Fig. 1. The main processes performed by the blockchain-based data-sharing system are as follows.

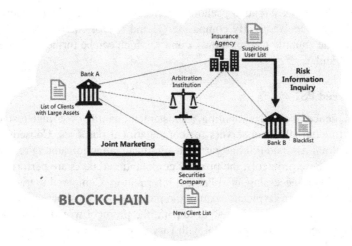

Fig. 1. Organizational architecture of the blockchain-based data-sharing system

1. Each financial institution performs an encryption operation on its respective list and uploads the hash of the entire ciphertext to the blockchain.
2. The inquirer initiates a data intersection request by means of a smart contract and engages in a private set intersection calculation with the service provider who consents to the request. The crucial outcomes obtained during the calculation process are hashed and uploaded to the blockchain to ensure secure record-keeping.
3. The inquirer may provide a reward to the service provider contingent upon the ultimate outcomes, or alternatively, express concerns to the arbitration institution.
4. The arbitration institution renders a decision on whether to punish the service provider, taking into consideration the on-chain proof.

3.2 BTPSI Algorithm

The schematic representation of the BTPSI algorithm is depicted in Fig. 2. Both the sender and the receiver collaboratively execute a two-round privacy algorithm, which includes pre-intersection and precise intersection off-chain. Additionally, they individually store crucial data obtained during the computation process on the blockchain respectively.

Considering the significant discrepancy in the sizes of the datasets belonging to the two parties, the first step entails conducting a pre-intersection process to acquire an approximate set of intersecting elements. Although the larger party's dataset has experienced a significant decrease in volume, the resulting intersection contains some data that is not present in the final intersection. To obtain accurate list data, further intersection operations are necessary.

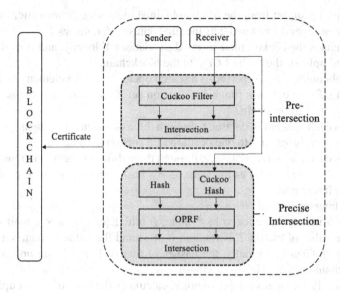

Fig. 2. Algorithm framework

Table 1. Definition of notations

Notation	Definition
A	Server Provider
B	Inquirer
$X(x_i \in X, 1 \le i \le n_1)$	Data set of Service Provider
$Y(y_i \in Y, 1 \le i \le n_2)$	Data set of Inquirer
$Z(z_i \in Z, 1 \le i \le n_3)$	Rough Data Intersection
$fingerprint(\cdot)$	Fingerprint function of the cuckoo filter
$Hash_i(\cdot)\ (0 \le i \le 3)$	Four hash functions
CF_Y	Cuckoo filter inserted into dataset Y
S	Size of stash in cuckoo hash
$C : \{0, 1\}^* \rightarrow \{0, 1\}^k$	Random Encoding Function
$H : [m] \times \{0, 1\}^k \rightarrow \{0, 1\}^l$	Association Robust Hash Function

The parameter definitions used in the BTPSI algorithm are shown in Table 1. It is assumed that both parties have previously participated in intersection, and the dataset of the service provider is significantly larger than that of the Inquirer. The protocol is primarily categorized into three sections: pre-intersection, previous intersection, and arbitration.

1. Pre-Intersection

a. B samples a subset from the results obtained in previous intersection requests to add the original dataset and gets the final dataset, denoted as Y.

b. B calculates the cuckoo filter inserted the dataset Y locally, and then sends CF_Y to A and uploads the hash of CF_Y to the blockchain.

c. After obtaining CF_Y, A performs a search operation on the elements in its dataset X with CF_Y to obtain a rough intersection set Z and generates the hash of Z to store it on the blockchain.

d. B maps each element in dataset Y through three different hash functions $Hash_1(\cdot)$, $Hash_2(\cdot)$, and $Hash_3(\cdot)$ to one position in a cuckoo hash table.

e. A maps each element in dataset Z through three different hash functions $Hash_1(\cdot)$, $Hash_2(\cdot)$, and $Hash_3(\cdot)$ as the same as B's hash functions to three positions in a cuckoo hash table.

2. Previous Intersection

a. For each item in the cuckoo hash table, B initializes two $m \times k$ matrices T and U. The value of matrix T is random 0 or 1, and the value of matrix U satisfies $u_i = t_i \oplus C(item_i)$. Finally, the hashes of matrix T and U are uploaded to the blockchain respectively;

b. A randomly initializes a k-bit vector v, calculates the hash of v, and uploads it to the blockchain for deposit;

c. The two parties perform k OT_2^1: the A gets the corresponding column of matrix T and U according to each bit value v_i of v, namely, A gets t^i when $v_i = 0$ and gets u^i when $v_i = 1$;

d. A combines all the received columns into a matrix Q of size $m \times k$, and uploads the hash of Q to the blockchain;

e. A calculates $F_{((C,v),(i,q_i))}(x) = H(i||q_i \oplus (C(x) \cdot v))$ for each data x in the same position in the cuckoo hash table. At last, A sends set $E = \{F_{((C,v),(i,q_i))}(x)|x \in X\}$ to B and records the hash of E on the blockchain

f. For each position in CF_Y, B calculates $H(i||t_i)$ and checks whether there is the same value as $H(i||t_i)$ in the set E. If so, the item stored at the position of index i in CF_Y is the intersection data.

3. Arbitration

a. After obtaining the final intersection, B can validate the correctness of results by introducing random data prior to the protocol. Specifically, if the aforementioned data are not included in the ultimate intersection, B possesses valid grounds to suspect that A is engaging in fraudulent activities.

b. Once the validity of final result has been confirmed, B can invoke the smart contract to automatically reward the service provider. If the calculated result is believed incorrect, B has the option to raise objections to an arbitration node by first depositing a specified amount.

c. The arbitration institution makes determinations on conflicts by relying on the relevant data stored on the blockchain. If it is determined that the service provider engages in fraudulent behavior during the calculation process, appropriate punitive measures will be enforced against the service provider through the utilization of the smart contract. Otherwise, the deposit of the inquirer will be retained as a punitive measure for engaging in frivolous litigation. Simultaneously, the outcomes of arbitration are recorded on the blockchain to facilitate the positive progression of the entire system.

The pseudocode for the inquirer and the service provider are described in Table 2 and Table 3, respectively.

Table 2. Blockchain-based two-round secure intersection protocol—Inquirer

---

Inquirer()

// Generate a cuckoo filter
foreach y in Y **do**
 CF.insert(y);
return CF_Y
Send CF_Y to Service Provider
Upload hash of CF_Y to Blockchain
//Cuckoo Hash
foreach y in Y **do**
 CuckooHash(y);
return $CH_Y[\ ]$
// OPRF
Generate random matrix $T[\ ][\ ]$
calculate matrix $U[\ ][\ ]$: $u_i = t_i \oplus C(y_i)$
input (t^i, u^i) into OT
Upload hash of T and U to Blockchain
// Intersection
foreach i in $CH_Y[\ ]$ **do**
 If $H(i||t_i)$ in set $E = \{F_{((c,v),(i,q_i))}(x)|x \in X\}$ **then**
  Add $CH_Y[i]$ to $X \cap Y$
return $X \cap Y$
// validate
Validate the correctness of $X \cap Y$
If valid
 Reward Service Provider
Else
 Apply for arbitration

Table 3. Blockchain-based two-round security negotiation protocol – Service Provider

Service Provider()
// Cuckoo filter intersection
Receive CF_Y from Inquirer
foreach x in X **do**
if CF_Y.lookup(x) = true **then**
add x to Z
return Z
//Cuckoo Hash
foreach x in Z **do**
CuckooHash(z);
return $CH_X[\ ]$
// OPRF
Generate random vector v
Take as input v_i to perform OT
receive q^i and make up matrix $Q[\ ][\ ]$
//PRF
foreach row q_i in Q **do**
add $H\big(i||q_i\oplus(C(x)\cdot v)\big)$ to $E[\ ]$
Send E to Inquirer
Upload hash of E to Blockchain

4 Performance and Security Evaluation

4.1 Performance Evaluation

A test system was developed using the Python programming language to assess the effectiveness of the BTPSI (Binary Tree-based Parallel Sorting Index) proposed in this research paper. The testing system comprises two primary components: the user system and the blockchain system, each serving distinct purposes. The pertinent configuration is displayed in Table 4.

Table 4. Test system development environment configuration

Item	Configuration
Operating System	Ubuntu Server 18.04.1 LTS 64bit
CPU	Intel(R) Core(TM) i7-9750H CPU @ 2.60 GHz
Memory Size	8 GB
Hard Disk Size	60 GB

The user system primarily assumes responsibility for facilitating communication between the two parties, enabling interaction, and storing data on the chain. Participants engage in a private set intersection, wherein the hashes of crucial data are generated and subsequently uploaded to the blockchain via the system.

The blockchain system is responsible for the management of the underlying blockchain network. It is also responsible for deploying a smart contract that implements the reward or punishment mechanism. Additionally, the system is tasked with storing crucial data related to the calculation process and recording all arbitration decisions for disputes. The WuTong chain functions as a blockchain system that has successfully implemented essential technologies, including cryptographic algorithms and smart contract engines.

According to the conclusion of [23], load factor α and target false positive rate ϵ are dependent on the number of entries per bucket b and the length of the fingerprint in bits f. The size of the cuckoo hash table is set to 1.2 times the number of elements, as suggested by the findings in [24]. In order to strike a trade-off between efficiency and safety, the experiment employs the parameter configurations as presented in Table 5.

Table 5. Parameter Settings.

Parameter	Description	Value
b	number of entries per bucket	2
f	fingerprint length in bits	4
c	ratio of the size of the cuckoo hash table to the size of Y	1.2

To replicate the scenario of a private set intersection on a special list, this system generates data in the list of both parties randomly, following the format of the ID number. Specifically, the formation of an ID number involves the concatenation of 18 decimal characters. These characters consist of a 6-digit AREA CODE, an 8-digit birth date, a 2-digit sequence code, a 1-digit gender code, and a 1-digit check code. The selection of the AREA CODE is conducted in a random manner, encompassing all codes assigned to the administrative divisions of the People's Republic of China. The range of birth dates is restricted to January 1st, 1960 to January 1st, 2022. The sequence code and gender code are represented by two and one decimal random numbers, respectively. The determination of the last digit of the check code is based on the content of the first 17 digits.

In order to assess the algorithm's performance across datasets of varying sizes, the experiments are conducted by testing different magnitudes of datasets for both the inquirer and the service provider. Specifically, the dataset sizes for the inquirer vary in scale, ranging from small to large, encompassing 1,000, 10,000, and 100,000 ID records. On the other hand, the dataset sizes for the service provider are set at a larger magnitude, comprising 100,000, 1 million, and 10 million ID numbers. The objective of this experimental design is to examine the performance of the algorithm when applied to datasets of varying scales. By progressively enlarging the dataset used by the inquirer, it becomes possible to investigate the influence of dataset size on the intersection of the inquirer. Simultaneously, the dataset sizes for the service provider are adjusted in order to facilitate a thorough evaluation.

The private set intersection protocol necessitates both data encryption and the successful transmission of data between the two parties. Therefore, the experiment aims to assess the efficacy of the protocol by considering two key factors: computing time and communication overhead. The experimental results are presented in Table 6 and Table 7. The former graph illustrates the running time of each step in the BTPSI algorithm, whereas the latter graph displays the size of communication bytes for each step. The total runtime and communication bytes are also presented in last column of Table 6 and Table 7 respectively.

Table 6. The running time of BTPSI

Data set size of Inquirer	Data set size of Service Provider	Pre-intersection Time (s)	Precise Intersection Time (s)	Total Time (s)
1k	10w	0.832013	8.278473	9.110476
1k	100w	2.770037	20.72564	23.49567
1k	1000w	22.03498	158.1204	180.1554
1w	10w	7.219221	21.75890	15.91597
1w	100w	9.172431	157.1833	30.93133
1w	1000w	28.69464	13.29266	185.8779
10w	10w	71.24536	13.29266	84.53802
10w	100w	73.10326	27.00179	100.1051
10w	1000w	91.86684	160.68830	252.5552

Table 7. The communication bytes size of BTPSI

Data set size of Inquirer	Data set size of Service Provider	Pre-intersection Communication Bytes (kb)	Precise Intersection Communication Bytes (kb)	Total Communication Bytes (kb)
1k	10w	0.046875	1686.047	1686.09388
1k	100w	0.046875	15766.16	15766.2069
1k	1000w	0.046875	155685.8	155685.847
1w	10w	0.046875	3092.023	3092.06988
1w	100w	0.046875	17192.85	17192.8969
1w	1000w	0.046875	157161.3	157161.347
10w	10w	0.046875	16631.961	16632.0079
10w	100w	0.046875	31009.02	31009.0669
10w	1000w	0.046875	171183.5	171183.547

The process of depositing data onto the WuTong chain is achieved through the utilization of the Restful API interface. Leveraging the inherent storage capabilities of the blockchain, the shared data is maintained in an immutable, tamper-proof, and traceable manner. The WuTong chain facilitates the secure storage of data through encryption techniques, allowing for authorized retrieval of data as needed. Nevertheless, the blockchain network is not optimally designed for the storage of large amounts of data, primarily due to the constraints of distributed ledger technology. At the same time, data confidentiality is a necessary requirement. The original data is stored locally, while only computed hash values are stored on the blockchain to ensure the integrity of the data.

4.2 Security Evaluation

In the specific scenario of financial institutions' special list query, the service provider may seek to falsify results by undermining the execution of the agreement. If the service provider fails to abide by the terms of the agreement, the intersection result returned to the inquirer must be incorrect or incomplete. Then the inquirer can apply to a third-party trusted institution for arbitration, which will trace the interaction during the request process. If the arbiter finds that the service provider has engaged in fraudulent behavior, it will determine that the settlement has failed and will punish the service provider. The high cost of engaging in malicious behavior reduces the likelihood of the service provider performing such actions to a certain extent.

In the context of querying a special list in financial institutions, the service provider may attempt to manipulate the outcomes by undermining the proper execution of the agreement. If the service provider does not adhere to the terms stipulated in the agreement, the outcome of the intersection returned to the inquirer may be inaccurate or insufficient. The inquirer has the option to seek arbitration from a reputable third-party institution, which will recall the entire interaction through the request records on the blockchain. If the arbiter determines that the service provider has been involved in fraudulent activities, it will impose penalties on the service provider. The elevated expenses associated with engaging in malicious behavior serve as a deterrent for service providers, thereby reducing the probability of them engaging in such actions to some degree.

If both parties have the intention to acquire the exclusive list data of the other party for the purpose of generating profit, they may endeavor to deduce the exclusive list data of the service provider from the information obtained during the privacy processing procedure. For the researcher, acquiring comprehensive data from the special list is unattainable due to the confidential nature of the OPRF seed, which is only known and safeguarded by the sender. If one desires to acquire the output of the Oblivious Pseudorandom Function (OPRF) through exhaustive enumeration of input data, it will be necessary to engage in multiple interactive communications with the server. However, this may raise suspicion with the service provider. In the context of server-side privacy processing, it is important to note that while the server possesses the OPRF seed, it remains unaware of the specific OPRF input associated with the data provided by the inquirer. Both individuals are incapable of acquiring the primary information from each other.

If both parties seek to obtain all special list data of the other party for profit reasons, they may attempt to infer the special list data of the service provider from the information received during the privacy processing process. For the inquirer, it is impossible to obtain all special list data because the OPRF seed is only known by the sender and kept confidential. If it wants to obtain the output of the OPRF by exhaustively enumerating input data, it must engage in multiple interactive communications with the server, which will raise suspicion with the service provider. For the server, in the privacy processing process, although it has the OPRF seed, it does not know the OPRF input corresponding to the data of the inquirer. Consequently, both of them are unable to learn the original data from the other one.

Furthermore, the blockchain records the hash values of key interaction data, indicating that the data stored on the chain becomes increasingly challenging to decrypt as it undergoes additional processing through a hash algorithm. In reality, in the event that malevolent nodes acquire the hash value of the data, they are unable to access any legitimate information regarding the plaintext of said data.

In consideration of the aforementioned, the BTPSI algorithm, as presented in this paper, successfully attains the security of intersection in scenarios where the server may exhibit dishonest behavior.

5   Conclusion

This paper presents the BTPSI algorithm, which introduces a pre-intersection step utilizing a cuckoo filter into the current OT-based PSI algorithm. The aim is to enhance the algorithm's performance in scenarios where datasets are unbalanced. At the same time, the BTPSI algorithm incorporates blockchain technology to guarantee the immutability of data records throughout the calculation process. Additionally, it has the capability to enforce penalties on participating parties engaging in malicious behavior through the utilization of smart contracts in the event of disputes.

The BTPSI algorithm, while capable of enhancing efficiency in cases of unbalanced PSI, still has potential for further improvement in the efficiency of balanced PSI. Furthermore, extending the proposed algorithm to an effective PSI scheme for three or even more participants is our next research direction.

Acknowledgments. This work was supported by National Key Research and Development Program under 2021YFC3340600.

References

1. Chen, M., Mao, S., Liu, Y.: Big data: a survey. Mob. Netw. Appl. **19**, 171–209 (2014)
2. Wylde, V., et al.: Cybersecurity, data privacy and blockchain: a review. SN Comput. Sci. **3**, 12 (2022)
3. Nakamoto, S.: Bitcoin: a peer-to-peer electronic cash system. Consulted (2008)
4. Tan, L., Yu, K., Shi, N., Yang, C., Lu, H.: Towards secure and privacy-preserving data sharing for covid-19 medical records: a blockchain-empowered approach. IEEE Trans. Netw. Sci. Eng. 1 (2021)
5. Piao, C., Hao, Y., Yan, J., Jiang, X.: Privacy preserving in blockchain-based government data sharing: a service-on-chain (SOC) approach. Inf. Process. Manage. **58**, 12 (2021)
6. Jiang, S., Cao, J., Wu, H., Chen, K., Liu, X.: Privacy-preserving and efficient data sharing for blockchain-based intelligent transportation systems. Inf. Sci.: Int. J. (2023)
7. Lu, Y., Huang, X., Dai, Y., Maharjan, S., Zhang, Y.: Blockchain and federated learning for privacy-preserved data sharing in industrial IoT. IEEE Trans. Industr. Inf. **16**, 4177–4186 (2020)

8. Sultana, T., Ghaffar, A., Azeem, M., Abubaker, Z., Javaid, N.: Data sharing system integrating access control based on smart contracts for IoT. In: in 14th International Conference on P2P, Parallel, Grid, Cloud and Internet Computing (2019)
9. Goyal, V., Pandey, O., Sahai, A., Waters, B.: Attribute-based encryption for fine-grained access control of encrypted data. In: Proceedings of the 13th ACM Conference on Computer and Communications Security, pp. 89–98. Association for Computing Machinery, Alexandria (2006)
10. Rouhani, S., Deters, R.: Blockchain based access control systems: State of the art and challenges. IEEE/WIC/ACM International Conference on Web Intelligence, pp. 423–428. Association for Computing Machinery, Thessaloniki (2019)
11. Wang, Z., Tian, Y., Zhu, J.: Data sharing and tracing scheme based on blockchain. In: 8th International Conference on Logistics, Informatics and Service Sciences (LISS), Toronto, ON, Canada, pp. 1–6 (2018)
12. Zheng, B.-K., et al.: Scalable and privacy-preserving data sharing based on blockchain. J. Comput. Sci. Technol. **33**, 557–567 (2018)
13. Feng, T., Pei, H., Ma, R., Tian, Y., Feng, X.: Blockchain data privacy access control based on searchable attribute encryption. Comput. Mater. Continua (2020)
14. Li, Y., Zhang, G., Zhu, J., Wang, X.: Privacy protection model for blockchain data sharing based on zk-SNARK. In: Zeng, J., Qin, P., Jing, W., Song, X., Lu, Z. (eds.) ICPCSEE 2021. CCIS, vol. 1452, pp. 229–239. Springer, Singapore (2021). https://doi.org/10.1007/978-981-16-5943-0_19
15. Wenge, O., Lampe, U., Müller, A., Schaarschmidt, R.: Data privacy in cloud computing - an empirical study in the financial industry (2014)
16. Freedman, M.J., Pinkas, B., Nissim, K., Hazay, C.: Efficient set intersection with simulation-based security. J. Cryptol.: J. Int. Assoc. Cryptol. Res. (2016)
17. De Cristofaro, E., Tsudik, G.: Practical private set intersection protocols with linear complexity. In: Sion, R. (ed.) FC 2010. LNCS, vol. 6052, pp. 143–159. Springer, Heidelberg (2010). https://doi.org/10.1007/978-3-642-14577-3_13
18. Freedman, M.J., Ishai, Y., Pinkas, B., Reingold, O.: Keyword search and oblivious pseudo-random functions. In: Kilian, J. (ed.) TCC 2005. LNCS, vol. 3378, pp. 303–324. Springer, Heidelberg (2005). https://doi.org/10.1007/978-3-540-30576-7_17
19. Jiang, H., Xu, Q.: Secure multiparty computation in cloud computing. J. Comput. Res. Dev. (2016)
20. Kerschbaum, F.: Collusion-resistant outsourcing of private set intersection. In: ACM Symposium on Applied Computing (2012)
21. Liu, F., Ng, W.K., Zhang, W., Giang, D.H., Han, S.: Encrypted set intersection protocol for outsourced datasets. In: Proceedings of the 2014 IEEE International Conference on Cloud Engineering, pp. 135–140. IEEE Computer Society (2014)
22. Kamara, S., Mohassel, P., Raykova, M., Sadeghian, S.: Scaling private set intersection to billion-element sets. In: Christin, N., Safavi-Naini, R. (eds.) FC 2014. LNCS, vol. 8437, pp. 195–215. Springer, Heidelberg (2014). https://doi.org/10.1007/978-3-662-45472-5_13
23. Fan, B., Andersen, D.G., Kaminsky, M., Mitzenmacher, M.D.: Cuckoo filter: practically better than bloom. In: Proceedings of the 10th ACM International on Conference on emerging Networking Experiments and Technologies, pp. 75–88. Association for Computing Machinery, Sydney (2014)
24. Pagh, R., Rodler, F.F.: Cuckoo hashing. J. Algorithms **51**, 122–144 (2004)

Novel Quaternion Orthogonal Fourier-Mellin Moments Using Optimized Factorial Calculation

Chunpeng Wang[1,2], Long Chen[1,2], Zhiqiu Xia[1,2(✉)], Jian Li[1,2], Qi Li[1,2], Ziqi Wei[3], Changxu Wang[4], and Bing Han[1,2]

[1] Key Laboratory of Computing Power Network and Information Security, Ministry of Education, Shandong Computer Science Center, Qilu University of Technology (Shandong Academy of Sciences), Jinan, China
sddxmb@126.com

[2] Shandong Provincial Key Laboratory of Computer Networks, Shandong Fundamental Research Center for Computer Science, Jinan, China

[3] Institute of Automation, Chinese Academy of Sciences, Beijing 100190, China

[4] Faculty of Computing, Universiti Teknologi Malaysia, Johor Bahru, Malaysia

Abstract. This paper provides an in-depth discussion on the application of quaternion orthogonal Fourier-Mellin Moments (QOFMM) in the field of digital image processing, and proposes a novel method of factorial operation aiming to optimize its computational efficiency and accuracy. In addition, this paper focuses on the importance of zero-watermark technology in the field of information security. QOFMM is an advanced feature mention technique, which is particularly applicable to the field of digital image processing and information security. In this technique, the factorial operation plays a key role. However, traditional factorial computation methods may encounter efficiency bottlenecks when dealing with large data, thus affecting the overall performance. To address this issue, this study proposes an innovative method for factorial operation, which performs factorial operation by improving the radial basis function computation strategy, aiming to reduce the computational complexity and enhance the computational accuracy. To verify the effectiveness of the proposed method, we compare it with existing methods of factorial computation. The experimental results show that the new method significantly improves both processing speed and computational capability. Overall, this paper provides a new QOFMM optimization strategy, which not only improves the computational efficiency and accuracy but also brings a new research direction to the field of digital image processing and zero-watermarking technology.

Keywords: Zero-watermarking · Quaternions · Orthogonal Fourier-Mellin moments · Factorial operations · Digital image processing · ISnformation security

1 Introduction

Image moments are an effective method for image feature extraction, and their geometric invariance makes them valuable in designing robust watermarking algorithms against geometric attacks. Common geometric invariant moments include Legendre moments

B. Ma et al. (Eds.): IWDW 2023, LNCS 14511, pp. 262–276, 2024.
https://doi.org/10.1007/978-981-97-2585-4_19

(LM) [1], Zernike moments (ZM) [1], pseudo-Zernike moments (PZM) [2], polar harmonic Fourier moments (PHFMs) [3], trinion fractional-order continuous orthogonal moments(TFrCOMs) [4], orthogonal Fourier-Mellin moments (OFMM) [5], Chebyshev-Fourier moments (CHFM) [6], radial harmonic Fourier moments (RHFM) [7], Bessel-Fourier moments (BFM) [8], polar harmonic transforms (PHT) [9], and Exponent moments (EM) [10]. These moments provide invariance to rotation, scaling, and other transformations, and are widely used to construct robust watermarking schemes, common schemes include: Zhang et al. proposed an image watermarking algorithm based on Legendre moments invariant moments, which can resist geometrical attacks such as rotation, scaling, and flipping [11]. Hosny et al. proposed robust steganographic image watermarking algorithms based on imitating Legendre moments [12]. Hossein Rahmani et al. proposed a semi-blind robust image watermarking scheme based on watermarking and the most perceptually important sub-image (MPISI) of the original image [13]. Xia et al. proposed a method for color medical image lossless watermarking that leverages chaotic systems and accurate quaternion polar harmonic transforms [14]. Deng et al. combined the technology with Harris detector to improve the robustness [15]. Sun et al. proposed an image watermarking algorithm based on a discrete wavelet transform (DWT) algorithm [16]. Ma et al. proposed a high-performance robust reversible data hiding algorithm that utilizes polar harmonic Fourier moments [17]. Shao et al. combined OFMM and chaotic mapping to realize dual image copyright protection [18]. Bi et al. proposed a blind image watermarking algorithm based on multi-band wavelet transform and empirical modal decomposition [19].

Domestic and foreign research on QOFMM, although some progress has been made, still faces some challenges in the application to practical image processing. In particular, how to combine modern image processing techniques, such as deep learning and big data analysis, to further improve its efficiency and robustness. However, at the same time, research institutes in Europe and the United States, have published several papers on the application of image moments in advanced image processing techniques, which not only explore the basic theory but also combine with modern image processing techniques such as deep learning and big data analysis, aiming to further improve the efficiency and robustness of the algorithms. In summary, image moments are widely used in designing robust watermarking algorithms against geometric attacks.

In this paper, QOFMM is used, which has a broad application prospect in the field of image processing by combining the advantages of quaternion [20] and OFMM [5]. This technique is used in the zero-watermarking algorithm to improve the robustness, and the experimental results verify the performance of the zero-watermarking algorithm designed based on the optimized QOFMM.

The contributions and innovations of this paper are as follows: we have optimized the radial basis function algorithm of QOFMM to save computational time and enhance computational power. By combining image moments with quaternion algebra, we have improved the performance of hypercomplex image moments. Specifically, we apply the quaternion framework in the computation of image moments to encapsulate the multidimensional nature of color image data, thereby extending traditional moment invariants into the hypercomplex number space. This combination makes use of the rich mathematical structure of quaternions, enabling the effective simultaneous representation of both spatial and color information of images. The scheme proposed in this paper can be applied to the zero-watermarking algorithm. Experimental results show that the scheme is robust to various common attacks and combinations of attacks.

The rest of the paper is organized as follows: Sect. 2 describes the definition and computation of QOFMM as well as experimental comparisons with traditional methods to verify the performance improvement. Section 3 introduces the zero-watermarking algorithm and the combined algorithm with the new QOFMM. Section 4 gives the experimental results to verify the performance enhancement of zero-watermark and the good invisibility and robustness of image watermarking based on QOFMM. Finally, Sect. 5 summarizes the full paper.

2 Quaternion Orthogonal Fourier-Mellin Moments

2.1 Definition of Orthogonal Fourier-Mellin Moments

The OFMM of order $n(n \geq 0)$ with repetition $m(|m| \geq 0)$ of an image $f(r, \theta)$ in a polar coordinate system is defined as follows:

$$\Phi_{nm} = \frac{1}{2\pi a_n} \int_0^{2\pi} \int_0^1 f(r, \theta) Q_n(r) \exp(-jm\theta) r dr d\theta \tag{1}$$

where n is a nonnegative integer, m is an integer, radial basis function $Q_n(r)$ is an n-th order polynomial of r expressed as:

$$Q_n(r) = \sum_{s=0}^{n} a_{ns} r^s \tag{2}$$

where $a_{ns} = \sum_{s=0}^{n} [(-1)^{(n+s)} \times \frac{(n+s+1)!}{(n-s)! \times s! \times (s+1)!}] \times r^s$ is the coefficient of the polynomial. It is weighted orthogonal in the range of $0 \leq r \leq 1$:

$$\int_0^1 Q_n(r) Q_K(r) r dr = a_n \delta_{nk} \tag{3}$$

where δ_{nk} is the Kronecker symbol and $a_n = 1/[2(n+1)]$ is the normalization factor, the basis functions of the orthonormal Fourier-Mellin moments $Q_n(r) \exp(jm\theta)$ are orthogonal in the unit circle according to the properties of the radial basis functions $Q_n(r)$ and the angular circular harmonic factors $\exp(jm\theta)$:

$$\int_0^{2\pi} \int_0^1 f(r, \theta) Q_n(r) \exp(-jm\theta) Q_K(r) \exp(-jl\theta) r dr d\theta = \frac{\pi}{n+1} \delta_{nk} \delta_{nl} \tag{4}$$

According to the theory of orthogonal function systems, the original image $f(r, \theta)$ can be reconstructed with a finite number of OFMM approximations. If the OFMM with the highest order $n_{\max}$ and the largest repetition $n_{\max}$ is known, then the image $f(r, \theta)$ can be reconstructed as follows:

$$f(r, \theta) \approx \sum_{n=0}^{n_{\max}} \sum_{m=-m_{\max}}^{m_{\max}} \Phi_{nm} Q_n(r) \exp(jm\theta) \tag{5}$$

2.2   Traditional Calculation of QOFMM

A quaternion [21] is a mathematical structure beyond the complex numbers, usually denoted by $q = a + bi + cj + dk$, where (a, b, c, d) is the real number and (i, j, k) is the imaginary part and satisfies the following relation:

$$i^2 = j^2 = k^2 = ijk = -1$$
$$ij = k = -ji$$
$$jk = i = -kj$$
$$ki = j = -ik$$

(6)

The multiplication of two imaginary units is not interchangeable, and quaternions are not multiplicative. Conjugate use of quaternions Eq. (7), is defined as follows:

$$q^* = a - bi - cj - dk \tag{7}$$

For color images, each channel (e.g., red, green, blue) can be represented by its corresponding function f_r, f_g, f_b. We can define a quaternion image function:

$$F(r, \theta) = f_r + f_g i + f_b j \tag{8}$$

For each color channel use the above formula to calculate the orthogonal Fourier-Mellin moments, and finally construct the QOFMM combining the results of each channel to obtain the quaternion form of the Fourier-Mellin moments. However, the most difficult part of the calculation is to compute the polynomial coefficients a_{ns}.

This involves very large factorial and exponential operations. The factorial is usually denoted as n! and is defined as the product of all positive integer values of n, i.e., $n! = n \times (n - 1) \times \dots \times 2 \times 1$. In particular, 0! is defined as 1. Especially in the computation of orthogonal Fourier-Mellin moments, the factorial plays an important role in the computation of various parameters. In traditional computational methods, the factorial is primarily computed by direct multiplication, which typically involves iterating from 1 to n using a loop and multiplying each value to obtain $n!$ Each multiplication involves a large number of operations whose computation increases dramatically as the number grows, and the true computational time and storage complexity exceeds that of a simple O(n). When dealing with very large numbers, such as 1000!, we need a huge amount of storage space to save the result, which can lead to a waste of storage resources. There is a risk of losing accuracy when we multiply large numbers consecutively, especially with limited computational resources, and it is difficult to apply it to more complex mathematical structures or algorithms.

2.3 Optimized Calculation of QOFMM

Start by going through the fact that any positive integer can be expressed as a product of its prime factors. For factorials, this means that we can represent them as a collection of prime factors rather than a single large number [22]. For example: $5! = 5 \times 4 \times 3 \times 2 \times 1 = 2^3 \times 3 \times 5$. Let's look at the formula a_{ns} and it's easy to see that in it the values of s accumulate from 0. For this formula, notice that there are multiple factorial combinations in the numerator and denominator. First, it is important to note that some combinations of factors can be eliminated directly. For example, for $(n + s + 1)!$ and $(n - s)!$, $(n - s)!$ is a subset of $(n + s + 1)!$ when $s \neq 0$, so they have many factors that can directly cancel each other out. And then, s! is a subset of $(s + 1)!$ a subset of s, the extra is just a multiplier of s + 1. With the above observations, we propose the following simplified computational steps:

For each value of s, calculating $(n + s + 1)! / (n - s)!$ actually only involves multiplying the numbers $s + 1$, $s + 2$, ..., $n + s + 1$, since the other parts are offset by $(n - s)!$. Because of $s!$ and $(s + 1)!$, we only need to multiply by $s + 1$ as an additional multiplier. This simplification provides us with computational convenience, as we no longer need to perform a full factorial calculation, but instead calculate the necessary multipliers directly. Using the Eratosthenes sieve, a pure, redundancy-free list of numbers is obtained. Finally, each factorial factor is decomposed and then saved then the remaining terms are finally performed by canceling the numerators and denominators. This optimization method not only reduces the total number of steps in the computation but also improves the accuracy of the computation, especially when the values of n and s are large. By removing unnecessary calculations, we also reduce the likelihood of errors, especially in the case of numeric overflow or underflow, and we only save prime numbers and their exponents, thus greatly reducing storage requirements.

2.4 Reconstruction Time Comparison

A series of detailed experiments are carried out for this purpose to truly represent the difference between the two methods in terms of reconstruction time. In the experiments, image data of different orders are selected for testing, and the average value of multiple experiments is selected by removing the CPU-influenced deviation excessive data in each group of experiments, focusing on their reconstruction time. Figure 1 shows the comparison results:

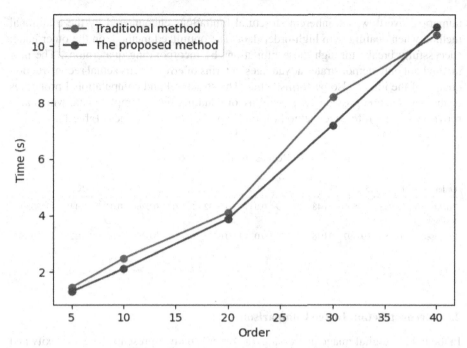

Fig. 1. Comparison of reconstruction times.

With the increase of image order, the reconstruction time of the two methods naturally rises, but the obvious difference is shown in that our proposed method outperforms the traditional computational methods in every stage, especially in middle to high-order processing. This means that the proposed method is more applicable and practical when faced with large amounts of data or complex images. In low-order reconstruction, although the time interval between the two methods is relatively small, the optimized QOFMM method starts to lead when the order is increased. Overall, our proposed method demonstrates significant superiority in both the quality and speed of image reconstruction. The experimental results depict the time optimization of our proposed method over the traditional computational methods.

2.5 Reconstruction Error Comparison

The above graphs clearly show how the reconstruction error varies with order as the order increases. Until the 21st order, the errors of both traditional and improved methods are quite stable, which indicates that the accuracy of both methods is quite reliable in low-order processing. However, after this, there is a clear divergence in the performance of the two. Especially at the 22nd and 23rd orders, the improved method still maintains a low error, while the error of the traditional method increases sharply and even becomes almost unacceptable at the 23rd order. Such a difference reflects the limitation of the traditional method in dealing with high-order images. Especially at the 22nd order, the reconstruction quality of the actual image is seriously impaired, although the reconstruction error is still within the acceptable range. This phenomenon is attributed to the numerical overflow and inherent structural limitations encountered by the traditional method when dealing with high-order data. The optimized method, on the other hand, successfully breaks through these limitations by incorporating quaternions. The new method not only demonstrates advantages in terms of error, but its actual reconstruction quality of the image also performs better. The structural and computational properties of the new method provide the possibility of handling more complex data, which also provides another reference method for image processing techniques (Table 1).

Table 1. Reconstruction error comparison

Order	1	5	7	10	12	13	15	20	21	22	23
Traditional method	0.0953	0.0597	0.048	0.0379	0.0349	0.0332	0.0302	0.0266	0.0269	0.4455	3490.656
Proposed method	0.0953	0.0597	0.048	0.0379	0.0349	0.0332	0.0302	0.0266	0.0259	0.0254	0.194

2.6 Reconstruction Image Comparison

In the field of digital image processing, the order usually represents the complexity and richness of the details of an image. With the advancement of technology, it has become a trend to process image data with high-order numbers because it can provide richer information and higher image quality.

Traditional method reaches their limit at the 21st order due to technical and theoretical limitations, and when processing images higher than this order, not only does the reconstruction error increases dramatically, but also the quality of the reconstructed image is seriously affected. Our proposed method employs the quaternion technique combined with a new factorial computation method, which can directly and efficiently process three-channel color images. In the following image Fig. 2, is the test image and Fig. 3 is the experimental comparison image. The figure demonstrates that the quality of the reconstructed image is significantly improved and shows better superiority in processing high-order image data. Our proposed method maintains low error and high quality in both 22nd and 23rd-order image reconstruction.

Fig. 2. Test images.

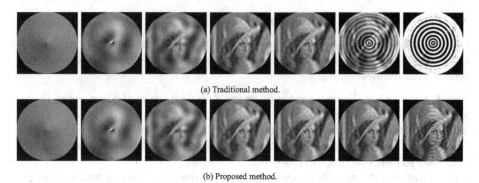

(a) Traditional method.

(b) Proposed method.

Fig. 3. Reconstruction image comparison ($n_{\max} = 1, 5, 10, 20, 21, 22, 23$).

3 The Proposed Zero-Watermarking Algorithm

Combined with the optimized QOFMM, we designed an efficient and robust zero-watermarking algorithm. The algorithm mainly includes two processes: the construction of zero-watermark and the verification of zero-watermark [23]. The addition of the improved QOFMM makes the entire zero-watermarking algorithm more efficient and reliable. The implementation process and specific steps of the algorithm are introduced in detail below. Figure 4 shows the overall flow chart:

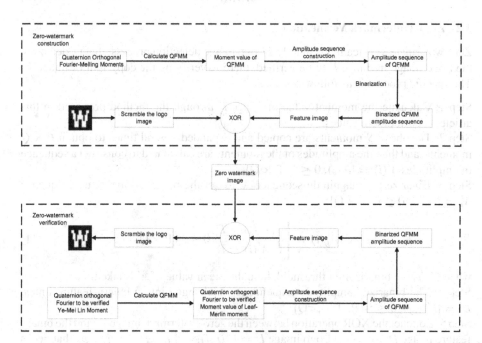

Fig. 4. Flowchart of proposed method.

3.1 Zero-Watermark Construction

Let $I = \{f(x, y), 0 \leq x, y \leq N\}$ be the original QOFMM image and $L = \{l(i, j), 0 \leq i \leq P, 0 \leq j \leq Q\}$ be the original binary Logo image, the zero-watermark construction process is as follows:

Step 1: Calculate the moment value of image I through the method proposed in this article, get S moment values.

Step 2: The obtained S moments are copied and expanded several times to obtain $P \times Q$ moments, and then the amplitudes of the moments are calculated to construct the amplitude sequence $A(i) = \{a(i), 0 \leq i < P \times Q\}$ of length $P \times Q$.

Step 3: Binarize the magnitude sequence A to get the binarized magnitude sequence $A_b = \{a_b(i), 0 \leq i < P \times Q\}$:

$$A_b(i) = \begin{cases} 1, A(i) \geq T \\ 0, A(i) \leq T \end{cases}, 0 \leq i < P \times Q \qquad (9)$$

where T is the binarization threshold, here the mean value of A is taken.

Step 4: The binarized amplitude sequence A_b is changed into a binary feature image $D = \{f(i, j), 0 \leq i < P, 0 \leq j < Q\}$.

Step 5: Embed the Logo image L into the binary feature image D to get the zero-watermark image $W = \{w(i, j), 0 \leq i < P, 0 \leq j < Q\}$ using the XOR operation.

$$W = XOR(L, D) \qquad (10)$$

3.2 Zero-Watermark Verification

Zero-watermark verification is mainly a Logo image detection process, that is, detecting the Logo image in image I' to be verified, thereby verifying the copyright of image I'. The specific process is as follows:

Step 1: Calculate the moment value of image I' through the method proposed in this article, get S moment values.

Step 2: The above S moments are copied and expanded several times to obtain $P \times Q$ moments, and then the amplitudes of the moments are calculated to construct a sequence of amplitudes $A'(i) = \{a'(i), 0 \leq i < P \times Q\}$.

Step 3: Binarize the magnitude sequence A' to get the binarized magnitude sequence $A'_b = \{a'_b(i), 0 \leq i < P \times Q\}$:

$$A'_b(i) = \begin{cases} 1, A'(i) \geq T \\ 0, A'(i) \leq T \end{cases}, 0 \leq i < P \times Q \qquad (11)$$

where T' is the binarization threshold, here the mean value of A' is taken.

Step 4: The binarized amplitude sequence A'_b is changed into a binary feature image $D' = \{f'(i, j), 0 \leq i < P, 0 \leq j < Q\}$.

Step 5: Conduct the XOR operation between the zero-watermark image W and the binary feature image D' to get the Logo image $L' = \{l'(i, j), 0 \leq i \leq P, 0 \leq j \leq Q\}$ that needs to be detected:

$$L' = XOR(W, D') \qquad (12)$$

4 Experiments

4.1 Simulation and Analysis

The match between the detected Logo image and the original Logo image is measured using the Bit Correct Rate (BCR), which is defined as the proportion of bits in the detected Logo image that match the original Logo image. It is a measure of the similarity of two images, where a higher BCR value indicates higher agreement. BCR is defined as:

$$BCR = \frac{C}{P \times Q} \times 100\% \tag{13}$$

where C is the correct number of bits in the detected Logo image and $P \times Q$ is the size of the Logo image. The value of the ratio ranges from 0 to 1. When the BCR value tends to 1, it means that the more similar the detected Logo image is to the original Logo image, which indicates that the zero-watermarking algorithm has better robustness. On the contrary, a lower BCR value means that there is a large difference between the two.

The simulation experiment was conducted in the Matlab R2016 environment. 15 256 × 256 color images were selected as the original images to be tested. The color image used for display in the text and a 16 × 16 binary watermarked image is shown in Fig. 5.

Fig. 5. Original image and watermarked image.

4.2 Robust Experiments

(1) **Filtering attacks**

Image filtering will result in a blurred edge of the watermarked image[24]. The watermarked image was tested under two types of filtering attacks including mean filtering, and Gaussian filtering. As can be seen from Fig. 6, our proposed algorithm has good robustness to filtering attacks.

(2) **Noise attacks**

Image noise is a common interference. To analyze the algorithm's ability to resist noise attacks, salt-and-pepper noise, and Gaussian noise were added to the watermarked images for testing. Figure 7 shows the watermarked images after the addition of two types of noises, the watermarks extracted from the watermarked images, and the corresponding BCR values. It can be seen that when the variance of Gaussian noise is 0.005 and the density of salt-and-pepper noise is 0.01, the algorithm proposed in this article can still effectively extract watermarks, indicating that the algorithm can effectively resist noise attacks.

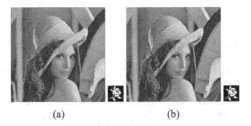

(a)   (b)

Fig. 6. Experimental results under image filtering attack with (a) mean filtering 3×3, BCR = 0.99414 and (b) Gaussian filtering 3×3, BCR = 0.99882.

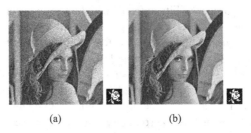

(a)   (b)

Fig. 7. Experimental results under image noise attack (a) Gaussian noise 0.005, BCR = 0.97345, (b) Salt and pepper noise, 0.01, BCR = 0.98925.

(3) **JPEG compression attacks**

JPEG compression is often seen as a potential attack on images. This compression may lead to the loss of image details, thus challenging the algorithm to correctly extract watermark information from the image. To address this issue, we tested the robustness of the proposed algorithm at different JPEG compression quality factors, i.e., 5, 10, 20, 30, and 40. Figure 8 shows the results of watermarked images after JPEG compression, the watermarks extracted from these images, and the associated BCR. Observing the results of Fig. 8, it is clear that at a quality factor of 5, the quality of the extracted watermark is somewhat affected. However, when the quality factor exceeds 20, the bit correct rate variance is all below 0.02, which strongly proves that our algorithm can effectively withstand JPEG compression attacks and ensure the accurate extraction of the watermark.

(a)   (b)   (c)   (d)   (e)

Fig. 8. Experimental results under image JPEG compression attack:(a) JPEG 5, BCR = 0.96416; (b) JPEG 10, BCR = 0.97792; (c) JPEG 20, BCR = 0.9806; (d) JPEG 30, BCR = 0.9945; (e) JPEG 40, BCR = 0.99667.

(4) **Geometric attacks**

Geometric attacks involve some form of spatial manipulation of an image that may result in changes to the geometric structure or layout of the image. These attacks include rotation, scaling, cropping, translation, flipping, and aspect ratio changes. The challenge for zero-watermarking algorithms in the face of geometric attacks is to accurately detect the watermark and prove ownership even after the image has undergone such operations. This paper applies vertical flipping and shearing attacks to the watermarked images. Observing the experimental results in Fig. 9 and successfully extracting the watermark logo demonstrate that the algorithm can withstand geometric attacks to a certain degree.

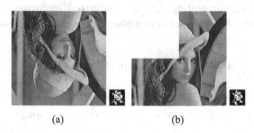

(a)   (b)

Fig. 9. Experimental results under image geometry attack:(a) Vertical flipping BCR = 0.95441 (b) Cropping, BCR = 0.95324.

(5) **Combination attacks**

A combination attack is a composite attack that involves multiple successive operations on an image. By combining different kinds of attacks such as filtering, noise, geometry, etc., the attacker tries to identify weaknesses in the algorithm and destroy the watermark information. For example, an image might first be rotated, then have Gaussian noise added, followed by scaling. The purpose of the combination attack is to try to increase the difficulty of detecting the watermark by mixing multiple attacks. For a watermarking algorithm, successfully defending against combinatorial attacks means that it is highly robust and secure. Figure 10 shows the experimental results after adding combined attacks.

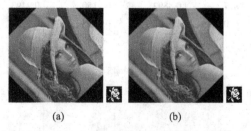

(a)   (b)

Fig. 10. Experimental results under image combination attack: (a) Rotation 45 + Salt and pepper noise, BCR = 0.97454. (b) Rotation 45 + Gaussian filter, BCR = 0.98097.

(6) **Comparison of various algorithms**

The method proposed in this paper enhances the representation and computational efficiency of color images. In the experiments, the optimized OFMM is applied to color image reconstruction and zero-watermarking algorithms, confirming the good stability of the method introduced in this paper. In future research, the method proposed in this paper can be applied to other areas of color image processing, such as image recognition, image retrieval, and image encryption.

During the course of the study, the researchers conducted several experiments to test the effectiveness of the proposed method. These experiments utilized the improved OFMM in tasks such as color image reconstruction and in the implementation of a zero-watermarking algorithm. The results from these experiments were promising, demonstrating that the optimized moments can maintain stable performance even when applied to complex image processing tasks (Table 2).

Table 2. Comparison of robustness (BCR)

Attacks	Proposed	Algorithm[18]	Algorithm[25]	Algorithm[26]	Algorithm[27]	Algorithm[28]
Median filtering 3 × 3	0.99414	0.9772	0.9818	0.9727	0.9658	0.9300
Gaussian filtering 3 × 3	0.99882	0.9835	0.9946	0.9629	0.9980	0.9290
Gaussian noise 0.005	0.97345	0.9705	0.9470	0.9609	0.9727	0.9370
Salt and pepper noise 0.01	0.98925	0.9641	0.9544	0.9424	0.9756	0.9248
JPEG 30	0.96416	0.9575	0.9462	0.9091	0.9222	0.9131
JPEG 40	0.99667	0.9872	0.9914	0.9932	0.9971	0.9610
Vertical flipping	0.95441	0.8798	0.9860	0.8555	0.9863	0.8540
Cropping	0.95324	0.8877	0.8892	0.9150	0.9493	0.8540
Rotation 45 + Salt and pepper noise	0.97454	0.9690	0.9585	0.9648	0.9590	0.9756
Rotation 45 + Gaussian filter	0.98097	0.9799	0.9971	0.9881	0.9980	0.9844

5 Conclusion

Quaternions provide a higher dimensional representation of data for color image processing and can directly process all three channels of a color image, thus providing a more accurate and efficient calculation. Zero watermark technology has proven to be advantageous in ensuring image integrity and data hiding compared to other watermark techniques, which are more difficult to detect and remove, providing a higher level of protection for images. The application of new optimization methods in image processing significantly improves the quality of image reconstruction and computational speed. Compared with the traditional methods, the new methods show significant advantages in both image reconstruction quality and speed, especially when dealing with image data of high-order numbers. The prime number decomposition and factorial computation play a key role in the algorithm We found that the optimized factorial computation method can handle large numbers more quickly and efficiently, and the combination with quaternions and Fourier-Mellin moments provides robust support for the whole algorithm.

Acknowledgment. This work was supported by Shandong Provincial Natural Science Foundation (ZR2023QF032, ZR2020MF054, ZR2023QF018, ZR2022LZH011); Taishan Scholar Program of Shandong (tsqn202306251); National Natural Science Foundation of China (62302249, 62272255, 62302248); Youth Innovation Team of Colleges and Universities in Shandong Province (2022KJ124); Ability Improvement Project of Science and Technology SMES in Shandong Province (2023TSGC0217, (2022TSGC2485); The Chunhui Plan Cooperative Scientific Research Project of Ministry of Education (HZKY20220482); National Key Research and Development Program of China (2021YFC3340600, 2021YFC3340602); Jinan "20 Universities" (2020GXRC056); Jinan " New 20 Universities" (20228016); QiLu First Talent Research Project (2023RCKY143), QiLu Integration Pilot Project of Science Education Industry (2023PX006, 2023PY060, 2023PX071); Key Research and Development Program of Shandong Academy of Science.

References

1. Teague, M.R.: Image analysis via the general theory of moments. J. Opt. Soc. Am. A **70**, 920–930 (1980)
2. Teh, C.-H., Chin, R.T.: On image analysis by the methods of moments. IEEE Trans. Pattern Anal. Mach. Intell. **10**, 496–513 (1988)
3. Wang, C., Wang, X., Xia, Z., Ma, B., Shi, Y.-Q.: Image description with polar harmonic Fourier moments. IEEE Trans. Circuits Syst. Video Technol. **30**, 4440–4452 (2020)
4. Wang, C., Ma, B., Xia, Z., Li, J., Li, Q., Shi, Y.-Q.: Stereoscopic image description with trinion fractional-order continuous orthogonal moments. IEEE Trans. Circuits Syst. Video Technol. **32**, 1998–2012 (2021)
5. Sheng, Y., Shen, L.: Orthogonal Fourier-Mellin moments for invariant pattern recognition. J. Opt. Soc. Am. A **11**, 1748–1757 (1994)
6. Ping, Z., Wu, R., Sheng, Y.: Image description with Chebyshev-Fourier moments. J. Opt. Soc. Am. A **19**, 1748–1754 (2002)
7. Wang, C., Xia, X.W.Z.: Geometrically invariant image watermarking based on fast radial harmonic Fourier moments. Signal Process. Image Commun. **45**, 10–23 (2016)

8. Xiao, B., Ma, J.-F., Wang, X.: Image analysis by Bessel-Fourier moments. Pattern Recogn. **43**, 2620–2629 (2010)
9. Yap, P.-T., Jiang, X., Chichung Kot, A.: Two-dimensional polar harmonic transforms for invariant image representation. IEEE Trans. Pattern Anal. Mach. Intell. **32**, 1259–1270 (2010)
10. Hu, H., Zhang, Y., Shao, C., Ju, Q.: Orthogonal moments based on exponent functions: exponent-Fourier moments. Pattern Recogn. **47**, 2596–2606 (2014)
11. Zhang, H., et al.: Affine Legendre moment invariants for Image watermarking robust to geometric distortions. IEEE Trans. Image Process. **20**, 2189–2199 (2011)
12. Hosny, K.M., Darwish, M.M.: Robust color image watermarking using invariant quaternion Legendre-Fourier moments. Multimedia Tools Appl. **77**, 24727–24750 (2018)
13. Rahmani, H., Moghaddam, M.E.: A semi-blind robust watermarking scheme based on most perceptually important region of images. In: 2010 International Conference on Electronics and Information Engineering, pp. V1–267–V1–271 (2010)
14. Xia, Z., Wang, X., Zhou, W., Li, R., Wang, C., Zhang, C.: Color medical image lossless watermarking using chaotic system and accurate quaternion polar harmonic transforms. Signal Process. **157**, 108–118 (2019)
15. Deng, C., Gao, X., Li, X., Tao, D.: A local Tchebichef moments-based robust image watermarking. Signal Process. **89**, 1531–1539 (2009)
16. Sun, X., Wang, Z., Zhang, D.: A watermarking algorithm based on MA and DWT. In: 2008 International Symposium on Electronic Commerce and Security, pp. 916–919. IEEE (2008)
17. Ma, B., Tao, Z., Ma, R., Wang, C., Li, J., Li, X.: A high-performance robust reversible data hiding algorithm based on polar harmonic Fourier moments. IEEE Trans. Circ. Syst. Video Technol. (2023) https://doi.org/10.1109/TCSVT.2023.3311483
18. Shao, Z., Shang, Y., Zhang, Y., Liu, X., Guo, G.: Robust watermarking using orthogonal Fourier-Mellin moments and chaotic map for double images. Signal Process. **120**, 522–531 (2016)
19. Bi, N., Sun, Q., Huang, D., Yang, Z., Huang, J.: Robust image watermarking based on multi-band wavelets and empirical mode decomposition. IEEE Trans. Image Process. **16**, 1956–1966 (2007)
20. Zhao, M., Jia, Z., Gong, D.: Improved two-dimensional quaternion principal component analysis. IEEE Access. **7**, 79409–79417 (2019)
21. Yan, C.-P., Pun, C.-M., Yuan, X.-C.: Quaternion-based image hashing for adaptive tampering localization. IEEE Trans. Inf. Forensics Secur. **11**, 2664–2677 (2016)
22. Yan, Z., Zhao, S.: Optimal fractions of three-level factorials under a baseline parameterization. Statist. Probab. Lett. **202**, 109902 (2023)
23. Wang, C., et al.: RD-IWAN: residual dense based imperceptible watermark attack network. IEEE Trans. Circuits Syst. Video Technol. **32**, 7460–7472 (2022)
24. Araghi, T.K., Manaf, A.A.: An enhanced hybrid image watermarking scheme for security of medical and non-medical images based on DWT and 2-D SVD. Futur. Gener. Comput. Syst. **101**, 1223–1246 (2019)
25. Liu, J., Li, J., Ma, J., Sadiq, N., Bhatti, U.A., Ai, Y.: A robust multi-watermarking algorithm for medical images based on DTCWT-DCT and Henon map. Appl. Sci. **9**, 700 (2019)
26. Kang, X., Zhao, F., Chen, Y., Lin, G., Jing, C.: Combining polar harmonic transforms and 2D compound chaotic map for distinguishable and robust color image zero-watermarking algorithm. J. Vis. Commun. Image Represent. **70**, 102804 (2020)
27. Xia, Z., et al.: Geometrically invariant color medical image null-watermarking based on precise quaternion polar harmonic Fourier moments. IEEE Access **7**, 122544–122560 (2019)
28. Niu, P., Wang, P., Liu, Y., Yang, H., Wang, X.: Invariant color image watermarking approach using quaternion radial harmonic Fourier moments. Multimed. Tools Appl. **75**, 7655–7679 (2016)

DNA Steganalysis Based on Multi-dimensional Feature Extraction and Fusion

Zhuang Wang⬤, Jinyi Xia⬤, Kaibo Huang⬤, Shengnan Guo⬤, Chenwei Huang,
Zhongliang Yang(✉), and Linna Zhou

School of Cyberspace Security, Beijing University of Posts and Telecommunications,
Beijing 100876, China
yangzl@bupt.edu.cn

Abstract. Steganalysis, as an adversarial technique to steganography, aims to uncover potential concealed information transmission, holding significant research implications and value in maintaining societal peace and stability. With the rapid development and application of DNA synthesis technology, an increasing number of information hiding technologies based on DNA synthesis have emerged in recent years. DNA, as a natural information carrier, boasts advantages such as high information density, robustness, and strong imperceptibility, making it a challenging target for existing steganalysis technologies to efficiently detect. This paper proposes a DNA steganalysis technique that integrates multi-dimensional features. It extracts short-distance and long-distance related features from the DNA long chain separately and then employs ensemble learning for feature fusion and discrimination. Experiments have shown that this method can effectively enhance the detection capability against the latest DNA steganography technologies. We hope that this work will contribute to inspiring more research on DNA-oriented steganography and steganalysis technologies in the future.

Keywords: DNA Steganalysis · Ensemble Learning · Convolutional Neural Network · Self-attention Mechanism

1 Introduction

Steganography primarily investigates techniques that securely and efficiently embed covert information into an alternate information carrier. While ensuring the confidentiality of the embedded content, it simultaneously obfuscates the act of transmitting crucial data. Such techniques play an instrumental role in safeguarding individuals' privacy during routine information communications. However, there's an inherent risk of technological misuse with steganography [22]. If malicious entities exploit it illicitly, such as for internal communications within terrorist organizations to evade scrutiny, it could pose a severe threat to societal security. Hence, studying countermeasures to steganography, specifically steganalysis, to discern if a regular communication medium contains steganographic content, is of paramount importance and value for ensuring cybersecurity and societal stability.

Z. Wang, J. Xia and K. Huang—These authors contributed equally to this work.

B. Ma et al. (Eds.): IWDW 2023, LNCS 14511, pp. 277–291, 2024.
https://doi.org/10.1007/978-981-97-2585-4_20

Steganography, based on the nature of the information carrier, can be categorized into two main classes: digital-medium-based steganography and physical-medium-based steganography. Digital-medium-based steganography primarily embeds secret information into digital media, such as digital images [19], digital audio [13], and network text [23], utilizing public cyberspace for covert communication. This category of technology, emerging in the late last century, has witnessed widespread adoption with the meteoric rise and ubiquity of the internet. Physical-medium-based steganography boasts a developmental trajectory spanning over a millennium, ranging from ancient Greek wax tablet transmissions to invisible inks and microfilms used during World Wars I and II. The advantage of physical-medium-based steganography lies in its versatility across application scenarios, unrestricted by networked environments, and its ability to communicate covertly through various physical media [18]. In recent years, with advancements in molecular biology techniques, there have even been emerging covert communication technologies based on large biomolecules, such as deoxyribonucleic acid (DNA) chains [1,4,5,7,9,11,12]. The advantages of this kind of technology are manifold. Firstly, DNA strands are exceedingly minute, colorless, and tasteless. Even a droplet of water can contain a plethora of DNA sequences. Thus, when using DNA-based steganography for transmitting large volumes of covert information in daily life, it is nearly imperceptible. Secondly, DNA sequences exhibit robust resilience, demonstrating strong adaptability to environmental fluctuations, allowing for the concealed information to be preserved stably over extended durations [8,14]. Thirdly, DNA is solely represented by the four nucleotide bases: A, T, G, and C. This ensures that the generated DNA sequences are indistinguishable from regular sequences. Given these attributes, in recent years, research into DNA-based steganography [4,7,11,12] and steganalysis [1,4,5,9] has increasingly garnered attention from researchers spanning interdisciplinary domains.

Steganalysis is used to capture the difference between the carrier before and after steganography to detect whether it contains secret information. Traditional DNA steganalysis models use some common manual statistical features extraction methods [1,4,5,9], such as Principal Component Analysis (PCA), Factor Analysis (FA), Cluster Analysis (CA) and so on. With the rapid advancement of deep learning technologies, the latest DNA steganography methods have incorporated neural network models designed for efficient sequence generation [7,11,12]. Coupled with the ever-maturing DNA synthesis techniques, researchers have the capability to produce increasingly indistinguishable DNA sequences. Traditional DNA steganalysis techniques are gradually finding effective detection to be challenging, even when employing more powerful feature extraction with deep neural network models, as these models still exhibit significant performance limitations. The primary reason is that these methods generally lack a biological understanding of DNA sequences and cannot effectively harness the complex biological characteristics inherent within, thus constraining their detection capabilities. For example, many related studies perceive DNA sequences as mere symbol chains concatenated by four types of deoxyribonucleotides. Yet in reality, extensive molecular biology research indicates that DNA sequences are not randomly assembled by four nucleotides but exhibit intricate interrelationships. Adjacent nucleotides, in groups of 3–5, might form combinations analogous to the concept of "words" in English letter sequences.

In this study, we have conducted a comprehensive and meticulous analysis and modeling of DNA sequences. We employed various segmentation methods to partition the DNA sequences into different lengths, obtaining sets of DNA sequences of distinct segment lengths. For these segmented sequence sets, convolutional neural networks and self-attention mechanism models were utilized for feature extraction. Subsequently, we adopted an ensemble learning approach based on random forests. Finally, the detection results from the trained models were integrated using RF for ensemble decision-making. Experimental results validate that our detection model can efficiently counteract the state-of-the-art DNA steganographic algorithms, substantially enhancing the capability of steganalysis for DNA sequences.

2   Related Work

2.1   DNA Steganography

In recent decades, advancements in genome editing and deoxyribonucleic acid (DNA) synthesis have rapidly progressed. Steganography oriented towards DNA has emerged as a focal point in the domain of physical carrier steganography. Compared to other physical carrier types, the utilization of DNA as a carrier boasts strong concealment and robustness. Presently, DNA steganography can be categorized into two primary types: sequence modification [10, 15, 16, 20] and sequence generation [7, 11, 12]. Sequence modification steganography refers to the process of replacing certain bases in a natural DNA strand or encoding secret information with base fragments and then inserting them into the natural DNA strand, thereby concealing the information. On the other hand, sequence generation steganography pertains to the synthesis of pseudo-sequences under specific generation constraints. Compared to the former, sequence generation steganography usually achieves a higher information hiding capacity since it is not limited by the original carrier during the information hiding process. Moreover, given that the communicating parties do not need to share the original carrier, it offers greater practicality in real-world scenarios. Owing to these reasons, there has been an increasing emergence of information-hiding methods based on DNA sequence generation in the past two years. For example, David et al. [11] proposed a DNA synthesis steganography algorithm restricted by non-start codons (NSCRS). This method examines the synthesized sequence at every step of base synthesis to adjust the candidate bases, thereby avoiding the generation of start codons. Simultaneously, the optimized bases are encoded, accomplishing the mapping of secret information to nucleotide sequences. Chun et al. [7] introduced a DNA synthesis steganography method constrained by codon usage ((CCRS)). Initially, start and stop codons are eliminated, followed by selecting 32 codons to encode secret information into nucleotide segments, guided by the distribution of codon usage in natural DNA strands. Subsequently, the remaining 28 codons are randomly inserted to ensure that the acquired pseudo-sequence exhibits a codon usage similar to that of natural DNA strands. Recently, Huang et al. [12] introduced an advanced strategy for DNA steganography. Initially, they conducted an in-depth analysis and modeling of the myriad of complex statistical characteristics present in natural DNA strands, subsequently utilizing an LSTM model to learn serial statistical properties. Upon obtaining an optimal sequence model that highly satisfies the statistical characteristics of natural DNA strands, they employed an Adaptive Dynamic Grouping (ADG) algorithm for information hiding. Compared to existing DNA steganography methods, this strategy can synthesize quasi-natural pseudo-sequences with higher statistical-imperceptibility, perceptual-imperceptibility, and more robust anti-steganography capabilities. Current DNA steganography methods do not fully leverage the advantages of neural network architectures. Future research may need to focus more on customized model designs and deep learning techniques tailored to DNA steganography to enhance the efficiency and robustness of DNA steganography methods.

2.2 DNA Steganalysis

The central premise of the DNA steganalysis task is to extract and analyze the statistical feature distribution differences between natural DNA sequences and DNA carriers containing covert information. Subsequently, one determines whether the input sample is a regular DNA specimen or a steganographic DNA carrier sample.

Early work on DNA steganalysis primarily revolved around traditional machine learning methods. For instance, in 2014, Beck M B [3] developed a software package wherein they primarily determined the presence of concealed information in DNA sequences by analyzing the frequency distribution of symbols therein. However, such methods generally lacked comprehensive feature extraction and analysis capabilities for DNA sequences, rendering them ill-equipped to tackle the increasingly sophisticated DNA steganography algorithms of today. With the swift evolution of deep neural network technologies in recent years, an increasing number of researchers have proposed DNA steganalysis algorithms based on deep neural networks. For example, in 2020, Ho Bae *et al.* [2] proposed a steganographic DNA sequence analysis method based on Recurrent Neural Networks (RNN). The proposed method learns the intrinsic distribution and uses classification scores to identify distribution changes, aiming to predict whether a sequence is an encoding sequence or a non-encoding sequence. In the most recent work by Huang *et al.* [12], to test the anti-steganalysis capability of the generated steganographic DNA, they implemented DNA steganalysis methods based on Convolutional Neural Networks (CNN), Random Forest (RF), and Recurrent Neural Networks (RNN), respectively.

While the aforementioned methods have, to some extent, augmented the capability of DNA steganalysis, they exhibit two primary shortcomings. Firstly, these approaches directly deploy existing neural network models for the task of DNA steganalysis without tailoring model designs to adapt to the intrinsic characteristics of DNA sequences. This may lead to insufficient feature extraction and analysis from the DNA sequences. Secondly, these techniques predominantly rely on the characteristics of a single neural network model, failing to harness the advantages offered by integrating diverse neural network architectures, thereby constraining their detection performance.

3 Proposed Method

The overall structure of the proposed model is depicted in Fig. 1, consisting of four main components: DNA sequence preprocessing, feature extraction, ensemble learning, and steganography determination. The purpose of DNA sequence preprocessing is to clean, standardize, and format the raw DNA sequences to ensure the data's quality and consistency, providing accurate input for the subsequent stages. Feature extraction is primarily concerned with extracting key and distinguishable features from the processed DNA sequences. Ensemble learning aims to amalgamate predictions from various models to enhance the accuracy and robustness of the steganography detection. Steganography determination is principally grounded in the prior preprocessing, feature extraction, and ensemble learning outcomes, ultimately ascertaining if the DNA sequence being examined harbors concealed information.

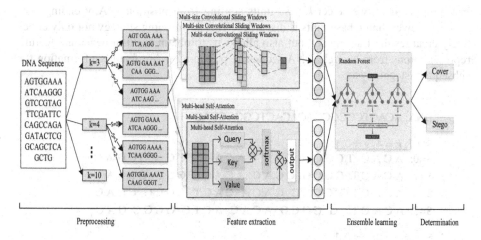

Fig. 1. The proposed model consists of four main components: DNA sequence preprocessing, feature extraction, ensemble learning, and steganography determination.

3.1   DNA Sequence Preprocessing

The preprocessing of DNA sequences aims to extract meaningful subsequences from long-chain DNA sequences for further analysis, allowing for a better understanding of their characteristics. The specific process is as follows. Firstly, our goal is to divide the original long DNA sequence S_{long} into a series of short sequences S_i. We then further split these short sequences into equal-length base units. Consequently, each base unit can be represented as a fixed-length subsequence, facilitating our statistical analysis and examination of its properties.

Given a long DNA sequence S_{long}, we partition it according to a predetermined short sequence length L_{short}, resulting in:

$$S_{long} = [S_1, S_2, ..., S_m],$$ (1)

where m represents the number of short sequences obtained from the partitioning.

For each short sequence S_i, it is further divided into a set of sub-sequences. It's note-worthy that different sub-sequence lengths k and different starting positions s provide us with DNA information from varied perspectives. Assuming we begin the partitioning of S_i from the starting position s into sub-sequences of length k, we obtain:

$$S_{iks} = [s_{iks1}, s_{iks2}, ..., s_{iksn}], \tag{2}$$

where, each s_{iksj} is a sub-sequence of length k partitioned from S_i starting from the position s.

Through this approach, the model can capture statistical features of DNA sequences at different scales, providing rich feature information for subsequent analysis. For instance, in Fig. 2, we select k base units starting from position s. Any ending portion with fewer than k base units is discarded. This partitioning strategy not only effectively enlarges the training set, but also allows us to obtain multi-dimensional feature representations from a single DNA sequence, providing richer feature information for subsequent steganalysis.

DNA: ACACTCGCGCGACACACTCGCGCGAC

SET 1: $k = 4$; $s = 0, 1, 2, 3$
S=0: A C A C   T C G C   G C G A   C A C A   C T C G   C G C G A C
S=1: A C A C T   C G C G   C G A C   A C A C   T C G C   G C G A C
S=2: A C A C T C   G C G C   G A C A   C A C T   C G C G   C G A C
S=3: A C A C T C G   C G C G   A C A C   A C T C   G C G C   G A C

SET 2: $k = 2$; $s = 0, 1$
S=0: A C   A C   T C   G C   G C   G A   C A   C A   C T   C G   C G   C G   A C
S=1: A C A   C T   C G   C G   C G   A C   A C   A C   T C   G C   G C   G A   C

Fig. 2. We select k base units starting from position s, any ending portion with fewer than k base units is discarded.

3.2 Feature Extraction

After tokenization and vectorization preprocessing, the DNA sequence, which origi-nally consisted of hundreds of consecutive base pairs, is transformed into a symbolic sequence linked by several tens of "words". The role of the feature extraction module is to use relevant neural network models to extract rich statistical features from these symbolic sequences to support subsequent steganalysis. To fully capture these statis-tical features, this paper primarily integrates two neural network models to perform feature extraction in parallel. They are the feature extraction module based on multi-size convolutional sliding windows and the feature extraction module based on self-attention mechanisms. The former mainly extracts short-range correlated features of different lengths from DNA symbol sequences to enhance the richness and diversity of features. The latter primarily captures long-range correlated features within DNA symbol sequences, effectively extracting important features from DNA sequences. Using these two methods allows for a more comprehensive extraction of various features from DNA sequences, thus enabling more accurate classification or identification of DNA steganographic regions.

Multi-size Convolutional Sliding Windows. The feature extraction module based on multi-size convolutional sliding windows, denoted as M_{CNN}, primarily serves to efficiently capture local base correlation features in DNA steganalysis. This enables precise localization and analysis of potential steganographic regions. By utilizing convolutional kernels of multiple sizes, this module can identify and extract short-distance correlation features of varying lengths within the DNA sequence, thus providing a wealth of information for subsequent steganalysis.

Firstly, through the model's embedding layer, the input base unit s_{iskj} is encoded into a continuous vector form, allowing the model to better understand and process the DNA sequence:

$$x_{iskj} = \text{Embedding}(s_{iskj}). \tag{3}$$

Here, the base unit s_{iskj} within the ith short sequence S_i is mapped to a fixed-dimensional word vector x_{iskj}, serving as the input for the subsequent convolutional operations.

Subsequently, utilizing multiple convolutional kernels of different sizes, we capture local base association features of varying lengths from the DNA sequence:

$$C_k = f(W_k \cdot x + b_k). \tag{4}$$

Here, C_k represents the local feature extracted by the kth convolutional kernel. This operation is adept at identifying and extracting potential steganographic traces or other anomalous associations present within the DNA sequence.

Next, we employ a max-pooling operation to reduce the dimensionality of the convolutional layer's output, ensuring the model's computational efficiency while simultaneously retaining the most prominent local features of the DNA sequence.

$$P_k = \max(C_k), \tag{5}$$

With this operation, we keep the most significant steganographic features or other local association features identified by the convolutional layer. This max-pooling step acts as a feature selector, emphasizing the most critical patterns that the convolutional layers detect in the sequence, which could be crucial hints or indicators of steganography.

Finally, through a fully connected layer, the features after pooling are transformed into the final decision, i.e., whether the DNA sequence fragment S_i might contain steganographic content.

$$Z = W_f \cdot P + b_f,$$
$$P_{CNN,S_{iks}} = \text{softmax}(W \cdot Z + b). \tag{6}$$

Here, $P_{CNN,S_{iks}}$ represents the probability judged by the model M_{CNN} regarding the presence of steganographic content in the DNA short sequence S_{isk}. The softmax function is utilized to normalize the outputs and convert them into probabilities, which provides clarity in decision-making. With this, the network has a clear output on how confident it is that the given DNA sequence fragment contains hidden messages or not. The higher the probability, the more likely it is that the fragment has steganographic content. Thus, for each input DNA short sequence S_i, by selecting a specific subsequence length k and starting position s, the M_{CNN} model will output a probability value $P_{CNN,S_{iks}}$, representing the likelihood that this sequence is a steganographic DNA sequence.

Multi-head Self-attention Mechanism. The feature extraction module based on the multi-head self-attention mechanism, M_{SA}, is mainly responsible for capturing long-distance correlation features in the DNA symbol sequence. This helps to reveal potential hidden information or anomalous patterns within the sequence.

Firstly, for each input DNA sequence fragment, we compute the Query, Key, and Value:

$$
\begin{aligned}
q_{iksj} &= x_{iksj} \cdot W_q \\
k_{iksj} &= x_{iksj} \cdot W_k \\
v_{iksj} &= x_{iksj} \cdot W_v
\end{aligned}
\tag{7}
$$

Here, for every position-specific division unit s_{iksj} in the input DNA sequence S_i, we calculate three vectors: Query (q), Key (k), and Value (v). Among them, the vector x_{iksj} represents the base unit s_{iskj} within the ith short sequence S_i. These three vectors bestow a new representation upon each position within the DNA sequence, thereby enabling the model to delve deeper into understanding and computing the relationships between various positions.

Secondly, by computing attention weights, different importance is allocated to various parts of the input DNA short sequence S_i, assisting the model in pinpointing potential steganographic regions more accurately. Given the same subsequence length k and starting position s, the attention weight between base positions i and j can be computed as follows:

$$
a_{ij} = \frac{\exp(q_i \cdot k_j / \sqrt{d_k})}{\sum_{l=1}^{N} \exp(q_i \cdot k_l / \sqrt{d_k})}.
\tag{8}
$$

By taking the dot product of the Query and Key, we obtain a value. This value is then scaled down by a factor of $\sqrt{d_k}$ to ensure stability. The softmax function σ is then applied to obtain normalized weights.

Then, using the weights a_{ij} we computed earlier, we weight the Value representations across the entire sequence, obtaining a new sequence representation z_i:

$$
z_i = \sum_{j=1}^{N} a_{ij} \cdot v_j.
\tag{9}
$$

The z_i is a new representation considering the context of the entire DNA sequence. It encompasses not only the information from its corresponding position but also from other positions, yielding a more enriched representation.

Finally, the extracted feature representation z_i is passed through a fully connected layer with weights W_f and bias b_f to obtain a score F. Subsequently, the sigmoid function σ is utilized to map this continuous score to a value between 0 and 1, yielding the steganography content probability $P_{SA,S_{iks}}$:

$$F = z \cdot W_f + b_f,$$
$$P_{SA,S_{iks}} = \sigma(F). \qquad (10)$$

Here, $P_{SA,S_{iks}}$ represents the probability given by the M_{SA} model that the DNA subsequence S_{iks} contains steganographic content.

Similarly, for each inputted DNA subsequence S_{isk}, the M_{SA} model will output a probability value indicating the likelihood that the sequence contains steganographic content.

3.3 Feature Integration

The feature extraction module based on multi-sized convolutional sliding windows, M_{CNN}, and the multi-head self-attention mechanism, M_{SA}, primarily focus on capturing short-distance and long-distance correlational features in DNA sub-sequences, respectively. By fusing the outputs from these two models, we harness the strengths of each while mitigating their limitations. As a result, the derived feature vector representation is both comprehensive and robust, providing a solid foundation for subsequent DNA steganography detection.

For each short sequence S_i, we process it through M_{CNN} and M_{SA} respectively, obtaining the predicted probabilities:

$$P_{CNN,S_{iks}} = M_{CNN}(S_{iks}), \qquad (11)$$
$$P_{SA,S_{iks}} = M_{SA}(S_{iks}). \qquad (12)$$

Then, for each short sequence S_i, the probabilities output by the M_{CNN} and M_{SA} models for various sub-sequence lengths k and starting positions s are integrated:

$$P_{CNN,S_i} = [P_{CNN,S_i,k_1,s_1}, P_{CNN,S_i,k_1,s_2}, \ldots, P_{CNN,S_i,k_n,s_m}], \qquad (13)$$
$$P_{SA,S_i} = [P_{SA,S_i,k_1,s_1}, P_{SA,S_i,k_1,s_2}, \ldots, P_{SA,S_i,k_n,s_m}]. \qquad (14)$$

Finally, the probability vectors P_{CNN,S_i} and P_{SA,S_i} are concatenated to construct a feature vector F_{S_i} for each short sequence S_i:

$$F_{S_i} = [P_{TextCNN,S_i}, P_{Self\text{-}Attention,S_i}]. \qquad (15)$$

Thus, each short sequence S_i corresponds to a feature vector F_{S_i}, collectively constituting the feature matrix F. For the comprehensive DNA sequence S_{long}, the feature matrix F integrates the advantages of the two neural network models, M_{CNN} and M_{SA}, achieving a holistic feature extraction from the DNA sequence. The core objective of this fusion process is to ensure that the model can identify not only short-range associated features but also capture the long-range associated features within the DNA symbol sequence.

Random Forest Model (M_{RF}). First, we define the set of decision trees in the random forest as $T_1, T_2, ..., T_{100}$, with the number of trees $T = 100$. For each decision tree T_i, the probability that it predicts the short sequence S_i as a positive class is denoted as $p_{T_i}(S_i)$. For the random forest RF, its predicted probability of the short sequence S_i being a positive class is the average of the predicted probabilities from all decision trees, $P_{RF}(S_i)$, as follows:

$$P_{RF}(S_i) = \frac{1}{T} \sum_{i=1}^{100} p_{T_i}(S_i). \tag{16}$$

Subsequently, we employ a decision threshold t to categorize the predicted probability $P_{RF}(S_i)$:

$$\text{Predict}(S_i) = \begin{cases} 1 & \text{if } P_{RF}(S_i) > t \\ 0 & \text{otherwise} \end{cases} \tag{17}$$

- When Predict $(S_i) = 1$, based on the current decision threshold t, the short sequence S_i is determined to contain steganographic information.
- When Predict $(S_i) = 0$, based on the current decision threshold t, the short sequence S_i is determined to be devoid of steganographic information.

To ensure optimal classification performance, we need to select an appropriate decision threshold t. Let t^* be the optimal threshold, defined as follows:

$$t^* = \arg \max_t \text{Accuracy}(t), \tag{18}$$

where Accuracy(t) represents the model's accuracy at threshold t. Thus, we can traverse the interval $[0, 1]$ to determine the optimal threshold t^*. Finally, given the short sequence S_i, and the computed feature matrix F, the random forest classifier M_{RF} is trained:

$$M_{RF}(F) \rightarrow \text{Prediction}, \tag{19}$$

here, the Prediction is the final predicted output from the random forest model M_{RF}, derived based on the aforementioned decision mechanism and threshold selection.

3.4   Steganographic Determination

In the final steganography determination, we utilize the Random Forest model M_{RF} for integrated steganographic feature detection. During the training phase, D_{train} is used to train the model M_{RF}. The training set D_{train} includes the feature matrix F_{train} and the corresponding label vector y_{train}. In the testing phase, D_{test} is used to evaluate the performance of the model M_{RF}. The testing set D_{test} includes the feature matrix F_{test} and the corresponding label vector y_{test}.

During the model training phase, we utilize F_{train} and y_{train} from the training set D_{train} to train the model M_{RF}:

$$M_{RF} \leftarrow \text{Train}(F_{train}, y_{train}). \tag{20}$$

We compute the output probabilities of the model M_{RF} on the training set and determine the optimal threshold t^* to maximize accuracy:

$$t^* = \arg\max_t \text{Accuracy}(t, F_{\text{train}}, y_{\text{train}}). \tag{21}$$

In the testing phase, utilizing the model M_{RF} and the optimal threshold t^*, for the test set D_{test}, if the output probability of the model is greater than t^*, i.e.:

$$\text{Accuracy, Recall, Precision, } F_1 \leftarrow \text{Evaluate}(M_{RF}, t^*, F_{\text{test}}, y_{\text{test}}). \tag{22}$$

Ultimately, if $\bar{P}_{S_i} > t^*$, we deduce that the DNA sequence contains steganographic content; conversely, if $\bar{P}_{S_i} \leq t^*$, we determine that the DNA sequence does not contain steganographic content. This procedure ensures that we can accurately and efficiently conduct steganographic detection in DNA sequences, providing researchers with specific and reliable detection results.

4 Experiments and Analysis

4.1 Experimental Setup

In the experimental section, we construct the DNA steganography dataset, taking reference from the work of Huang et al. [12]. They randomly picked five long chains from the European Bioinformatics Institute (EBI) database[1], as shown in Table 1.

Table 1. DNA CHAINS OF NON-CODING DNA FROM EBI DATABASE

Accession	Organism	Base Num.
GCA_002863745	Acinetobacter pittii	1 MB
GCA_009497935	Bacillus paralicheniformis	846 kB
GCA_018219195	Saccharomyces cerevisiae	4 MB
GCA_001417925	Xylella fastidiosa	482 kB
GCA_00400647	Brevibacterium frigoritolerans	797 kB

To effectively analyze the statistical properties of DNA strands, it is essential to segment them into shorter sequences that are multiples of the base units consisting of several bases. Following Bornholt's recommendation [6], the sequence length for one-step DNA synthesis is limited to 200 bases to ensure synthesis accuracy. Hence, we first uniformly split the original DNA sequences into sequences comprised of 198 bases. Based on this, we generated a large number of steganographic DNA sequences using the dynamic grouping encoding-based DNA steganography algorithm proposed by Huang et al. [12]. In the end, we obtained a total of 30,696 original DNA sequences and 142,748 steganographic DNA sequences. To enable the model to learn from DNA information of different dimensions, during the preprocessing phase, we further split the

---

[1] https://www.ebi.ac.uk/.

aforementioned DNA sequences based on fragment length (k) and starting segmentation position (s). Herein, the value range for k is {2, 3, 4, 5, 6, 8, 10}; for each k, the value range for S is [0, k).

For the input DNA sequences, we map them into a uniform 128-dimensional feature vector. In the feature extraction module based on multi-scale convolutional sliding windows, we employ 64 convolutional kernels, with sizes of 3, 4, and 5. For the feature extraction module based on the multi-head self-attention mechanism, we use 4 attention heads and the number of layers in the model is set to 1. During the model training process, we adopt a learning rate of 0.001 and set the batch size to 64. The algorithm is implemented in Pytorch. Training acceleration is achieved using a GeForce GTX 2060 GPU and CUDA 11.1, with a total of 30 training iterations.

4.2 Results and Discussion

In order to effectively evaluate the detection performance of the proposed DNA steganography detection model, we replicate several DNA steganography detection models used in the paper by Huang *et al.* [12], specifically those based on CNN [17], LSTM [2], and self-Attention (SA) model [21].

We use several evaluation indicators commonly used in classification tasks to evaluate the performance of our model, which are accuracy (Acc), precision (P), recall (R) and F1-score. The conceptions and formulas are described as follows:

- Accuracy measures the proportion of true results (both true positives and true negatives) among the total number of cases examined

$$Accuracy = \frac{TP + TN}{TP + FN + FP + TN}. \tag{23}$$

- Precision measures the proportion of positive samples in the classified samples.

$$Precision = \frac{TP}{TP + FP}. \tag{24}$$

- Recall measures the proportion of positives that are correctly identified as such.

$$Recall = \frac{TP}{TP + FN}. \tag{25}$$

- F1-score is a measure of a test's accuracy. It considers both the precision and the recall of the test. The F1 score is the harmonic average of the precision and recall, where an F1-score reaches its best value at 1 and worst at 0.

$$F_1 - score = \frac{2 \times Precision \times Recall}{Precision + Recall}. \tag{26}$$

TP (True Positive) represents the number of positive samples that are predicted to be positive by the model, FP (False Positive) indicates the number of negative samples predicted to be positive, FN (False Negative) illustrates the number of positive samples predicted to be negative and TN (True Negative) represents the number of negative samples predicted to be negative.

To validate the performance of model integration in a single dimension for DNA steganalysis, we respectively tested the DNA steganography detection performance of the model proposed in this paper and other relevant models under three scenarios with fixed S = 0 and k = 3, 5, 10. The results are shown in Table 2. From Table 2, we can draw the following conclusions. First, under different scenarios, the detection performance of the proposed model is superior to previous models. This indicates that our model can fully utilize the biological characteristics of DNA sequences, better extract their feature expression, and therefore has better detection performance. Secondly, we find that although both feature extraction modules can bring a certain detection performance gain, the detection effect after integration is better. This shows that our model integrates the advantages of each neural network model and can exert the detection capability of the model to a greater extent.

Table 2. When $s = 0$ and $k = 3, 5, 10$ are fixed, a comparison of the DNA steganalysis performance between the model proposed in this paper and other relevant models is presented.

		CNN [17]	LSTM [2]	SA [21]	Ours (Ensemble)		
					w/o M_{SA}	w/o M_{CNN}	Final
k = 3	Accuracy	90.62%	85.06%	75.22%	93.96%	88.67%	**94.60%**
	Recall	91.31%	99.64%	76.51%	97.11%	94.05%	97.25%
	Precision	97.36%	85.00%	92.05%	95.61%	92.34%	96.25%
	F1 score	94.12%	91.45%	83.55%	96.36%	93.19%	**96.75%**
k = 5	Accuracy	89.63%	86.68%	86.78%	94.05%	94.02%	**94.12%**
	Recall	90.17%	98.95%	86.92%	97.18%	97.50%	97.22%
	Precision	97.02%	86.69%	96.68%	95.66%	95.34%	95.69%
	F1 score	93.47%	92.41%	91.54%	96.42%	96.41%	**96.45%**
k = 10	Accuracy	84.94%	82.01%	81.56%	93.28%	92.29%	**93.37%**
	Recall	84.45%	99.79%	91.71%	96.74%	95.96%	96.72%
	Precision	96.84%	82.12%	86.65%	95.16%	94.76%	95.30%
	F1 score	90.22%	90.10%	89.11%	95.94%	95.35%	**96.00%**

In order to validate the effectiveness of integrating models across different dimensions for DNA steganalysis, we tested the performance metrics of models with a fixed single dimension s = 0 and integrated multi-dimensions, with results shown in Table 3. From Table 3, we can observe that the detection performance when combining multi-dimensional features is superior to that in a single dimension. Whether utilizing SA or CNN, or integrating both model types, dimensional integration enhances various metrics of the model. This indicates that integrating information across different DNA dimensions is beneficial for steganalysis. At the same time, for SA and CNN, which have two different mechanisms, integrated usage of both offers certain advantages over using either in isolation, whether in single or multiple dimensions. This suggests that integrating SA and CNN, and fully exploiting the short and long-range correlation characteristics of DNA sequences, provides substantial assistance in DNA steganalysis.

Table 3. Comparison of Model Performance between Fixed Single Dimension $s = 0$ and Multi-Dimensional Integration.

	Ours (Single-dimensional)			Ours (Multi-dimensional)		
	w/o M_{SA}	w/o M_{CNN}	Ensembled	w/o M_{SA}	w/o M_{CNN}	Ensembled
Accuracy	94.08%	93.96%	94.44%	94.45%	94.98%	94.93%
Recall	96.97%	96.74%	97.28%	97.17%	97.46%	97.38%
Precision	95.87%	95.97%	96.03%	96.13%	96.48%	96.51%
F1 score	96.42%	96.35%	96.65%	96.65%	96.97%	96.94%

5 Conclusion

This paper proposes a DNA steganalysis technique that integrates multi-dimensional features. It extracts short-distance and long-distance related features from long DNA chains, respectively, and employs ensemble learning for feature fusion and identification. Experiments demonstrate that this method can effectively enhance the detection capability against the latest DNA steganography techniques. We hope that this work will inspire further research into DNA-oriented steganography and steganalysis technologies.

Acknowledgments. This work was supported in part by the National Key Research and Development Program of China under Grant 2022YFC3303301 and in part by the National Natural Science Foundation of China under Grant 62172053 and Grant 62302059.

References

1. Arita, M.: Comma-free design for DNA words. Commun. ACM **47**(5), 99–100 (2004)
2. Bae, H., Min, S., Choi, H.S., Yoon, S.: DNA privacy: analyzing malicious DNA sequences using deep neural networks. IEEE/ACM Trans. Comput. Biol. Bioinf. **19**(2), 888–898 (2020)
3. Beck, M.B., Desoky, A.H., Rouchka, E.C., Yampolskiy, R.V.: Decoding methods for DNA steganalysis. In: 6th International Conference on Bioinformatics and Computational Biology (BICoB) (2014)
4. Beck, M.B.: A forensics software toolkit for DNA steganalysis (2015)
5. Berthold, O., Federrath, H., Köpsell, S.: Web MIXes: a system for anonymous and unobservable internet access. In: Federrath, H. (ed.) Designing Privacy Enhancing Technologies. LNCS, vol. 2009, pp. 115–129. Springer, Heidelberg (2001). https://doi.org/10.1007/3-540-44702-4_7
6. Bornholt, J., Lopez, R., Carmean, D.M., Ceze, L., Seelig, G., Strauss, K.: A DNA-based archival storage system. In: Proceedings of the Twenty-First International Conference on Architectural Support for Programming Languages and Operating Systems, pp. 637–649 (2016)
7. Chun, J.Y., Lee, H.L., Yoon, J.W.: Passing go with DNA sequencing: delivering messages in a covert transgenic channel. In: 2015 IEEE Security and Privacy Workshops, pp. 17–26. IEEE (2015)
8. Clelland, C.T., Risca, V., Bancroft, C.: Hiding messages in DNA microdots. Nature **399**(6736), 533–534 (1999)

9. Danezis, G.: Statistical disclosure attacks: traffic confirmation in open environments. In: Gritzalis, D., De Capitani di Vimercati, S., Samarati, P., Katsikas, S. (eds.) SEC 2003. IFIPAICT, vol. 122, pp. 421–426. Springer, Boston (2003). https://doi.org/10.1007/978-0-387-35691-4_40

10. Guo, C., Chang, C.C., Wang, Z.H.: A new data hiding scheme based on DNA sequence. Int. J. Innov. Comput. Inf. Control **8**(1), 139–149 (2012)

11. Haughton, D., Balado, F.: BioCode: two biologically compatible algorithms for embedding data in non-coding and coding regions of DNA. BMC Bioinform. **14**(1), 1–16 (2013)

12. Huang, C., et al.: DNA synthetic steganography based on conditional probability adaptive coding. IEEE Trans. Inf. Forensics Secur. **18**, 4747–4759 (2023)

13. Huang, Y.F., Tang, S., Yuan, J.: Steganography in inactive frames of VoIP streams encoded by source codec. IEEE Trans. Inf. Forensics Secur. **6**(2), 296–306 (2011)

14. Ivanova, N.V., Kuzmina, M.L.: Protocols for dry DNA storage and shipment at room temperature. Mol. Ecol. Resour. **13**(5), 890–898 (2013)

15. Khalifa, A., Atito, A.: High-capacity DNA-based steganography. In: 2012 8th International Conference on Informatics and Systems (INFOS), p. BIO-76. IEEE (2012)

16. Khalifa, A., Elhadad, A., Hamad, S.: Secure blind data hiding into pseudo DNA sequences using playfair ciphering and generic complementary substitution. Appl. Math. **10**(4), 1483–1492 (2016)

17. Kim, Y.: Convolutional neural networks for sentence classification, pp. 1746–1751 (2014). https://doi.org/10.3115/v1/D14-1181. https://aclanthology.org/D14-1181

18. Sarkar, A., Madhow, U., Manjunath, B.: Matrix embedding with pseudorandom coefficient selection and error correction for robust and secure steganography. IEEE Trans. Inf. Forensics Secur. **5**(2), 225–239 (2010)

19. Tang, W., Li, B., Tan, S., Barni, M., Huang, J.: CNN-based adversarial embedding for image steganography. IEEE Trans. Inf. Forensics Secur. **14**(8), 2074–2087 (2019)

20. Taur, J.S., Lin, H.Y., Lee, H.L., Tao, C.W.: Data hiding in DNA sequences based on table lookup substitution. Int. J. Innov. Comput. Inf. Control **8**(10), 6585–6598 (2012)

21. Vaswani, A., et al.: Attention is all you need. In: Advances in Neural Information Processing Systems, vol. 30 (2017)

22. Wiseman, S.: Stegware–using steganography for malicious purposes (2017)

23. Yang, Z.L., Zhang, S.Y., Hu, Y.T., Hu, Z.W., Huang, Y.F.: VAE-Stega: linguistic steganography based on variational auto-encoder. IEEE Trans. Inf. Forensics Secur. **16**, 880–895 (2020)

VStego800K: Large-Scale Steganalysis Dataset for Streaming Voice

Xuan Xu⬤, Shengnan Guo, Zhengyang Fang, Pengcheng Zhou, Zhongliang Yang$^{(\boxtimes)}$, and Linna Zhou

School of Cyberspace Security, Beijing University of Posts and Telecommunications, Beijing 100876, China
yangzl@bupt.edu.cn

Abstract. In recent years, more and more steganographic methods based on streaming voice have appeared, which poses a great threat to the security of cyberspace. In this paper, in order to promote the development of streaming voice steganalysis technology, we construct and release a large-scale streaming voice steganalysis dataset called VStego800K. To truly reflect the needs of reality, we mainly follow three considerations when constructing the VStego800K dataset: large-scale, real-time, and diversity. The large-scale dataset allows researchers to fully explore the statistical distribution differences of streaming signals caused by steganography. Therefore, the proposed VStego800K dataset contains 814,592 streaming voice fragments. Among them, 764,592 samples (382,296 cover-stego pairs) are divided as the training set and the remaining 50,000 as testing set. The duration of all samples in the data set is uniformly cut to 1 s to encourage researchers to develop near real-time speech steganalysis algorithms. To ensure the diversity of the dataset, the collected voice signals are mixed with male and female as well as Chinese and English from different speakers. For each stegano-graphic sample in VStego800K, we randomly use two typical streaming voice steganography algorithms, and randomly embed random bit with embedding rates of 10%–40%. We tested the performance of some latest steganalysis algorithms on VStego800K, with specific results and analysis details in the experimental part. We hope that the VStego800K dataset will further promote the development of universal voice steganalysis technology. The description of VStego800K and instructions will be released here: https://github.com/YangzlTHU/VStego800K.

Keywords: VStego800K · Voice Steganalysis · Dataset

1 Introduction

The rapid development of network technology not only brings convenience to people's communication, but also makes people increasingly worried about the privacy and security of communication in public cyberspace. Covert communication technology mainly studies how to embed privacy information into other normal carriers, hide the existence of privacy information and thus protect its security. However, this technology can also

B. Ma et al. (Eds.): IWDW 2023, LNCS 14511, pp. 292–303, 2024.
https://doi.org/10.1007/978-981-97-2585-4_21

be used by hackers, terrorists, and other law breakers for malicious intentions [1]. There-fore, studying and developing effective steganalysis techniques becomes an increasingly promising and challenging task.

There are various media forms of carrier that can be used for information hiding, including image [2, 37], audio [3, 4], text [5–7] and so on [8]. In recent years, with the popularity and development of the Internet, communication based on streaming media has been greatly developed. Streaming media refers to continuous time-based media that uses streaming technology in the Internet, such as audio, video, or multimedia files. Voice over IP (VoIP) [9] is one of the most popular streaming communication service in the Internet. Therefore, with these emerging communication channels, more and more VoIP-based covert communication systems have appeared in recent years [3, 10–16].

Usually we can model a covert communication system as a "Prisoners' Problem" [17]. In this model, Alice needs to transmit the secret message m, which from the secret message space, to Bob. Alice selects a suitable cover c from the cover space C and embeds the secret message m into the cover c under the guidance of the hidden key k, which is from the key space K. The cover c becomes a stego carrier s after embedding the covert information m, and large number of stego carriers constitute the stego space S. The information embedding process can be expressed by the embedding function f(), that is:

$$Emb = C \times K \times M \rightarrow S, f(c, k, m) = s \tag{1}$$

Generally speaking, this mapping function will inevitably affect the probability dis-tributions differences of normal carriers and steganographic carriers. For Alice and Bob, their main purpose is to ensure the successful transmission of information without arous-ing Eve's suspicion, so they need to reduce the difference in statistical distribution of carriers before and after steganography as much as possible.

Steganalysis technology is the countermeasure technology of steganography. Its main purpose is to detect whether covert information is contained in the information carrier being transmitted in cyberspace. It can help identify potential network attacks in cyberspace and maintain cyberspace security. Any steganalysis can be described by a map F: Rd $\rightarrow$ {0,1}, where F = 0 means that carrier is detected as cover, while F = 1 means that carrier is detected as stego. Therefore, steganalysis researchers usually con-struct a variety of corresponding statistical features, and based on these features to find the differences in the statistical distribution between cover and stego carriers [21–30].

Compared with steganalysis methods for static carrier, the steganalysis methods for streaming media are more demanding and thus more challenging. Firstly, since the signal streams are transmitted online in real time, the detection algorithm should also be efficient enough. This involves two requirements. On the one hand, the steganalysis model needs to be able to perform high-performance detection on a speech signal which is as short as possible. Therefore, once Alice and Bob are found to be transmitting secret information, Eve can terminate the communication within the shortest time after they establish the communication. On the other hand, the steganalysis model needs to have a sufficiently fast judgment ability. That is, when inputted a speech signal, it is required to complete the process of feature extraction, analysis and judgment in the shortest time. Secondly, a very important characteristic of covert communication based on VoIP is that, for both communicator parties, the capacity of the carrier is completely controllable and can be expanded arbitrarily. This means that Alice and Bob can spread the secret information that needs to be transmitted in a sufficiently long speech signal to achieve a covert communication with low embedding rate. Therefore, from the perspective of steganalysis, it has always been an important research goal to achieve efficient and high-performance steganalysis for streaming voice signals with short duration and low embedding rate.

Therefore, in the process of constructing the VStego800K dataset, we focused on these two challenges. In general, the motivation for constructing this dataset comes from the following aspects. Firstly, in order to achieve higher performance steganalysis technology, researchers usually need to analyze the statistical distribution differences between a large number of normal samples and steganographic samples [21–30]. Especially with the development of deep learning technology, some voice steganalysis methods based on deep neural network have a growing demand for data [26–30]. However, as far as we know, there is currently no general large-scale steganalysis dataset for streaming media. This makes it difficult to compare various voice steganalysis algorithms uniformly, and greatly limits the development of voice steganalysis technology. Therefore, we constructed a large-scale streaming voice steganalysis dataset VStego800K, which includes 382,296 pairs of cover and stego voice fragments. Secondly, considering the real-time requirements and challenges of voice steganalysis, we encourage researchers to develop more and more efficient real-time streaming voice steganalysis. Therefore, for the samples in the dataset, we uniformly cut them to a duration of 1 s. In other words, we hope that researchers can identify whether the voice contains secret information within 1 s when Alice and Bob to establish streaming voice communication, and achieve almost real-time steganalysis. Thirdly, considering the requirements and challenges of effective steganography detection for real-time streaming voice signals at low embedding rates, we set the secret information embedding rate of steganographic voice samples in the dataset to a minimum of 10% and a maximum of 40%. The percentage here means, when conducting a% embedding rate steganography, we embedded each frame with a% probability. Finally, we noticed that many previous methods of voice steganalysis are highly targeted. Mainly for a specific speech steganography algorithm, specific duration and even specific embedding rate. However, in a real network environment, Eve usually does not know the steganography algorithm Alice may adopt, or even the specific identities of the two sides of the communication. Therefore, in order to encourage researchers to develop more robust voice steganalysis algorithms, we have introduced a lot of general designs in the process of constructing dataset. Two widely used steganography algorithms, which are Complementary Neighbor Vertices (CNV) [32] steganography and Pitch Modulation Steganography (PMS) [16], are used in the dataset for information hiding. The voice samples are mixed with male voice and female voice, and the language types are mixed with Chinese and English. We hope that the large-scale streaming voice steganalysis dataset VStego800K can help researchers develop more general, efficient and robust audio steganalysis algorithms to maintain the security of public cyberspace.

In the remainder of this paper, Sect. 2 introduces related streaming voice steganalysis methods. Section 3 introduces the detailed information of the VStego800K dataset, including data collection and preprocessing, information embedding algorithms. The Following part, Sect. 4, describes the steganalysis benchmarks we use and their performance on VStego800K dataset. Finally, conclusions are drawn in Sect. 5.

2 Related Works

There has been much effort in steganalysis of digital audio [18–21]. The most common way is to directly extract statistical features from the audio and conduct classification subsequently. For example, in 2007, C Kraetzer et al. [20, 22] proposed to extract and analyze the Mel-frequency features of VoIP signals, and then used SVM as classifier to perform steganalysis. In 2011, Y. Huang et al. [23] proposed to use Regular Singular (RS) algorithm [24] to extract VoIP signals in each sliding window to determine whether secret information is hidden in the speech stream. In 2012, S. Li et al. [25] used the Markov model to calculate the transition probabilities between frames in each sliding window and then used SVM to classify these features. The main difference between the above-mentioned models lies in the features extraction module.

In recent years, with the development of deep neural network technology, there have been more and more steganalysis methods that use various neural network models to extract code word correlation features in VoIP stream signals. In 2018, Z. Lin et al. [26] used an Long Short-Term Memory (LSTM) Neural Nework to extract the temporal correlations of codewords. Z. Yang et al. [27] proposed a VoIP steganalysis method based on multi-channel convolution sliding window. In order to analyze the correlations between frames and different neighborhood frames in a VoIP signal. After that, H. Yang et al. [28] further improved the feature extraction module. They indicated a proper way to take advantage of two main deep learning architectures, which are convolutional neural network (CNN) and recurrent neural network (RNN) and proposed a novel CNN-LSTM model to extract correlation features in VoIP streams. In 2019, H. Yang and Z. Yang [29] proposed a Hierarchical Representation Network based on convolutional neural networks and attention mechanisms to tackle the steganalysis of QIM steganography in low-bitrate speech signal. Futher, they found that by mapping the quantized codewords into dense semantic space, and then compared the knowledge distillation training framework can achieve both high detection efficiency and high accuracy steganalysis performance [30].

All these works have made important contributions to the development of voice steganalysis technology. In this paper, we reproduced the most typical methods and conducted experiments on VStego800K. Experimental results show that their performances are still unsatisfactory under the conditions of hybrid steganography algorithms, short speech signal duration and low information embedding rate. We will provide these steganalysis models and codes at the same time we release this dataset to encourage subsequent researchers to develop more efficient streaming speech steganalysis algorithms.

3   VStego800K Dataset

In this section, we will introduce in details of the construction process of VStego800K, including source voice collection, voice preprocessing and information hiding. Finally, we give the overall distribution characteristics of VStego800K.

3.1   Source Voice Collection and Preprocessing

All voice samples in the VStego800K dataset are voice fragments intercepted from public streaming media services after obtaining user authorization. The streaming media voice in VStego800K uniformly adopts the G.729 coding standard, with a mixture of male and female as well as Chinese and English from different speakers. Among them, the total duration of Chinese pronunciation is about 43 h, and that of English is about 72 h. For data preprocessing, we first uniformly cut the obtained encoded streaming voice into fragments with the same duration of 1 s, and obtained a total of 814,592 voice samples. Among them, 764,592 samples (382,296 cover-stego pairs) are divided as the training set and the remaining 50,000 as testing set.

3.2   Information Hiding

Currently, steganographic methods which are based on VoIP streams can be roughly divided into two strategies. The first strategy conducts information hiding mainly by introducing Quantization Index Modulation (QIM) algorithm [31] to segment and encode the codebook in the process of speech quantization [15, 32]. The second strategy integrates information hiding into pitch period prediction process [16, 33, 34].

In order to construct a universal and practical streaming voice steganalysis dataset, we have selected the most typical steganography algorithms for these two steganographic strategies, namely Complementary Neighbor Vertices (CNV) [32] steganography and Pitch Modulation Steganography (PMS) [16]. In the information embedding process, we randomly selected one of these two steganographic algorithms and used them to embed the random bits stream into all the samples in the training set and the random half of the test set. For each steganographic samples, the embedding rate was randomly set to be 10%, 20%, 30%, and 40%. Here, the definition of embedding rate is, when conducting a% embedding rate steganography, we embedded each frame with a% probability. A brief introduction of CNV and PMS methods are as follows.

CNV Steganography. Complementary Neighbor Vertices (CNV) algorithm is a steganography method based on QIM strategy and graph theory proposed by Xiao et al. [32], which has been widely used in the field of streaming voice covert communication. The QIM method divides the codebook into two parts, each representing '0' and '1', respectively. Instead of randomly partitioning the codebook, CNV algorithm further considers the relationship between codewords. It guarantees that every codeword in the opposite part to its nearest neighbor, and the distortion is limited by a bound. CNV algorithm embeds secret message in the field of vector quantization index of LPC coefficients, getting the benefit that the distortion due to QIM is lightened adaptively by the rest of the encoding procedure. As a result, the improper random partitions were prevented and the speech quality was guaranteed.

PMS Steganography. PMS steganography [16] mainly hides secret information in streaming voice by modifying the adaptive code topic delay (ACD) parameter. It first predicts the open-loop pitch and closed-loop pitch of the voice signal. Then for each frame, two pitch periods are calculated by using the first two subframes and the last two subframes, respectively. Then, For subframes 0 and 2, the closed loop pitch lag is selected around the appropriate open loop pitch lag in the range $\pm$ 1 and coded using 7 bits. For subframes 1 and 3, the closed loop pitch lag is coded differentially using 2 bits. In this way we can realize the embedding of secret information. PMS steganography algorithm can obtain high-quality speech, prevent the detection of steganalysis, and has good compatibility with standard speech codecs, and will not cause further delay due to data embedding and extraction. Therefore, PMS steganography has been widely used in streaming voice covert communication.

3.3 Overall Details of VStego800K

After these above operations, the overall characteristics of VStego800K are shown in Table 1. In addition, it is worth noting that we will also provide a steganography toolkit (https://github.com/YangzlTHU/VoIP-Stega) for future researchers when we public this VStego800K dataset. It contains tools and instructions on how to make these steganographic samples, so researchers can adjust the duration and embedding rate of steganography samples according to their needs.

Table 1. The overall characteristics of VStego800K.

VStego800K	Traing Set	Test Set
Number of Samples (Cover: Stego)	382,296: 382,296	25,000: 25,000
Number of Different Steganography (Cover: CNV: PMS)	382,296: 191,147: 191,149	25,000: 12,500: 12,500

4 VStego800K Dataset

4.1 Benchmark Methods and Evaluation Metrics

To evaluate the quality of VStego800K and provide benchmark results for researchers who subsequently use this dataset, we tested five latest and widely used VoIP steganalysis methods on VStego800K, which are CCN [35], SS-QCCN [29], RNN-SM [26], Fast VoIP Steganalysis (FVS) [30] and SFFN [36]. Details of these methods are introduced as follows.

- The SS-QCCN [29] model is a targeted steganalysis method for the QIM steganography. SS-QCCN is a strong version of the quantization codeword correlation network (QCCN). QCCN can be constructed from adjacent speech frames by using the split vector quantization codewords. Furthermore, SS-QCCN is obtained by pruning the weaker edges of QCCN. In this method, the principal component analysis (PCA) is utilized to reduce the feature dimension and the support vector machine (SVM) is adopted as the classifier.
- The codebook correlation network (CCN) [35] is a targeted steganalysis method for PMS. In CCN, conditional probability is adopted to quantify the correlation between the coefficients of an adaptive codebook. Similar with SS-QCCN, PCA is used to reduce the feature dimension and SVM is applied to conduct classification in this method.
- The RNN-based steganalysis model (RNN-SM) [26] consists of a two-layer LSTM-based codeword correlation model and a fully-connected layer-based feature classification model. For full RNN-SM, all outputs from the second layer of LSTM are forwarded to the final output node. But for pruned RNN-SM, the outputs at the end of the sequence from the second layer of LSTM are forwarded. As discussed in [26], pruned RNN-SM has more efficiency and other benefits compared with full RNN-SM. Thus, we use pruned RNN-SM in our experiments.
- The Fast steganalysis method (FSM) [30] maps vector quantization code-words into a semantic space to better exploit the correlations in code-words. In order to achieve high detection efficiency, only one hidden layer was utilized to extract the correlations between these code-words. Based on the extracted correlation features, FSM model uses the softmax classifier to categorize the input stream carriers. To boost the performance, FSM model incorporates a simple knowledge distillation framework into the training process.
- The Steganalysis Feature Fusion Network (SFFN) [36] aims to detect the existence of the confidential messages hidden in the frames of streaming media with multiple kinds of orthogonal steganographic methods. SFFN consists of three sub-networks, i.e., a feature learning network, a feature fusion network and a classification network. With the three sub-networks, SFFN can effectively extract steganalysis features for the steganographic methods and can fuse the features to make credible prediction.

We use several evaluation indicators commonly used in classification tasks to evaluate the performance of our model, which are precision (P), recall (R), F1-score (F1) and accuracy (Acc). The conceptions and formulas are described as follows:

- Accuracy measures the proportion of true results (both true positives and true negatives) among the total number of cases examined

$$Acc = \frac{TP + TN}{TP + FN + FP + TN} \tag{2}$$

- Precision measures the proportion of positive samples in the classified samples.

$$P = \frac{TP}{TP + FP} \tag{3}$$

- Recall measures the proportion of positives that are correctly identified as such.

$$R = \frac{TP}{TP + FN} \quad (4)$$

- F1-score is a measure of a test's accuracy. It considers both the precision and the recall of the test. The F1 score is the harmonic average of the precision and recall, where an F1 score reaches its best value at 1 and worst at 0.

TP (True Positive) represents the number of positive samples that are predicted to be positive by the model, FP (False Positive) indicates the number of negative samples predicted to be positive, FN (False Negative) illustrates the number of positive samples predicted to be negative and TN (True Negative) represents the number of negative samples predicted to be negative. All these indicators are the higher the better.

4.2 Detection Results of Benchmark Methods

For the five benchmark steganalysis algorithms, SS-QCCN [29] and CCN [35] algorithms are mainly for specific streaming voice steganography, and RNN-SM [26], FSM [30] and SFFN [36] algorithms are general steganalysis algorithms based on deep neural networks. Therefore, in the experiment process, for the SS-QCCN [29] algorithm, we used the samples which conduct information hiding by CNV algorithm for training, and then evaluated the performance in the complete test set; for the CCN [35] algorithm, we used the samples which conduct information hiding by PMS algorithm for training, and then evaluated the performance in the complete test set; for the latter three steganographic models based on neural networks, we extracted the steganographic features of CNV and PMS respectively and sent them to the model for training after stitching, we finally evaluated the performance in the complete detection set. Table 2 records the test performance of each benchmark steganalysis model on test set.

Table 2. The overall performance of each steganalysis methods.

Steganalysis Method	Accuracy	Precision	Recall	F1
SS-QCCN [29]	0.6117	0.6595	0.4617	0.5432
CCN [35]	0.5542	0.5544	0.5517	0.5531
RSM [26]	0.5174	0.5103	0.8605	0.6407
FSM [30]	0.7094	0.7085	0.7115	0.7100
SFFN [36]	0.7048	0.7206	0.6689	0.6938

From the results in Table 2, we can get the following conclusions. Firstly, the VStego800K dataset is very challenging, and the detection accuracy of the best model is only about 70%. It indicates that the previous models are still not satisfactory under the conditions of hybrid steganography algorithms, short speech signal duration, and low information embedding rate. Secondly, we found that even RSM is a neural network-based model, the detection performance of RSM lags far behind FSM and SFFN, or even worse than CCN and SS-QCCN. We think that there are many reasons for this result. The RNN-SM model [26] is mainly based on the LSTM network. Generally, we think that LSTM is better at modeling the long-distance dependence of sequence signals, and its detection ability for short-duration sequence samples is weak; the RNN-SM model realizes steganalysis by analyzing the correlation between sequence frames. For samples with low embedding rate, the steganography frames are very sparse. In this case, it is difficult to realize steganalysis through the correlation between frames. Thirdly, we found that the detection results of SS-QCCN [29] and CCN [35] are not satisfactory. We believe that except that they are based on non-neural networks and therefore have weak signal feature extraction capabilities, the most critical reason is that they are all for specific steganographic algorithms. When samples are mixed with different steganography algorithms, it is difficult for them to extract the steganographic features of other steganography algorithms and thus to achieve effective detection.

Table 3. The steganalysis results of benchmark methods for different embedding rates in VStego800K.

Steganalysis Embedding Rates		10%	20%	30%	40%
SS-QCCN [29]	Acc	0.6792	0.6933	0.7189	0.7149
	R	0.3274	0.4224	0.5279	0.5533
CCN [35]	Acc	0.5510	0.5436	0.5582	0.5495
	R	0.5267	0.4917	0.5651	0.6138
RSM [26]	Acc	0.3029	0.3157	0.3087	0.3186
	R	0.8390	0.8595	0.8687	0.8737
FSM [30]	Acc	0.6683	0.7026	0.7232	0.7381
	R	0.5057	0.6847	0.7893	0.8568
SFFN [36]	Acc	0.6865	0.7171	0.7411	0.7601
	R	0.4610	0.6265	0.7340	0.8349

We further analyzed the impact of different embedding rates on the test results. We calculate the performance of each steganalysis method for voice samples with different embedding rates in the test set. The results are shown in Table 3. The steganalysis results of benchmark methods for different embedding rates in VStego800K.. From the results in Table 3, firstly, we can easily find a very obvious change rule, that is, as the embedding rate increases, the detection performance of each detection model is gradually improved. For example, for the FSM algorithm [30], when the embedding rate is 0.1, the detection accuracy is only 0.6683. When the embedding rate is increased to 0.4, the detection is also improved to 0.7381. This trend can be explained. Embedding additional information in the original voice signal is equivalent to introducing noise into the original signal, which will inevitably change the statistical distribution characteristics of the original signal carrier. The higher the embedding rate, the more extra information is embedded, which will cause this statistical distribution to become larger and therefore easier to be detected. Secondly, from Table 3, it is easy to see that the reason why the overall detection accuracy of RSM model is not good is that it predicts a large number of non-steganography samples as steganographic samples, resulting in a large number of false detection rate, so the average detection accuracy rate decreases. Thirdly, we find that SFFN has the best detection performance at any embedding rate. The main reason is that SFFN uses two sub-models to extract the feature distributions of the two steganography algorithms, and then implements the steganalysis after feature fusion. Compared with other models to extract features uniformly, SFFN seems to be more effective.

The above experimental results show that even though voice steganalysis technology has achieved rapid development in recent years, when faced with multiple steganography algorithms, short duration, low embedding rate, etc., their detection performance still cannot meet actual needs. We also released the code, model and test code used in the experiment (https://github.com/YangzlTHU/VStego800K). We hope that this paper can help those in the field of voice steganalysis to develop more efficient speech steganalysis algorithms in the future to maintain the security of public cyberspace.

5 Conclusion

In this paper, in order to promote the development of streaming voice steganalysis technology, we construct and release a large-scale streaming voice steganalysis dataset called VStego800K. To truly reflect the needs of reality, we mainly follow three considerations when constructing the VStego800K dataset: large-scale, real-time, and diversity. The large-scale dataset allows researchers to fully explore the statistical distribution differences of streaming signals caused by steganography. Therefore, the proposed VStego800K dataset contains 814,592 streaming voice fragments. Among them, 764,592 samples (382,296 cover-stego pairs) are divided as the training set and the remaining 50,000 as testing set. The duration of all samples in the data set is uniformly cut to 1 s to encourage researchers to develop near real-time speech steganalysis algorithms. To ensure the diversity of the dataset, the collected voice signals are mixed with male and female as well as Chinese and English from different speakers. For each steganographic sample in the dataset, we randomly use two typical streaming voice steganography algorithms, and randomly embed random bit with embedding rates of 10%–40%. We hope that the VStego800K dataset will further promote the development of universal voice steganalysis technology.

Acknowledgement. This work was supported in part by the National Key Research and Development Program of China under Grant 2018YFB2101501 and the National Natural Science Foundation of China (No.61862002, No.U1705261 and No.U1936208).

References

1. Theohary, C.A.: Terrorist Use of the Internet: Information Operations in Cyberspace. DIANE Publishing (2011)

2. Yang, Z., Wang, Ke., Ma, S., Huang, Y., Kang, X., Zhao, X.: Istego100k: large-scale image steganalysis dataset. In: Wang, H., Zhao, X., Shi, Y., Kim, H.J., Piva, A. (eds.) IWDW 2019. LNCS, vol. 12022, pp. 352–364. Springer, Cham (2020). https://doi.org/10.1007/978-3-030-43575-2_29

3. Yang, Z., Peng, X., Huang, Y.: A sudoku matrix-based method of pitch period steganography in low-rate speech coding. In: Lin, X., Ghorbani, A., Ren, K., Zhu, S., Zhang, A. (eds.) SecureComm 2017. LNICSSITE, vol. 238, pp. 752–762. Springer, Cham (2018). https://doi.org/10.1007/978-3-319-78813-5_40

4. Yang, Z., Du, X., Tan, Y., Huang, Y., Zhang, Y.J.: Aag-stega: automatic audio generation-based steganography. arXiv preprint arXiv:1809.03463 (2018)

5. Yang, Z.L., Guo, X.Q., Chen, Z.M., Huang, Y.F., Zhang, Y.J.: RNN-stega: linguistic steganography based on recurrent neural networks. IEEE Trans. Inf. Forensics Secur. 14(5), 1280–1295 (2018)

6. Yang, Z.L., Zhang, S.Y., Hu, Y.T., Hu, Z.W., Huang, Y.F.: VAE-Stega: linguistic steganography based on variational auto-encoder. IEEE Trans. Inf. Forensics Secur. 16, 880–895 (2020)

7. Yang, Z., Zhang, P., Jiang, M., Huang, Y., Zhang, Y.-J.: Rits: real-time interactive text steganography based on automatic dialogue model. In: Sun, X., Pan, Z., Bertino, E. (eds.) ICCCS 2018. LNCS, vol. 11065, pp. 253–264. Springer, Cham (2018). https://doi.org/10.1007/978-3-030-00012-7_24

8. Johnson, N.F., Sallee, P.A.: Detection of hidden information, covert channels and information flows. Wiley Handbook of Science and Technology for Homeland Security, pp. 1–37 (2008)

9. Goode, B.: Voice over internet protocol (VoIP). Proc. IEEE 90(9), 1495–1517 (2002)

10. Hamdaqa, M., Tahvildari, L.: ReLACK: a reliable VoIP steganography approach. In: 2011 Fifth International Conference on Secure Software Integration and Reliability Improvement, pp. 189–197. IEEE (2011)

11. Tian, H., Zhou, K., Jiang, H., Huang, Y., Liu, J., Feng, D.: An adaptive steganography scheme for voice over IP. In: 2009 IEEE International Symposium on Circuits and Systems, pp. 2922–2925. IEEE (2009)

12. Xu, E., Liu, B., Xu, L., Wei, Z., Zhao, B., Su, J.: Adaptive VoIP steganography for information hiding within network audio streams. In: 2011 14th International Conference on Network-Based Information Systems, pp. 612–617. IEEE (2011)

13. Ballesteros L.D.M., Moreno A.J.M.: Highly transparent steganography model of speech signals using efficient wavelet masking. Exp. Syst. Appl. 39(10), 9141-9149 (2012)

14. Huang, Y.F., Tang, S., Yuan, J.: Steganography in inactive frames of VoIP streams encoded by source codec. IEEE Trans. Inf. Forensics Secur. 6(2), 296–306 (2011)

15. Tian, H., Liu, J., Li, S.: Improving security of quantization-index-modulation steganography in low bit-rate speech streams. Multimedia Syst. 20(2), 143–154 (2014)

16. Huang, Y., Liu, C., Tang, S., Bai, S.: Steganography integration into a low-bit rate speech codec. IEEE Trans. Inf. Forensics Secur. 7(6), 1865–1875 (2012)

17. Simmons, G.J.: The prisoners' problem and the subliminal channel. In: Chaum, D. (eds) Advances in Cryptology, pp. 51–67. Springer, Boston (1984). https://doi.org/10.1007/978-1-4684-4730-9_5

18. Liu, Q., Sung, A.H., Qiao, M.: Temporal derivative-based spectrum and mel-cepstrum audio steganalysis. IEEE Trans. Inf. Forensics Secur. 4(3), 359–368 (2009)

19. Paulin, C., Selouani, S.A., Hervet, E.: Audio steganalysis using deep belief networks. Int. J. Speech Technol. 19(3), 585–591 (2016)

20. Kraetzer, C., Dittmann, J.: Mel-cepstrum based steganalysis for VoIP steganography. In: Security, Steganography, and Watermarking of Multimedia Contents IX, vol. 6505, p. 650505. International Society for Optics and Photonics (2007)

21. Wang, J., Huang, L., Zhang, Y., Zhu, Y., Ni, J., et al.: An effective steganalysis algorithm for histogram-shifting based reversible data hiding. Comput. Mater. Continua **64**(1), 325–344 (2020)

22. Yang, C., Wang, J., Lin, C., Chen, H., Wang, W.: Locating steganalysis of LSB matching based on spatial and wavelet filter fusion. Comput. Mater. Continua **60**(2), 633–644 (2019)

23. Huang, Y.F., Tang, S., Zhang, Y.: Detection of covert voice-over Internet protocol communications using sliding window-based steganalysis. IET Commun. **5**(7), 929–936 (2011)

24. Fridrich, J., Goljan, M., Du, R.: Detecting LSB steganography in color, and gray-scale images. IEEE Multimedia **8**(4), 22–28 (2001)

25. Li, S.B., Tao, H.Z., Huang, Y.F.: Detection of quantization index modulation steganography in G. 723.1 bit stream based on quantization index sequence analysis. J. Zhejiang Univ. SCI. C, **13**(8), 624–634 (2012)

26. Lin, Z., Huang, Y., Wang, J.: RNN-SM: fast steganalysis of VoIP streams using recurrent neural network. IEEE Trans. Inf. Forensics Secur. **13**(7), 1854–1868 (2018)

27. Yang, Z., Yang, H., Hu, Y., Huang, Y., Zhang, Y.J.: Real-time steganalysis for stream media based on multi-channel convolutional sliding windows. arXiv preprint arXiv:1902.01286 (2019)

28. Yang, H., Yang, Z., Huang, Y.: Steganalysis of voip streams with cnn-lstm network. In: Proceedings of the ACM Workshop on Information Hiding and Multimedia Security, pp. 204–209 (2019)

29. Yang, H., Yang, Z., Bao, Y., Huang, YongFeng: Hierarchical representation network for steganalysis of qim steganography in low-bit-rate speech signals. In: Zhou, J., Luo, X., Shen, Q., Xu, Z. (eds.) ICICS 2019. LNCS, vol. 11999, pp. 783–798. Springer, Cham (2020). https://doi.org/10.1007/978-3-030-41579-2_45

30. Yang, H., Yang, Z., Bao, Y., Liu, S., Huang, Y.: Fast steganalysis method for voip streams. IEEE Signal Process. Lett. **27**, 286–290 (2019)

31. Chen, B., Wornell, G.W.: Quantization index modulation: A class of provably good methods for digital watermarking and information embedding. IEEE Trans. Inf. Theory **47**(4), 1423–1443 (2001)

32. Xiao, B., Huang, Y., Tang, S.: An approach to information hiding in low bit-rate speech stream. In: IEEE GLOBECOM 2008–2008 IEEE Global Telecommunications Conference, pp. 1–5. IEEE (2008)

33. Nishimura, A.: Data hiding in pitch delay data of the adaptive multi-rate narrow-band speech codec. In: 2009 fifth International Conference on Intelligent Information Hiding and Multimedia Signal Processing, pp. 483–486. IEEE (2009)

34. Janicki, A.: Pitch-based steganography for Speex voice codec. Secur. Commun. Netw. **9**(15), 2923–2933 (2016)

35. Li, S.B., Jia, Y.Z., Fu, J.Y., Dai, Q.X.: Detection of pitch modulation information hiding based on codebook correlation network. Chinese J. Comput. **37**(10), 2107–2117 (2014)

36. Hu, Y., Huang, Y., Yang, Z., Huang, Y.: Detection of heterogeneous parallel steganography for low bit-rate VoIP speech streams. Neurocomputing **419**, 70–79 (2021)

37. Wang, J., Yang, C., Wang, P., Song, X., Lu, J.: Payload location for JPEG image steganography based on co-frequency sub-image filtering. Int. J. Distrib. Sens. Netw. **16**(1), 1550147719899569 (2020)

Linguistic Steganalysis Based on Clustering and Ensemble Learning in Imbalanced Scenario

Shengnan Guo[ID], Xuekai Chen[ID], Zhuang Wang[ID], Zhongliang Yang[✉][ID], and Linna Zhou[ID]

School of Cyberspace Security, Beijing University of Posts and Telecommunications, Beijing 100876, China
yangzl@bupt.edu.cn

Abstract. With the rapid development of the Internet, more and more methods of text steganography have emerged. However, these methods are easily abused in public networks for malicious purposes, which poses a great threat to cyberspace security. At present, a large number of text steganalysis methods have been proposed to game with text steganography. However, existing methods typically assume a balanced class distribution. In reality, stego texts are far less than cover texts. How to accurately detect stego texts in massive texts becomes a challenge. In this paper, we propose a text steganalysis method based on an under-sample method and ensemble learning in imbalanced scenarios. Specifically, we introduce the thinking of clustering to under-sample the majority class samples (cover texts) based on the detection difficulty of the samples, in order to select samples with rich information. Ensemble learning is then used to ensemble the detection results of multiple base classifiers and guide the sampling process. We designed several experiments to test the detection performance of the proposed model. Experimental results show that the proposed model can effectively compensate for the deficiencies of existing methods, even in highly imbalanced datasets, the model can still detect stego texts effectively.

Keywords: Linguistic Steganalysis · Clustering · Ensemble Learning

1 Introduction

In recent years, information exchange through the Internet has become a necessary part of people's daily lives. Along with this growing trend, there is also a growing concern about privacy and security in cyberspace. In order to safeguard the privacy and security of people's information transmission in the public network environment, researchers use the core technology in the covert communication system called steganography. Steganography is a technique that focuses on how to embed secret information securely and efficiently into other information carriers to protect the security of information content while concealing the act of transmitting important information. However, once steganography is illegally abused in social networks, it can cause terrible damage to personal privacy, network security, and even social security. Therefore, the technology used to combat steganography, namely steganalysis technology, must be designed to detect these steganographic carriers in public media.

© The Author(s), under exclusive license to Springer Nature Singapore Pte Ltd. 2024
B. Ma et al. (Eds.): IWDW 2023, LNCS 14511, pp. 304–318, 2024.
https://doi.org/10.1007/978-981-97-2585-4_22

Steganalysis, as the counter-technique of steganography, is used to capture the difference between the carrier before and after steganography to detect whether it contains secret information. The carrier used to hide secret information can be image [20,21], audio [8,29], text [32,34] and so on [9]. Among them, whether in the era of paper media or digital media, the text carrier is the most commonly used communication form in daily life. In recent years, the development of social media has brought an explosion of massive data to the Internet. A large number of users comment, reply, chat, etc. On social platforms, these interactions generate a large number of social texts. However, this situation also brings opportunities for criminals to abuse text steganography. Once the stego texts generated by steganography methods flow into the social networks, it is very difficult to detect them in the huge nature texts, which brings challenges to the imbalanced text steganalysis.

Traditional text steganalysis algorithms [23,31] typically assume a balanced class distribution. However, the reality may not be consistent with this assumption. In the field of information hiding, the proportion of stego texts in massive texts is extremely small, so text steganalysis in real scenarios has the challenge of extreme data imbalance. When one class of data is significantly more prevalent than another in the dataset, the model will learn the prior information of the proportion of texts in the training set, so that the actual prediction will focus on the majority class. Specifically, the stego texts belonging to the minority class may be falsely treated as cover texts, thereby blurring the classification boundary. The result is often that the prediction effect of the stego texts is not satisfactory, which deviates from the original intention of text steganalysis to learn how to distinguish whether the text contains secret information or not. So research on imbalanced text steganalysis methods is of great value and significance to the public security of cyberspace in real-world scenarios.

The current imbalanced detection methods can be roughly divided into three big types. The first type is at the data level, which focuses on balancing the dataset by resampling the original dataset [7,10]. The second type is mainly at the algorithm level [4,6]. The method of this type mainly corrects its classifier's preference for most classes by making adaptive modifications to the existing balanced learning algorithm. The third type is hybrid methods, that is, the combination of the above two methods [5,12]. The most common is to use the data-level method to resample a balanced dataset, and then obtain a strong classifier through the ensemble learning method.

However, in the context of text steganalysis, the above methods are not always satisfactory. Most of the existing text steganography methods mainly adopt the generative steganographic framework of the neural networks language model with conditional probability coding. That is, the secret information is embedded when the stego texts are generated. If we synthesize samples to make the dataset balance, then the steganography method used to generate the existing stego texts needs to be known in advance, which is impossible. Therefore, it is not feasible to synthesize stego texts to achieve over-sample. When using under-sample methods, information loss may occur because of a large reduction in the cover texts, and meanwhile, it is not sure that the selected texts are useful. The use of the algorithmic level methods is usually applicable to specific tasks and cannot be generalized to different situations. Compared with the previous two types of methods, the combination of resampling and ensemble learning is effective to some extent. In order to make up for the information loss caused by under-sampling, ensemble learning can be introduced to learn as much information as possible from the majority class of texts in many classifiers.

In this paper, aiming at the data imbalance between stego texts and cover texts, we propose a text steganalysis method combining clustering-based under-sample and ensemble learning. In the under-sample stage, we design a dynamic clustering method according to the distribution of detection difficulty in the dataset. We then select texts from these sub-clusters to form an information-rich subset. Finally, this subset will be merged with the stego samples to form a new dataset. As the number of iterations increases, the class distribution gradually changes from imbalanced to balanced. Ensemble learning can dynamically calculate the detection difficulty of the original dataset through the iteration of the base classifier, and use it to guide the sampling process. We experiment with the proposed method on three stego datasets, and the experimental results demonstrate that the proposed model outperforms previous models and achieves excellent performance in imbalanced scenarios.

The rest of this paper is organized as follows. Section 2 introduces the related work about general text steganalysis methods and above-mentioned hybrid methods for imbalanced detection. Section 3 presents the proposed method in detail. Section 4 introduces the experimental setup and discusses the experimental result. Finally, conclusions are drawn in Sect. 5.

2   Related Work

2.1   Text Steganalysis

The idea of the text steganalysis task is to extract and analyze the feature differences between the stego texts and the cover texts in the feature space. Early methods [2, 16, 26] using manual features usually extract some artificial features that are considered to be designed, such as word frequency [26] and synonym frequency [24], and then analyze whether the text contains secret information through the statistical differences of these features before and after steganography. For example, in 2011, Chen *et al.* [2] introduced the contextual clustering to estimate contextual fitness of texts and showed how to distinguish stego texts from cover texts. In 2014, Xiang *et al.* [24] used the feature vectors of different attribute pairs with relative frequency differences to implement steganalysis. In 2016, S. Samanta *et al.* [16] used Bayesian estimation and correlation coefficients to calculate the statistical similarity between cover and stego texts.

These manual features are effective for steganography that only involves shallow statistical features. However with the continuous advancement of machine learning and natural language processing, neural networks have shown increasingly strong feature extraction and expression capabilities. Accordingly, text steganalysis gradually began to use complex text features to detect stego texts. Among them, in 2019, Yang *et al.* [30] proposed a fast and efficient deep neural networks (DNN) model to extract the correlation between words in stego texts. In 2020, Yang *et al.* [31] analyzed three kinds of word correlation patterns in the texts and used convolutional sliding windows (CSW) with various sizes to extract text features, and then implemented steganalysis through analyzing the distribution differences of these features in the high-level feature space.

In 2021, Wu *et al.* [23] extracted features from the directed graph through the graph convolutional network, so that the texts can make full use of global information to obtain better self-representation. In 2022, Li *et al.* [11] presented the concept of text word entropy for the first time, and they fully mined and utilized the text word relation features implicit in the context to classify stego texts.

Although the existing text steganalysis methods have strong performance, almost all text steganalysis methods limit the research object to the condition that the dataset is in a balanced distribution by default, while ignoring the reality that the stego texts are less than the cover texts. For imbalanced datasets, the above text steganalysis methods may not be applicable and some targeted methods are needed to deal with the imbalanced distribution of datasets.

2.2 Imbalanced Detection Methods

Among some simple data-level and algorithm-level methods, ensemble learning is considered to be a mainstream technique to solve the information loss caused by resampling in imbalanced detection methods, because they can significantly improve the detection performance of classifiers. There are two basic types of ensemble learning: Bagging and Boosting. These two types of algorithms are widely used in imbalanced scenarios in combination with under-sample, such as UnderBagging [1], RUSBoost [17], EasyEnsemble [12], EUSBoost [5], etc. These algorithms mainly include three modules, which are resampling, training weak classifiers, and model ensemble. In 2015, Sun *et al.* [19] proposed a novel ensemble method, which first divides the imbalanced dataset into multiple balanced datasets, then uses multiple base classifiers for training, and finally ensembles all base classifiers to obtain the final model. In 2018, Sun *et al.* [18] tried to introduce evolutionary under-sample (EUS) into the bagging framework, and they designed a novel fitness function so that the EUS-based bagging ensemble method EUS-Bag can train a set of highly accurate and diverse base classifiers. In 2020, Liu *et al.* [14] introduced the histogram distribution of training and validation errors as the meta-state for ensemble training. They proposed an ensemble framework based on meta-learning, the main idea of which is to design a solution for adaptive under-sampling in iterative ensemble training. In 2022, Zhou *et al.* [33] proposed an effective meta-framework that uses self-paced sampling to gradually transform the class distribution from imbalanced to balanced to alleviate the severe information loss problem and assign appropriate weights to solve the problem of noisy data.

The above approach may compensate for the general under-sampling that loses important information. However, when highly imbalanced, these methods still suffer from information loss and are even prone to overfitting, which degrades the detection performance. The proposed method will optimize these problems. In the experimental part, we will analyze the experimental results in detail.

3   The Proposed Method

The overall framework of the proposed text steganalysis method is shown in Fig. 1. The proposed model consists of two parts, namely under-sampling and model ensemble. In the under-sampling part, for the input texts, the model first uses the dynamic clustering method to divide the cover texts into multiple sub-clusters according to their detection difficulty and then selects texts proportionally from each sub-cluster. Finally, we combine the sampling texts with the stego texts to form a dataset with the updated imbalanced ratio and use the new dataset to train the base classifier. In the model ensemble part, in each iteration, we ensemble the trained base classifiers into the ensemble classifier to guide the under-sampling of the next iteration.

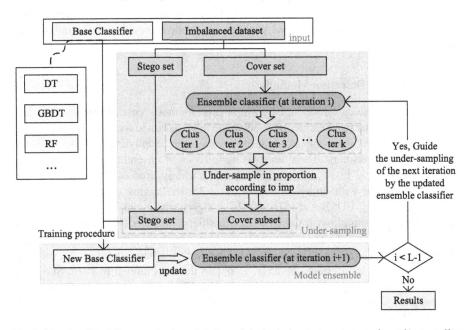

Fig. 1. The details of the proposed model. It mainly includes two parts: one is under-sampling based on clustering, which is used to select the informative subset. The other is the model ensemble, which ensembles the results of the trained i base classifiers and guides the under-sampling of the next iteration. Finally, after the iteration is completed, output the result of the final Ensemble classifier.

We now introduce the notations used in this paper. Let $X\{S;C\}$ denotes the original training set, where S denotes stego texts and C denotes cover texts. S and C are then defined as: $S = \{(x,y)|y = 1\}$, $C = \{(x,y)|y = 0\}$. Assuming that f is a trained classifier, we use $f(x)$ to indicate the probability that the classifier outputs x as a stego text. (The closer $f(x)$ is to 0, the more likely x is a cover text and the closer to 1, the more likely x is a stego text). We use $F_k(x)$ to denote an ensemble classifier consisting of k base classifiers. Let N_s and N_c denote the numbers of S and C, then the imbalance ratio $IR = \frac{N_c}{N_s}$. The value of IR is much greater than 1 for highly imbalanced datasets.

3.1   Imbalanced Ratio

When there are very few stego samples, the size of the balanced dataset selected by simple under-sampling for the cover set is also small, which will affect the performance of the base classifier, make it overfit and cause serious information loss.

Therefore, we consider using the sampling scheduler function [22] to dynamically resize the IR. The sampling scheduler function $h(l)$ is a function that returns a value between 0 and 1, where l indicates that it is currently in the l-th iteration. The IR at the l-th iteration is calculated by $IR_l = IR_o^{h(l)}$, where IR_o represents the original imbalanced ratio. $h(l)$ can be defined in different ways, it can be a convex function indicating the learning speed from slow to fast, a concave function indicating the learning speed from fast to slow, or a linear function indicating the constant learning speed. Here we choose the linear function as the sampling scheduler function $h_{linear}(l)$:

$$h_{linear}(l) = 1 - \frac{l}{L}, \tag{1}$$

where L represents the total number of iterations. In the beginning, when $l = 0$, $h(0) = 1$, $IR = IR_o$. As iteration increases, $h(l)$ gradually decreases, and the class distribution gradually changes from imbalanced to balanced. At the last iteration, $h(L) = 0$, $IR = 1$, that is, the dataset becomes balanced.

Because the number of stego texts is limited, each iteration is only for the C set, and the S set is unchanged. The C set at the i-th iteration is expressed as follows:

$$C_l = \lfloor IR_l \times N_s \rfloor = \left\lfloor N_c^{1-\frac{l}{L}} \times N_s^{\frac{l}{L}} \right\rfloor, \tag{2}$$

3.2   Dynamic Clustering

In order to select informative samples, we take a fancy to clustering methods which can help to select more diverse data among cover class data.

The proposed method adopts a dynamic clustering algorithm. Unlike the multi-dimensional input, the input of our proposed method is the detection difficulty of the texts, which simplify the computational complexity. The model first counts all cover texts as a large cluster and then divides them into multiple sub-clusters depending on the degree of dispersion of texts' detection difficulty. Due to the use of ensemble learning, the quantity and composition of sub-clusters undergo dynamic changes, that is, dynamic clustering.

Before clustering, we use the symbol D to denote the detection difficulty function, then the detection difficulty of the sample (x, y) with respect to the ensemble classifier F is represented by the function $D(x, y, F)$, where we use the absolute error.

When calculating the degree of dispersion of the texts, we use the coefficient of variation to measure it. The coefficient of variation is the ratio of the standard deviation to the mean, so it is more sensitive to the mean of the detection difficulty. When the mean is between 0 and 1, the coefficient of variation will be larger than the standard deviation, which can more obviously express the degree of dispersion of the data in the cluster, and its calculation is as follows:

$$C_v(c) = \frac{std(D_O)}{mean(D_O)},\qquad(3)$$

where $C_v(O)$ represents the coefficient of variation of the cluster, $std(D_O)$ represents the standard deviation of the detection probability in the cluster, and $mean(D_O)$ represents the average value of the predicted probability in the cluster.

In the process of molecular clustering, in each cluster, two sets of data are separated, resulting in two sub-clusters when they both have minimal internal differences and data fluctuations. We design formula (4) to measure whether both the left and right clusters have the smallest difference inside. If both $C_v(O_r)$ and $C_v(O_l)$ are small enough, then the value of $S(O_l, O_r)$ will be small, and it can also show that the data fluctuations in these two clusters are very small.

$$S(O_l, O_r) = \frac{N_{O_l} \times C_v(O_l) + N_{O_r} \times C_v(O_r)}{N_S},\qquad(4)$$

where N_S represents the number of samples in the cluster to be split, N_{O_l} represents the number of samples in the left cluster, and N_{O_r} represents the number of samples in the right cluster.

The specific process of the algorithm can be described as follows. First, according to the ensemble classifier F, each text obtains the predicted value, from which we get the detection difficulty of the texts. Sort the detection difficulty to form a large cluster, and traverse the detection difficulty of each sample from small to large. When traversing each text, the subset whose detection difficulty is smaller than the text is called the left cluster O_l, and the rest form the right cluster O_r. Then calculate the coefficients of variation $C_v(O_l)$ and $C_v(O_r)$ of these two sub-clusters respectively. According to formula (4), $S(O_l, O_r)$ can be obtained. After one round of traversal, $minS(O_l, O_r)$ can be obtained, which determines where to divide the large cluster into two sub-clusters. It is worth noting that we set a parameter num, if a cluster has just been divided into sub-clusters with fewer than num texts, we will not split this cluster. Subsequently, the obtained two sub-clusters will undergo the above operations respectively to continuously generate sub-clusters until the number of sub-clusters reaches N_O.

$$N_O = max(int(IR), c),\qquad(5)$$

where c is a constant, which is used to prevent the number of sub-cluster from being too small due to low IR, resulting in the generation of a sampling subset that is not rich in information.

In each iteration, with the change of the ensembled classifier F, the detection difficulty of the texts is different, and the clustering result of the proposed dynamic clustering algorithm will be updated accordingly, which provides effective and positive guidance for subsequent under-sampling.

3.3   Under Sampling

Based on the above analysis, the coefficient of variation can reflect the compactness of the texts in the cluster. The smaller the coefficient of variation of a sub-cluster, the more concentrated the detection difficulty of the texts in this subcluster, and the better the clustering quality of the subcluster. When under-sampling in each subcluster, we develop the thinking of the importance of the subclusters *imp* and sample in proportion based on the importance of sub-clusters for detection.

In the importance measurement of sub-clusters, considering both the quality of sub-clusters and the detection difficulty of texts, the final expression of *imp* is as follows:

$$imp_k = \frac{\sum_i^{N_{O_k}} D_i}{C_v(O_k) + 1},\tag{6}$$

where N_{O_k} represents the number of texts in the k-th sub-cluster, $\sum_i^{N_{O_k}} D_i$ represents the sum of detection difficulties of all texts in the sub-cluster, and $C_v(O_k)$ represents the coefficient of variation of the k-th sub-cluster. To avoid the situation where the coefficient of variation is zero, we added one to the denominator.

After obtaining the *imp* of each sub-cluster, the sampling amount in each sub-cluster is calculated according to the proportion of *imp* of each sub-cluster.

It is worth noting that if the number of texts in a certain sub-cluster is larger than the amount of sampling required in the sub-cluster, under-sampling will be performed; otherwise, then all texts in the sub-cluster are selected, and the remaining texts are averaged to other sub-clusters for sampling. According to the sampling amount, the sampled subset C_{new} is obtained, and the new training set X_{new} can be obtained by combining C_{new} with the original S set.

3.4 Ensemble

Each time we get an updated X_{new}, we will use it to train a new base classifier. Then at the beginning of each iteration, we will use the combination strategy of the average method to ensemble multiple base classifiers. Specifically, every time a weak classifier is trained, we average it with the previous classifiers and combine them into an ensemble classifier. The ensemble classifier after the l-th iteration is expressed as follows:

$$F_l(x) = \frac{1}{l} \sum_{i=1}^{l} f_i(x),\tag{7}$$

4   Experiments and Analysis

4.1   Datasets and Evaluation Indicators

In this paper, all experiments are carried out on three natural text corpora including IMDB, Twitter, and News. To obtain the stego texts, we use the datasets generated by the generative steganography model RNN-Stega [28] with different conditional probabilistic coding methods (Huffman Coding (HC) and Arithmetic Coding (AC)). Among them, the embedding rate of stego texts is measured by the number of embedded bits per word (*bpw*). The average *bpw* is about 5 for HC and about 7 for AC. And we arbitrarily select 75% of the samples for training and validation and 25% of the samples for testing in every dataset.

We use several evaluation indicators commonly used in imbalanced classification tasks to evaluate the performance of our model, namely Area Under Precision-Recall Curve (AUCPRC), F1-score, Matthews correlation coefficient (MCC), and G-mean. Their concepts and formulas are described as follows:

– AUCPRC measures the area under the precision-recall curve. The larger the value of AUCPRC, the better the performance of the model.

$$AUCPRC = AreaUnderPrecision - RecallCurve, \tag{8}$$

– F1-score is the harmonic mean of precision and recall. The closer its value is to 1, the better the model is.

$$F1 - score = 2 \cdot \frac{Recall \times Precision}{Recall + Precision}, \tag{9}$$

– MCC essentially describes the correlation coefficient between the predicted results and the real results.

$$MCC = \frac{TP \times TN - FP \times FN}{\sqrt{(TP + FP)(TP + FN)(TN + FP)(TN + FN)}}, \tag{10}$$

– G-mean is the geometric mean of Specificity and Recall.

$$G - mean = \sqrt{Recall \cdot Specificity}, \tag{11}$$

where TP (True Positive) indicates the number of positive samples predicted by the model as positive, FP (False Positive) indicates the number of negative samples predicted by the model as positive, FN (False Negative) indicates the number of positive samples predicted by the model as negative, TN (True Negative) indicates the number of negative samples predicted by the model as negative.

4.2 Baselines

In order to objectively reflect the effect of the proposed model, we selected several most representative imbalanced detection methods: RUS, Under [13], Easy [12], Cascade [12], MESA [14], and DAPS [33]. RUS (Random Under-sampling) refers to randomly deleting multiple samples from the majority class samples so that the distribution of the dataset is balanced. Under [13] uses multiple balanced datasets extracted from bootstrap under-sampling to train multiple weak classifiers at the same time, and finally uses voting to determine the final result. Easy [12] randomly under-samples the majority samples to obtain balanced subsets, and finally ensembles all AdaBoost base classifiers to obtain the final model. The basic architecture of Cascade [12] is the same as Easy. The difference is that misclassified samples are selectively returned to the original samples space. MESA [14] can adaptively resample the training set in iterations to obtain multiple classifiers and form a cascade ensemble model. DAPS [33] uses self-paced learning to gradually transform the class distribution from imbalanced to balanced and uses instance weighting to mine valuable instances.

In addition, we also used several state-of-the-art steganalysis algorithms: TS-FCN [30], TS-CSW [31], BERT-LSTM-ATT [35], R-BiLSTM-C [15], DENSE [25], and SeSy [27] to compare with our proposed model. TS-FCN [30] proposes a fast and efficient text steganalysis method, which uses neural networks to extract the correlations between words. TS-CSW [31] first analyzes the three word correlation patterns existing in the discourse and then utilizes multi-size convolution sliding windows (CSW) to extract text features. BERT-LSTM-ATT [35] uses BERT [3] to extract the contextualized association relationships between the words and BiLSTM to fuse the context information after outputting the embedded layer. R-BiLSTM-C [15] uses the BILSTM structure to obtain the long-term semantic information of texts and the asymmetric convolution kernels with different sizes extract the local relationship between words. Considering the presence of low-level features, DENSE [25] introduces two components of dense connections and feature pyramid to fuse the low-level features in the feature vector. SeSy [27] comprehensively considers the grammar structures and semantic features of the texts.

4.3   Performance of Our Proposed Model

We first test the effectiveness of the proposed model and the baseline models on three different corpora with an imbalance ratio of 400. Table 1 shows the F1-Score, AUCPUC value, MCC value, and G-mean value obtained by all models using the Decision Tree model as a classifier. Overall, our model outperforms other baseline models in all metrics. Especially in the News dataset with HC, our model can directly improve the F1 score by up to 92% when Cascade's result is only 2.5%.

For the baseline methods, we can find that Under, RUSBoost, Cascade, and Easy suffer from information loss as resampling may lose samples with important information, which is more obvious when the imbalance rate is extremely high. In contrast, the improved sampling strategies of MESA and DAPS will make them achieve relatively better performance, and DAPS is even better. This may be because DAPS also adopts the method of gradually changing the dataset from imbalanced to balanced in iterations when training the model, which can effectively alleviate the problem of information loss, unlike other methods that directly change the dataset to balanced when resampling. But the detection ability of these baseline models is still not as good as the proposed method. We introduce the idea of clustering when resampling, and make the ensemble training also participate in the guided sampling. By gradually reducing the number of samples of the cover class in iterations, the effect of the base classifier is getting better and better, especially in highly imbalanced text steganalysis task.

4.4   Compared with Other Text Steganalysis Methods

We further compare the proposed method with several text steganalysis methods and use AUCPRC as the evaluation metric. In the actual experiment, we used a dataset with an imbalance ratio of 40 at the beginning, but all the steganalysis baseline methods had

Table 1. Comparisons of the proposed model with other imbalanced detection methods on different evaluation indicators with an imbalanced ratio of 400.

Dataset		Metric	RUS	Under	Cascade	Easy	MESA	DAPS	Ours
IMDB	hc	AUCPRC	0.286	0.153	0.032	0.153	0.604	0.714	**0.769**
		F1-score	0.302	0.286	0.061	0.286	0.67	0.833	**0.87**
		MCC	0.381	0.388	0.171	0.388	0.584	0.845	**0.877**
		G-mean	0.991	0.943	0.961	0.943	0.983	0.999	**1.0**
	ac	AUCPRC	0.668	0.625	0.016	0.164	0.651	0.601	**0.701**
		F1-score	0.662	0.769	0.032	0.282	0.733	0.75	**0.824**
		MCC	0.734	0.79	0.118	0.402	0.766	0.774	**0.836**
		G-mean	0.739	0.749	0.622	0.694	0.74	0.775	**0.837**
News	hc	AUCPRC	0.668	0.714	0.05	0.159	0.811	0.8	**0.87**
		F1-score	0.662	0.833	0.025	0.274	0.878	0.889	**0.945**
		MCC	0.734	0.845	0.048	0.396	0.859	0.87	**0.943**
		G-mean	0.739	0.899	0.549	0.693	0.847	0.894	**0.945**
	ac	AUCPRC	0.541	0.112	0.149	0.109	0.685	0.582	**0.712**
		F1-score	0.707	0.202	0.37	0.196	0.714	0.762	**0.842**
		MCC	0.715	0.332	0.475	0.326	0.811	0.762	**0.843**
		G-mean	0.724	0.547	0.594	0.793	0.758	0.794	**0.894**
Twitter	hc	AUCPRC	0.735	0.588	0.579	0.5	0.755	0.821	**0.896**
		F1-score	0.829	0.741	0.75	0.667	0.888	0.906	**0.951**
		MCC	0.909	0.766	0.76	0.706	0.899	0.868	**0.948**
		G-mean	0.889	0.861	0.848	0.749	0.892	0.884	**0.947**
	ac	AUCPRC	0.772	0.588	0.017	0.417	0.875	0.81	**0.9**
		F1-score	0.759	0.741	0.033	0.588	0.905	0.9	**0.947**
		MCC	0.799	0.766	0.121	0.644	0.896	0.9	**0.949**
		G-mean	0.791	0.829	0.339	0.66	0.888	0.949	**0.949**

results of zero, so we gradually decreased the imbalance rate to test the effect of the model. In order to present the most obvious experimental results, we finally selected the imbalance ratio from 5 to 20 with a double increase for comparative experiments. The full results are shown in Table 2.

It can be observed from Table 2 that our model still achieves the best results when the imbalance ratio is relatively small and is hardly affected by changes in low imbalance ratios. The baseline models are all worse due to the decrease of the imbalance ratio, especially most methods have an AUCPRC value close to zero at an imbalance ratio of 20, which indicates that they have a high probability of misclassifying all the samples of the stego class. The reason may be that these models only focus on the semantic features of the samples while ignoring their quantitative distribution characteristics, which leads to them being only suitable for balanced datasets, and they will largely fail when they are in imbalanced scenarios. Compared with the general steganalysis methods based on

Table 2. Comparisons of the proposed model with other representative text steganalysis methods under different imbalanced ratio.

Dataset	ratio	TS-CSW	BERT-LSTM-ATT	R-BiLSTM-C	TS-FCN	DENSE	Sesy	Ours
IMDB	5	0.615	0.578	0.589	0.599	0.598	0.629	**0.999**
	10	0.388	0.327	0.322	0.357	0.4	0.62	**0.995**
	20	0.123	0.08	0.05	0.05	0.09	0.537	**0.987**
News	5	0.77	0.759	0.758	0.772	0.761	0.379	**0.991**
	10	0.647	0.623	0.623	0.644	0.631	0.342	**0.975**
	20	0.461	0.336	0.233	0.253	0.455	0.299	**0.946**
Twitter	5	0.423	0.399	0.423	0.424	0.426	0.419	**0.999**
	10	0.202	0.143	0.09	0.161	0.177	0.273	**0.986**
	20	0.05	0.05	0.05	0.05	0.05	0.224	**0.974**

balanced datasets, the proposed model is more competitive and practical in dealing with imbalanced steganalysis problems.

4.5 Performance Under Different Number of Base Classifiers

To examine the effect of the number of base classifiers on the performance of ensemble methods, we conduct tests with varying the number of base classifiers between 10 and 100 in steps of 10. All experiments are performed on the News dataset with an imbalance ratio of 400. In order to intuitively reflect the experimental results, we plot the F1-score curves of each model under different numbers of base classifiers, as shown in Fig. 2.

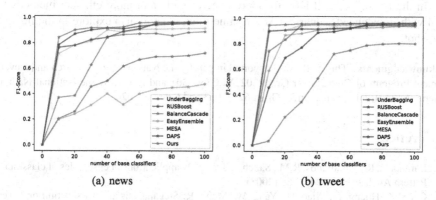

(a) news (b) tweet

Fig. 2. Comparisons of the proposed model with other imbalanced detection methods under different number of base classifiers.

We can see that although the proposed model is trained with few stego samples, it still achieves reasonable results. First, as the number of base classifiers increases, the performance of each model gradually improves. The reason may be that when there are fewer base classifiers, the results in the base classifiers are prone to randomness, so the result of the final ensemble classifier will be inconsistent. Well, the increase of base classifiers can alleviate this drawback. Another reason is that in MESA and DAPS, the number of base classifiers corresponds to the number of iterations of the ensemble model. An increase in the number of iterations within a certain range means that the base classifier can learn more and more meaningful samples, thereby improving the performance of the ensemble classifier. Then as the number of base classifiers continues to increase, other methods are less stable, and the proposed model can be the fastest to keep the F1-Score stable. It shows that compared with other baseline models, the proposed model only needs the least number of base classifiers to achieve the best results, which can effectively save the computational cost of steganalysis, which has very important practical significance.

5 Conclusion

In reality, stego samples are much smaller than cover samples, and the problem of data imbalance is involved in text steganalysis. However, most of the existing text steganalysis methods consider the class balance scenario. To address the data imbalance problem in the field of text steganalysis, in this paper, we propose a text steganalysis method based on clustering under-sampling and ensemble learning. It uses the clustering method to divide texts into several sub-clusters according to their detection difficulty, and then under-samples from the separated sub-clusters. The sampled texts are predicted by base classifiers, and these base classifiers are ensembled into the ensemble classifier. We use extensive experiments to validate our model in multiple aspects. Experimental results show that the model can still perform well in highly imbalanced datasets. We hope that this work can inspire methods for imbalanced text steganalysis. In the future, we will look for more generalizable and more efficient methods to deal with imbalanced scenarios. At the same time, we will also explore in the field of few-shot.

Acknowledgments. This work was supported in part by the National Key Research and Development Program of China under Grant 2022YFC3303301 and in part by the National Natural Science Foundation of China under Grant 62172053 and Grant 62302059.

References

1. Barandela, R., Valdovinos, R.M., Sánchez, J.S.: New applications of ensembles of classifiers. Pattern Anal. Appl. **6**, 245–256 (2003)
2. Chen, Z., Huang, L., Miao, H., Yang, W., Meng, P.: Steganalysis against substitution-based linguistic steganography based on context clusters. Comput. Electr. Eng. **37**(6), 1071–1081 (2011)
3. Devlin, J., Chang, M.W., Lee, K., Toutanova, K.: BERT: pre-training of deep bidirectional transformers for language understanding. arXiv preprint arXiv:1810.04805 (2018)
4. Freund, Y.: Boosting a weak learning algorithm by majority. Inf. Comput. **121**(2), 256–285 (1995)

5. Galar, M., Fernández, A., Barrenechea, E., Herrera, F.: EUSBoost: enhancing ensembles for highly imbalanced data-sets by evolutionary undersampling. Pattern Recogn. **46**(12), 3460–3471 (2013)
6. Gao, L., Zhang, L., Liu, C., Wu, S.: Handling imbalanced medical image data: a deep-learning-based one-class classification approach. Artif. Intell. Med. **108**, 101935 (2020)
7. He, H., Bai, Y., Garcia, E.A., Li, S.: ADASYN: adaptive synthetic sampling approach for imbalanced learning. In: 2008 IEEE International Joint Conference on Neural Networks (IEEE World Congress on Computational Intelligence), pp. 1322–1328. IEEE (2008)
8. Huang, Y.F., Tang, S., Yuan, J.: Steganography in inactive frames of VoIP streams encoded by source codec. IEEE Trans. Inf. Forensics Secur. **6**(2), 296–306 (2011)
9. Johnson, N.F., Sallee, P.A.: Detection of hidden information, covert channels and information flows. In: Wiley Handbook of Science and Technology for Homeland Security, pp. 1–37 (2008)
10. Laurikkala, J.: Improving identification of difficult small classes by balancing class distribution. In: Quaglini, S., Barahona, P., Andreassen, S. (eds.) AIME 2001. LNCS, vol. 2101, pp. 63–66. Springer, Heidelberg (2001). https://doi.org/10.1007/3-540-48229-6_9
11. Li, S., Wang, J., Liu, P.: Detection of generative linguistic steganography based on explicit and latent text word relation mining using deep learning. IEEE Trans. Dependable Secure Comput. **20**(2), 1476–1487 (2022)
12. Liu, X.Y., Wu, J., Zhou, Z.H.: Exploratory undersampling for class-imbalance learning. IEEE Trans. Syst. Man Cybern. Part B (Cybern.) **39**(2), 539–550 (2008)
13. Liu, Y., Chawla, N.V., Harper, M.P., Shriberg, E., Stolcke, A.: A study in machine learning from imbalanced data for sentence boundary detection in speech. Comput. Speech Lang. **20**(4), 468–494 (2006)
14. Liu, Z., Wei, P., Jiang, J., Cao, W., Bian, J., Chang, Y.: MESA: boost ensemble imbalanced learning with meta-sampler. In: Advances in Neural Information Processing Systems, vol. 33, pp. 14463–14474 (2020)
15. Niu, Y., Wen, J., Zhong, P., Xue, Y.: A hybrid R-BILSTM-C neural network based text steganalysis. IEEE Sig. Process. Lett. **26**(12), 1907–1911 (2019)
16. Samanta, S., Dutta, S., Sanyal, G.: A real time text steganalysis by using statistical method. In: 2016 IEEE International Conference on Engineering and Technology (ICETECH), pp. 264–268. IEEE (2016)
17. Seiffert, C., Khoshgoftaar, T.M., Van Hulse, J., Napolitano, A.: RUSBoost: a hybrid approach to alleviating class imbalance. IEEE Trans. Syst. Man Cybern.-Part A Syst. Hum. **40**(1), 185–197 (2009)
18. Sun, B., Chen, H., Wang, J., Xie, H.: Evolutionary under-sampling based bagging ensemble method for imbalanced data classification. Front. Comput. Sci. **12**, 331–350 (2018)
19. Sun, Z., Song, Q., Zhu, X., Sun, H., Xu, B., Zhou, Y.: A novel ensemble method for classifying imbalanced data. Pattern Recogn. **48**(5), 1623–1637 (2015)
20. Tang, W., Li, B., Tan, S., Barni, M., Huang, J.: CNN-based adversarial embedding for image steganography. IEEE Trans. Inf. Forensics Secur. **14**(8), 2074–2087 (2019)
21. Wang, Y., Zhang, W., Li, W., Yu, X., Yu, N.: Non-additive cost functions for color image steganography based on inter-channel correlations and differences. IEEE Trans. Inf. Forensics Secur. **15**, 2081–2095 (2019)
22. Wang, Y., Gan, W., Yang, J., Wu, W., Yan, J.: Dynamic curriculum learning for imbalanced data classification. In: Proceedings of the IEEE/CVF International Conference on Computer Vision, pp. 5017–5026 (2019)
23. Wu, H., Yi, B., Ding, F., Feng, G., Zhang, X.: Linguistic steganalysis with graph neural networks. IEEE Sig. Process. Lett. **28**, 558–562 (2021)
24. Xiang, L., Sun, X., Luo, G., Xia, B.: Linguistic steganalysis using the features derived from synonym frequency. Multimedia Tools Appl. **71**, 1893–1911 (2014)

25. Yang, H., Bao, Y., Yang, Z., Liu, S., Huang, Y., Jiao, S.: Linguistic steganalysis via densely connected LSTM with feature pyramid. In: Proceedings of the 2020 ACM Workshop on Information Hiding and Multimedia Security, pp. 5–10 (2020)

26. Yang, H., Cao, X.: Linguistic steganalysis based on meta features and immune mechanism. Chin. J. Electron. **19**(4), 661–666 (2010)

27. Yang, J., Yang, Z., Zhang, S., Tu, H., Huang, Y.: SeSy: linguistic steganalysis framework integrating semantic and syntactic features. IEEE Sig. Process. Lett. **29**, 31–35 (2021)

28. Yang, Z.L., Guo, X.Q., Chen, Z.M., Huang, Y.F., Zhang, Y.J.: RNN-Stega: linguistic steganography based on recurrent neural networks. IEEE Trans. Inf. Forensics Secur. **14**(5), 1280–1295 (2018)

29. Yang, Z., Du, X., Tan, Y., Huang, Y., Zhang, Y.J.: AAG-Stega: automatic audio generation-based steganography. arXiv preprint arXiv:1809.03463 (2018)

30. Yang, Z., Huang, Y., Zhang, Y.J.: A fast and efficient text steganalysis method. IEEE Sig. Process. Lett. **26**(4), 627–631 (2019)

31. Yang, Z., Huang, Y., Zhang, Y.J.: TS-CSW: text steganalysis and hidden capacity estimation based on convolutional sliding windows. Multimedia Tools Appl. **79**, 18293–18316 (2020)

32. Zhang, S., Yang, Z., Yang, J., Huang, Y.: Provably secure generative linguistic steganography. arXiv preprint arXiv:2106.02011 (2021)

33. Zhou, F., et al.: Dynamic self-paced sampling ensemble for highly imbalanced and class-overlapped data classification. Data Min. Knowl. Disc. **36**(5), 1601–1622 (2022)

34. Ziegler, Z.M., Deng, Y., Rush, A.M.: Neural linguistic steganography. arXiv preprint arXiv:1909.01496 (2019)

35. Zou, J., Yang, Z., Zhang, S., ur Rehman, S., Huang, Y.: High-performance linguistic steganalysis, capacity estimation and steganographic positioning. In: Zhao, X., Shi, Y.Q., Piva, A., Kim, H.J. (eds.) IWDW 2020. LNSC, vol. 12617, pp. 80–93. Springer, Cham (2021). https://doi.org/10.1007/978-3-030-69449-4_7

Author Index

B. Ma et al. (Eds.): IWDW 2023, LNCS 14511, pp. 319–320, 2024.
https://doi.org/10.1007/978-981-97-2585-4